BUILDING COST MANUAL

1981 Square Foot Costs for Residential, Commercial, Industrial, Agricultural and Military Structures

Edited By
Gary Moselle

Craftsman Book Company, 542 Stevens Avenue, Solana Beach, California 92075

ISBN: 0-91046077-9

Contents Of This Manual

Explanation of the Cost Tables **4**
Location Factor Table **7**
Construction Cost Index **9**

Residential Structures Section **10**
Single Family Residences 10
Multi-Family Residences 16
Motels 20
Additional Costs for Residences 24
Multi-Family and Motel Garages 28
Cabins and Rescreational Dwellings 29
Conventional Recreational Dwellings 35
"A-Frame" Cabins 35
Additional Costs for Recreational Dwellings 39
Typical Lives for Residential Structures 40

Commercial Structures Section **41**
Urban Stores, Masonry or Concrete 43
Urban Stores, Wood or Wood and Steel 61
Suburban Stores, Masonry or Concrete 56
Suburban Stores, Wood or Wood and Steel 61
Supermarkets, Masonry or Concrete 70
Supermarkets, Wood or Wood and Steel 72
Small Food Stores, Masonry or Concrete 74
Small Food Stores, Wood Frame 76
Discount Houses, Masonry or Concrete 76
Discount Houses, Wood or Wood and Steel 80
Banks and Savings Offices, Masonry or Concrete 82
Banks and Savings Office, Wood Frame 87
Department Stores, Reinforced Concrete 93
Department Stores, Masonry or Concrete 96
Department Stores, Wood Frame 99
General Office Buildings, Masonry or Concrete 102
General Office Buildings, Wood Frame 110
Medical-Dental Buildings, Masonry or Concrete 118
Medical-Dental Buildings, Wood Frame 126
Convalescent Hospitals, Masonry or Concrete 134
Convalescent Hospitals, Wood Frame 136
Furneral Homes 138
Self Service Restaurants 140
Coffee Shop Restaurants 143
Conventional Restaurants 146
"A-Frame" Restaurants 148
Theaters, Masonry or Concrete 150
Theaters, Wood Frame 156
Mobile Home Parks 161
Service Stations, Wood, Masonry or Steel 163
Service Stations, Porcelain Finished Steel 165
Service Stations, Ranch or Rustic 167
Additional Costs for Service Stations 169
Service Garages, Masonry or Concrete 173
Service Garages, Wood Frame 178
Auto Service Centers, Masonry or Concrete 183

Industrial Structures Section **187**
Warehouses 189
Light Industrial Buildings 190
Factory Buildings 191
Internal Offices 192
External Offices 192
Steel Buildings 194
Alternate Costs for Steel Buildings 195
Commercial and Industrial Building Lives 200
Additional Commercial and Industrial Costs 201
Display Fronts 208
Signs 211
Yard Improvements 212

Agricultural Structures Section **214**
General Purpose Barns 215
Hay Storage Barns 216
Feed Barns 217
Shop Buildings 218
Machinery and Equipment Sheds 219
Small Sheds 220
Pole Barns 221
Low Cost Dairy Barns 222
Stanchion Dairy Barns 223
Walk-Through Dairy Barns 224
Modern Herringbone Barns 225
Miscellaneous Dairy Costs 226
Poultry Houses, Conventional 227
Poultry Houses, Modern Type 228
Poultry Houses, High Rise Type 229
Poultry Houses, Deep Pit Type 230
Poultry House Equipment 231
Labor Camps 232
Miscellaneous Agricultural Structures 233
Typical Lives for Agricultural Buildings 233

Military Structures Section **234**
Index **237**

Explanation of the Cost Tables

This manual provides construction or replacement costs for a wide variety of residential, commercial, industrial, agricultural and military buildings. For your convenience and to avoid possible errors, all the cost and reference information you need for each building type is listed with the primary cost figures for that building. After reading this and the following two pages you should be able to turn directly to any building type and make an error-free estimate or appraisal.

The costs are per square foot of floor area for the basic building and additional costs for optional or extra components that differ from building to building. Building shape, floor area, design elements, materials used, and overall quality influence the basic structure cost. These and other cost variables are isolated for the building types. Components included in the basic square foot cost are listed with each building type. Instructions for using the basic building costs are included above the cost tables. These instructions include a list of components that may have to be added to the basic cost to find the total cost for your structure.

The figures in this manual are intended to reflect the amount that would be paid by the end user of a building as of mid 1981.

The costs in this manual are the builder's selling price plus the design cost. The following costs are excluded: land value, site improvements other than the building itself, the expense of overcoming unusual site conditions, the owner-entrepreneur's profit, interest expense to the land owner.

Building Quality
Structures vary widely in quality and the quality of construction is the most significant variable in the finished cost. For estimating purposes the structure should be placed in one or more quality classes. These classes are numbered from 1 which is the highest quality generally encountered. Each section of this manual has a page describing typical specifications which define the quality class. Each number class has been assigned a word description (such as best, good, average or low) for convenience and to help avoid possible errors.

The quality specifications do not reflect some design features and construction details that can make a building both more desirable and more costly. When substantially more than basic design elements are present, and when these elements add significantly to the cost, it is appropriate to classify the quality of the building as higher than would be warranted by the materials used in construction.

Many structures do not fall into a single class and have features of two quality classes. The tables have "half classes" which apply to structures which have some features of one class and some features of a higher or lower class. Classify a building into a "half class" when the quality elements are fairly evenly divided between two classes. Generally quality elements do not vary widely in a single building. For example, it would be unusual to find a top quality single family residence with minimum quality roof cover. The most weight should be given to quality elements that have the greatest cost. For example, the type of wall and roof framing or the quality of interior finish are more significant than the roof cover or bathroom wall finish. Careful evaluation may determine that certain structures fall into two distinct classes. In this case the cost of each part of the building should be evaluated separately.

Building Shapes
Shape classification considers any cost differences that arise from variations in building outline. Shape classification considerations vary somewhat with different building types. Where the building shape often varies widely between buildings and shape has a significant effect on the building cost, basic building costs are given for several shapes. Use the table that most closely matches the shape of the building you are evaluating. If the shape falls near the division between two basic building cost tables, it is appropriate to average the square foot cost from those two tables.

Area of Buildings
The basic building cost tables reflects the fact that larger buildings generally cost less per square foot than smaller buildings. The cost tables are based on square foot areas which include the following:

1. All floor area within and including the exterior walls of the main building.
2. Inset areas such as vestibules, entrances or porches outside of the exterior wall but under the main roof.
3. Any enclosed additions, annexes or lean-tos with a square foot cost greater than three-fourths of the square foot cost of the main building.

Select the basic building cost listed below the area which falls closest to the actual area of your building. If the area of your building falls nearly mid-way between two listed building areas, it is appropriate to average the square foot costs for the listed areas.

Wall Heights
Building costs are based on the wall heights given in the instructions for each building cost table. Wall height for the various floors of a building are computed as follows: The basement is measured from the bottom of floor slab to the bottom of the first floor slab or joist. The main or first floor extends from the bottom of the first floor slab or joist to the top of the roof slab or ceiling joist. Upper floors are measured from the top of the floor slab or floor joist to the top of the roof slab or ceiling joist. These measurements may be illustrated as follows:

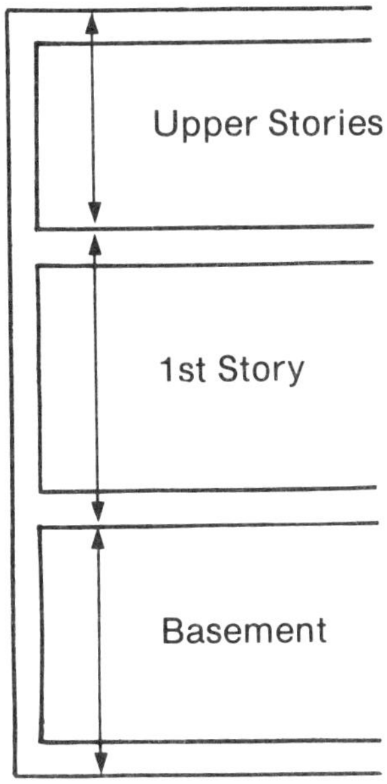

Square foot costs of most building design types must be adjusted if the actual wall height differs from the listed wall height. Wall height adjustment tables are included for buildings requiring this adjustment. Wall height adjustment tables list square foot costs for a foot of difference in perimeter wall height of buildings of various areas. The amount applicable to the actual building area is added or deducted for each foot of difference from the basic wall height.

Buildings such as residences, medical-dental buildings, funeral homes and convalescent hospitals usually have a standard eight foot ceiling height except in chapels or day room areas. If a significant cost difference exists due to a wall height variation, this factor should be considered in establishing the quality class.

Other Adjustments
A common wall exists when two buildings share one wall. Common wall adjustments are made by deducting the in-place cost of the exterior wall finish plus one-half of the in-place cost of the structural portion of the common wall area.

If an owner has no ownership in a wall, the in-place cost of the exterior wall finish plus the in-place cost of the structural portion of the wall should be deducted from the total building costs. Suggested common wall and no wall ownership costs are included for many of the building types.

Some square foot costs include the cost of expensive veneer finishes on the entire perimeter wall. When these buildings butt against other buildings, adjustments should be made for the lack of this finish. Where applicable, linear foot cost deductions are provided.

The square foot costs in this manual are based on composite costs of total buildings including usual work room or storage areas. They are intended to be applied on a 100% basis to the total building area even though certain areas will have inferior or no interior finish. Only in rare instances will it be necessary to modify the square foot cost of a portion of a building.

Multiple story buildings usually share a common roof structure and cover, a common foundation and common floor or ceiling structures. The costs of these components are included in the various floor levels as follows:

The first or main floor includes the cost of a floor structure built at ground level, foundation costs for a one-story building, a complete ceiling and roof structure, and a roof cover. The basement includes the basement floor structure and the difference between the cost of the first floor structure built at ground level and its cost built over a basement. The second floor includes the difference between the cost of a foundation for a one-story building and the cost of a foundation for a two story building and the cost of the second story floor structure.

Location Adjustments
The figures in this manual are intended as national averages for metropolitan areas of the United States. Use the information on page 7 to adapt the basic building costs to any area listed. Frequently building costs outside metropolitan areas are 2 to 6 percent lower if skilled, productive, lower cost labor is available in the area. The factors on page 7 can be applied to nearly all the square foot costs and some of the "additional" costs in this book.

Depreciation
Depreciation is the loss in value of a structure from all causes and is caused primarily by three forms of obsolescence: (1) physical (2) functional, and (3) economic.

Physical obsolescence is the deterioration of building components such as paint, carpets or roofing. This deterioration may be partially curable. The tables on pages 40, 200 and 233 consider only typical physical obsolescence. Individual judgments will have to be made of functional and economic obsolescence.

Functional obsolescence is due to a deficiency or inadequacy in some characteristic of the building, such as too few bathrooms for the number of bedrooms, or some excess, such as a 10 foot ceiling in a residence. This obsolescence may be curable. The tables do not include functional obsolescence considerations.

Economic obsolescence is caused by factors not directly concerning the structure, but rather, by adverse environmental factors, resulting in loss of desirability. Examples include the obsolescence of a store in an area of declining economic activity, or obsolescence resulting from governmental regulation changing the zoning of an area. Because this kind of obsolescence is particularly difficult to measure, it is not included in the tables.

''Effective age'' considers all forms of depreciation. It may be less than chronological age, if recently remodeled or improved, or more than the actual age, if deterioration is particularly bad. Though effective age is not considered in the physical life tables, it may yield a better picture of a structure's life than the actual physical age. Once the effective age is determined, considering physical, functional and economic deterioration, use the percent good tables on pages 40 or 200 to determine the present value of a depreciated building. Present value is the result of multiplying the replacement cost (found by using the cost tables) by the appropriate percent good.

Limitations

This manual will be a useful reference for anyone who has to develop budget estimates or replacement costs for buildings. Anyone familiar with construction estimating understands that even very competent estimators with complete working drawings, full specifications and precise labor and material costs can disagree on the cost of a building. Frequently exhaustive estimates for even relatively simple structures can vary 5% or more. The range of competitive bids on some building projects is as much as 10%. Estimating costs is not an exact science and there is room for legitimate disagreement on what the ''right'' cost is. This manual can **not** help you do in a few minutes what skilled estimators may not be able to do in several hours. This manual **will** help you determine a reasonable replacement or construction cost for most buildings. It is not intended as a substitute for judgment or as a replacement for sound professional practice, but should prove a valuable aid to developing an informed opinion of value.

Location Factors

The cost factors below are geographical adjustments for the structure costs in this manual. Multiply the factor listed by the building costs to find the cost for any area listed.

Alabama	**.86**
Birmingham	.88
Dothan	.85
Florence	.86
Huntsville	.85
Mobile	.92
Montgomery	.85
Tuscaloosa	.85
Alaska	**1.48**
Anchorage	1.48
Fairbanks	1.57
Juneau	1.44
Ketchikan	1.44
Sitka	1.47
Arizona	**.98**
Casa Grande	.99
Douglas	.98
Flagstaff	.99
Kingman	.99
Phoenix	.95
Safford	1.00
Tucson	1.02
Yuma	.94
Arkansas	**.85**
Fayetteville	.85
Fort Smith	.85
Jonesboro	.83
Little Rock	.88
Texarkana	.85
California	**1.12**
Arrowhead	1.17
Bakersfield	1.13
Barstow	1.15
Big Bear	1.17
El Cajon	1.11
Eureka	1.15
Fresno	1.07
Inoykern	1.18
Lancaster	1.15
Los Angeles	1.14
Modesto	1.05
Mojave	1.17
Needles	1.07
Oakland	1.09
Ojai	1.07
Oxnard	1.13
Paso Robles	1.13
Piru	1.07
Placerville	1.07
Redding	1.06
Ridgecrest	1.12
Sacramento	1.05
San Bernardino	1.13
San Diego	1.11
San Francisco	1.28
San Jose	1.11
Santa Ana	1.14
Santa Barbara	1.15
Santa Maria	1.15
Santa Rosa	1.15
South Lake Tahoe	1.10
Tehachapi	1.15
Ventura	1.13
Victorville	1.15
Yreka	1.06
Colorado	**.95**
Colorado Springs	.92
Denver	.94
Durango	1.00
Grand Junction	.94
Greeley	.95
Pueblo	.94
Tierra Amarilla	.93
Connecticut	**.96**
Bridgeport	1.11
Hartford	1.09
New Haven	1.08
New London	1.10
New Milford	1.08
Norwich	1.09
Windham	1.10
Delaware	**1.00**
Dover	1.00
Wilmington	1.01
Florida	**.87**
Jacksonville	.83
Key West	.91
Miami	.91
Orlando	.88
Pensacola	.81
Tampa	.88
Georgia	**.83**
Albany	.85
Atlanta	.86
Augusta	.84
Brunswick	.81
Columbus	.84
Macon	.84
Rome	.82
Savannah	.81
Valdosta	.83
Guam	**1.39**
Agana	1.39
Hawaii	**1.66**
Hilo	1.67
Honolulu	1.52
Kauai	1.73
Kona	1.71
Maui	1.66
Idaho	**1.05**
Boise	1.02
Idaho Falls	1.05
Lewiston	1.05
McCall	1.05
Pocatello	1.06
Twin Falls	1.05
Illinois	**1.01**
Belleville	1.02
Chicago	1.19
East St. Louis	1.02
Moline	1.01
Springfield	.99
Indiana	**.99**
Bloomington	.92
Evansville	.92
Fort Wayne	.90
Gary	1.00
Hammond	1.03
Indianapolis	1.10
Lafayette	1.10
South Bend	.96
Terre Haute	1.02
Iowa	**.96**
Bettendorf	.96
Council Bluffs	.93
Davenport	.97
Des Moines	.95
Dubuque	.96
Mason City	.96
Sioux City	.96
Waterloo	.96
Kansas	**.91**
Garden City	.89
Kansas City	.96
Pittsburg	.87
Salina	.88
Topeka	.95
Wichita	.91
Kentucky	**.95**
Ashland	.97
Bowling Green	.93
Corbin	.95
Covington	.98
Frankfort	.92
Hopkinsville	.92
Lexington	.95
Louisville	.95
Middlesboro	1.03
Murray	.93
Newport	.97
Owensboro	.95
Paducah	.93
Louisiana	**.91**
Alexandria	.85
Baton Rouge	.95
Buras	.93
Houma	.91
Lafayette	.92
Lake Charles	.96
Marshall	.84
Monroe	.86
New Orleans	.93
Shreveport	.91
Maine	**1.03**
Augusta	1.05
Bangor	1.01
Brunswick	1.04
Lewiston	1.03
Portland	1.04
Waterville	1.02
Maryland	**.95**
Baltimore (City)	.99
Baltimore (Suburban)	.95
Hagerstown	.94
Salisbury	.96
Waldorf	.93
Massachusetts	**1.11**
Boston	1.14
Fall River	1.10
Worchester	1.11
Michigan	**1.00**
Ann Arbor	1.00
Battle Creek	.98
Benton Harbor	1.04
Detroit	.96
Flint	1.00
Grand Rapids	.96
Jackson	1.03
Lansing	1.07
Marquette	1.05
Mt. Pleasant	1.00
Muskegon	.94
Saginaw	.96
Traverse City	1.03
Ypsilanti	1.02
Minnesota	**1.00**
Duluth	1.05
Mankato	1.00
Minneapolis	1.03
Rochester	.99
St. Cloud	.98
Worthington	.94
Mississippi	**.86**
Corinth	.87
Greenville	.85
Greenwood	.86
Gulfport	.86
Hattiesburg	.86
Jackson	.86
Southaven	.86
Missouri	**.91**
Cape Girardeau	.94
Columbia	.96
Joplin	.89
Kansas City	.96
Kirksville	.96
Rolla	.88
Sedalia	.92
Springfield	.91
St. Joseph	.92
St. Louis	.99
Montana	**.96**
Billings	.96
Bozeman	.92
Butte	1.01
Glasgow	.94
Great Falls	1.01
Havre	.94
Helena	1.01
Kalispell	.94
Miles City	.94
Missoula	.94
Nebraska	**.97**
Grand Island	1.01
Lincoln	.95
Norfolk	.99
North Platte	.93
Omaha	.97
Scottsbluff	.97
Nevada	**1.08**
Las Vegas	1.06
Reno	1.11
New Hampshire	**1.04**
Concord	1.07
Dover	1.01
Keene	1.03
Manchester	1.03
Nashua	1.04
Portsmouth	1.05
New Jersey	**1.12**
Asbury Park	1.17
Atlantic City	1.06
Burlington	1.09

Location Factors

Camden	1.09
Freehold	1.17
Gloucester	1.07
Newark	1.19
North Bergen	1.19
Trenton	1.12
Vineland	1.10
New Mexico	**.91**
Alamogordo	.87
Albuquerque	.89
Artesia	.88
Carlsbad	.88
Clovis	.87
Farmington	.94
Fort Sumner	.91
Gallup	.94
Hobbs	.87
Las Cruces	.87
Las Vegas	.91
Los Alamos	.93
Raton	.91
Ruidoso	.93
Santa Fe	.91
Silver City	.93
Socorro	.87
Taos	.99
New York	**.98**
Albany	.98
Binghamton	.98
Buffalo	.98
Elmira	1.02
Jamestown	.95
Nassau County	1.02
New York City (Inner)	1.23
New York City (Metro)	1.06
Orange County	.87
Plattsburgh	.93
Poughkeepsie	.97
Rochester	.96
Rockland County	.88
Suffolk County	.91
Syracuse	.98
Westchester County	.96
North Carolina	**.84**
Asheville	.87
Charlotte	.86
Durham	.83
Elizabeth City	.87
Fayetteville	.82
Greensboro	.84
Greenville	.82
Raleigh	.83
Wilmington	.83
Winston-Salem	.84
North Dakota	**1.00**
Bismark	1.03
Dickinson	1.01
Fargo	.96
Grand Forks	1.00
Minot	1.00
Ohio	**1.04**
Akron	1.07
Athens	1.00
Cincinnati	1.16
Cleveland	1.08
Columbus	1.01
Dayton	1.17
Findlay	.98
Lima	.99
Lorain	1.07
Mansfield	1.00
Newark	.98
Springfield	1.00
Toledo	1.08
Troy	.97
Youngstown	1.04
Zanesville	1.02
Oklahoma	**.89**
Ada	.89
Ardmore	.92
Bartlesville	.90
Enid	.89
Guymon	.92
Lawton	.86
McAlester	.88
Muskogee	.92
Oklahoma City	.86
Shawnee	.89
Stillwater	.89
Tulsa	.88
Woodward	.90
Oregon	**1.09**
Bend	1.07
Coos Bay	1.10
Eugene	1.03
Medford	1.06
Ontario	1.15
Pendleton	1.16
Portland	1.09
West Salem	1.06
Pennsylvania	**1.04**
Allentown	1.03
Altoona	1.03
Bellefonte	1.05
Erie	1.05
Harrisburg	1.03
Johnstown	1.03
Lancaster	1.01
Philadelphia	1.10
Pittsburg	1.07
Reading	1.02
Scranton	1.02
Wellsboro	1.14
York	1.01
Puerto Rico	**.87**
Arecibo	.84
Mayaguez	.84
Old San Juan	1.00
Ponce	.84
San Juan	.84
Rhode Island	**1.12**
Providence	1.12
South Carolina	**.87**
Aiken	.87
Anderson	.85
Beaufort	.88
Charleston	.93
Columbia	.86
Florence	.84
Greenville	.86
Greenwood	.87
Myrtle Beach	.88
North Augusta	.91
Orangeburg	.86
Rock Hill	.87
Spartanburg	.88
South Dakota	**1.04**
Aberdeen	1.02
Pierre	1.06
Rapid City	1.05
Sioux Falls	1.02
Tennessee	**.88**
Chattanooga	.90
Clarksville	.82
Columbia	.86
Jackson	.96
Johnson City	.84
Kingsport	.84
Knoxville	.86
Memphis	.91
Nashville	.84
Oak Ridge	.85
Union City	.97
Texas	**.87**
Abilene	.90
Amarillo	.88
Austin	.84
Beaumont	.92
Bryan	.94
Corpus Christi	.84
Dallas	.87
Del Rio	.82
Eagle Pass	.85
El Campo	.92
El Paso	.88
Fort Worth	.89
Harlingen	.83
Houston	.90
Junction	.84
Laredo	.83
Lubbock	.85
Lufkin	.94
Midland	.83
Odessa	.83
San Angelo	.91
San Antonio	.85
Sherman	.87
Texas City	.91
Tyler	.86
Victoria	.84
Waco	.86
Wichita Falls	.93
Utah	**.95**
Cedar City	.98
Logan	.94
Moab	.97
Provo	.92
Salt Lake City	.92
Vernal	.95
Vermont	**1.03**
Bennington	1.04
Brattleboro	1.04
Burlington	1.04
Montpelier	1.02
Rutland	1.03
Virgin Islands	**.98**
St. Croix	.97
St. Thomas	.99
Virginia	**.85**
Bristol	.85
Charlottesville	.96
Chesapeake	.84
Danville	.82
Eastern Shore	.82
Fredericksburg	.85
Hampton	.85
Harrisonburg	.83
Lynchburg	.81
Newport News	.85
Norfolk	.86
Portsmouth	.84
Richmond	.96
Roanoke	.82
Suffolk	.82
Virginia Beach	.84
Warrenton	.85
Washington	**1.08**
Aberdeen	1.08
Bellingham	1.09
Cheney	1.03
Kennewick	1.10
Longview	1.09
Olympia	1.09
Port Angeles	1.09
Pullman	1.07
Seattle	1.08
Spokane	1.01
Yakima	1.12
Washington, D.C.	**1.04**
Washington, D.C.	1.04
West Virginia	**.94**
Beckley	.94
Bluefield	.94
Charleston	.97
Fairmount	.95
Huntington	.95
Martinsburg	.90
Parkersburg	.96
Point Pleasant	.92
Wisconsin	**1.06**
Eau Claire	1.07
Green Bay	1.01
Madison	1.07
Milwaukee	1.09
Reedsville	1.03
Superior	1.11
Wausau	1.03
Wyoming	**.97**
Casper	.97
Cheyenne	.93
Cody	1.00
Gillette	.98
Jackson	1.00
Laramie	.94
Powell	1.00
Riverton	.96
Rock Springs	.99

Building Cost Historical Index

Use this table to find the approximate current dollar building cost when the actual cost is known for any year since 1921. Multiply the figure listed for the building type and year by the known cost. The result is the estimated 1981 construction cost.

Year	Masonry Buildings	Concrete Buildings	Steel Buildings	Wood-Frame Buildings	Agricultural Buildings	Year
1921	11.25	9.58	9.93	10.08	10.19	1921
1922	11.59	9.79	11.55	10.30	10.29	1922
1923	12.25	0.44	10.17	9.50	9.71	1923
1924	11.59	11.05	12.43	10.30	10.40	1924
1925	11.83	9.65	12.72	10.74	10.73	1925
1926	11.98	11.71	11.74	10.62	10.62	1926
1927	12.22	11.94	12.89	10.74	10.85	1927
1928	12.35	11.94	13.03	10.84	10.95	1928
1929	12.22	11.83	13.18	10.86	10.62	1929
1930	12.50	12.45	12.72	11.23	11.08	1930
1931	13.22	13.52	13.65	12.06	12.31	1931
1932	15.00	14.89	15.26	13.01	13.44	1932
1933	14.40	14.55	13.83	12.68	12.92	1933
1934	13.40	13.19	13.34	12.06	12.45	1934
1935	13.40	12.87	13.17	11.76	12.31	1935
1936	13.25	12.74	13.17	11.37	11.58	1936
1937	11.83	11.71	11.43	10.30	10.51	1937
1938	11.49	11.66	11.35	10.30	10.40	1938
1939	11.36	11.60	11.21	9.99	10.40	1939
1940	11.25	11.46	10.92	9.88	10.10	1940
1941	10.51	10.61	10.46	8.98	9.34	1941
1942	9.86	9.87	9.82	8.30	8.40	1942
1943	9.45	9.46	9.39	7.91	7.89	1943
1944	8.93	9.17	9.03	7.26	7.54	1944
1945	8.46	8.82	8.75	6.85	7.09	1945
1946	7.25	7.74	7.76	6.12	6.11	1946
1947	6.17	6.36	6.79	4.92	4.96	1947
1948	5.40	5.46	6.04	4.49	4.49	1948
1949	5.44	5.40	6.01	4.56	4.66	1949
1950	5.16	5.15	5.90	4.35	4.33	1950
1951	4.82	4.85	5.34	4.05	4.02	1951
1952	4.75	4.75	5.24	3.99	3.98	1952
1953	4.59	4.58	5.01	3.90	3.89	1953
1954	4.50	4.44	4.94	3.84	3.88	1954
1955	4.33	4.23	4.73	3.68	3.72	1955
1956	4.10	4.03	4.35	3.53	3.55	1956
1957	3.96	3.87	4.16	3.50	3.49	1957
1958	3.85	3.73	3.97	3.48	3.52	1958
1959	3.74	3.60	3.88	3.37	3.32	1959
1960	3.65	3.56	3.79	3.29	3.25	1960
1961	3.58	3.53	3.76	3.23	3.22	1961
1962	3.49	3.42	3.66	3.18	3.18	1962
1963	3.44	3.35	3.60	3.13	2.90	1963
1964	3.34	3.28	3.55	3.02	3.04	1964
1965	3.26	3.22	3.43	2.97	2.95	1965
1966	3.09	3.11	3.29	2.83	2.87	1966
1967	3.01	2.99	3.09	2.69	2.74	1967
1968	2.88	2.80	2.94	2.52	2.62	1968
1969	2.72	2.68	2.84	2.43	2.47	1969
1970	2.60	2.55	2.70	2.31	2.36	1970
1971	2.41	2.35	2.49	2.16	2.20	1971
1972	2.25	2.16	2.33	2.00	2.03	1972
1973	2.07	2.04	2.06	1.84	1.91	1973
1974	1.85	1.86	1.91	1.72	1.77	1974
1975	1.68	1.65	1.72	1.61	1.55	1975
1976	1.57	1.58	1.65	1.54	1.79	1976
1977	1.45	1.47	1.54	1.43	1.39	1977
1978	1.35	1.37	1.42	1.31	1.27	1978
1979	1.23	1.24	1.26	1.18	1.17	1979
1980	1.10	1.11	1.12	1.08	1.07	1980
1981 (estimated)	1.00	1.00	1.00	1.00	1.00	1981

RESIDENTIAL STRUCTURES SECTION

Section Contents

Structure Type	Page
Single Family Residences	10
Multi-Family Residences	16
Motels	20
Additional Costs for Residential Structures	20
Cabins and Recreational Dwellings	29
Additional Costs for Recreational Dwellings	39
Typical Lives for Residential Structures	40

The figures in this section include all costs associated with normal construction:

Foundations as required for normal soil conditions. Excavation for foundations, piers, and other foundation components given a fairly level construction site. Floor, wall, and roof structures. Interior floor, wall, and ceiling finishes. Exterior wall finish and roof cover. Interior partitions as described in the quality class. Finish carpentry, doors, windows, trim, etc. Electric wiring and fixtures. Rough and finish plumbing as described in applicable building specifications. Built-in appliances as described in applicable building specifications. All labor and materials including supervision. All design and engineering fees, if necessary. Permits and fees. Utility hook-ups. Contractors' contingency, overhead and profit.

The square foot costs do not include heating and cooling equipment or the items listed in the section "Additional Costs for Residential Structures" which appears on pages 24 to 28. The costs of the following should be figured separately and added to the basic structure cost: porches, basements, balconies, exterior stairways, built-in equipment beyond that listed in the quality classifications, fireplaces, garages and carports.

Single Family Residences

Single family residential structures vary widely in quality and the quality of construction is a significant factor in the finished cost. For estimating purposes the structure should be placed in one or more classes from 1 to 4. Class 1 construction is top quality work and Class 4 is the minimum required under most building codes. Many structures do not fall into a single class and have features of two quality classes. The tables have 3 "half classes" which apply to structures which have some features of two quality classes. Some structures may fall into two distinct classes and the square foot area of each part should be computed separately.

The shape of the outside perimeter is an important consideration in estimating the total construction cost. Generally, the more complex the shape, the more expensive the structure per square foot of floor area. The shape classification of multiple story or split level homes is based on the outline formed by the outer-most exterior walls, including the garage area, regardless of the varying level. Most structures have 4, 6, 8 or 10 corners, as illustrated in the example. Small insets not requiring a change in the roof shape can be ignored when determining the shape.

These figures can be applied to nearly all single family dwellings built using conventional methods and common materials. There is no simplified method of evaluating the cost of the relatively small number of highly decorative, starkly original or exceptionally well-appointed residences. It is safe to assume, however, that the cost of these highly customized homes will exceed the figures that appear on the following pages.

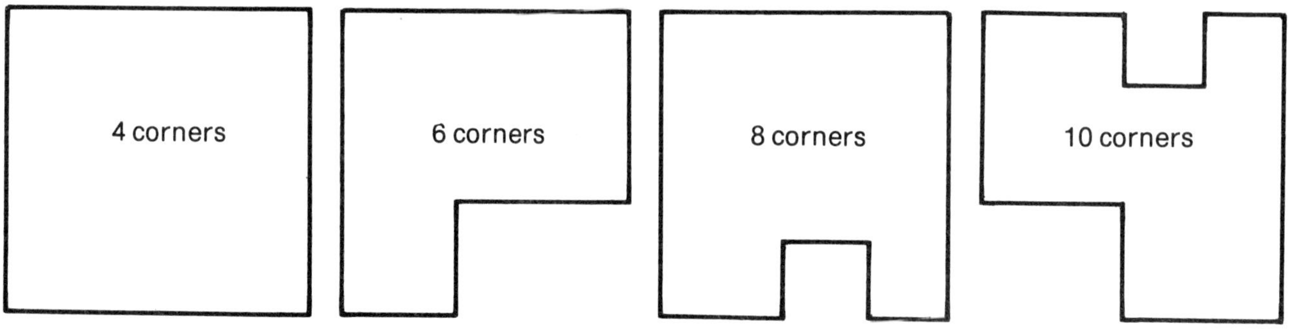

Single Family Residences Quality Classification

	Class 1 Best Quality	Class 2 Good Quality	Class 3 Average Quality	Class 4 Low Quality
Exterior Finish	Good quality wood siding, stucco or masonry, with good trim.	Good quality hardboard or plywood siding, some masonry veneer or stucco.	Hardwood or shingle siding or minimum quality stucco.	Composition siding or minimum quality stucco.
Windows	Good quality wood casement or aluminum.	Average quality aluminum or wood.	Minimum quality wood or aluminum.	Aluminum sliding.
Roofing	Heavy shake, slate or tile with a 24" to 36" overhang.	Shingle or built-up with rock cover. Less than 24" overhang.	Minimum shingle or built-up with rock cover or good quality composition roof.	Minimum quality built-up with gravel surface or composition shingle. Simple or small open cornice.
Flooring	Terrazzo or tile entry. Carpet or good quality hardwood in most rooms. Sheet vinyl in kitchen, family and breakfast rooms.	Average quality tile or hardwood entry. Average quality hardwood or carpet in most rooms. Resilient tile in kitchen, family and breakfast rooms.	3/8" hardwood or low cost carpet in most rooms, linoleum or vinyl-asbestos tile in remaining rooms.	Asphalt or vinyl-asbestos tile in most rooms, some low cost carpet.
Interior Finish	Gypsum board with heavy texture or plaster. Some wallpaper or vinyl covered walls or wood paneling.	Taped and textured gypsum wallboard, some wallpaper and plywood paneling in family room.	Taped and textured gypsum wallboard, some wallpaper.	Minimum quality wallboard.
Bathrooms	2½ bathrooms, tile floor, ceramic tile shower over tub, glass shower door, 4' to 6' tile or cultured marble pullman in 2 baths.	2 bathrooms, sheet vinyl floor, some average quality tile, 4' to 6' tile or laminated plastic pullman in each bath.	2 bathrooms, usually back to back, resilient tile floor, plastic coated hardboard or fiberglass around shower and over tub. 4' laminated plastic pullman in each bath.	2 bathrooms back to back, vinyl-asbestos tile, plastic shower door, fiberglass enclosures.
Kitchen	20 linear feet of hardwood veneer base cabinets, 20 linear feet of ceramic tile or good quality laminated plastic counter. Built-in oven, range, dishwasher, garbage disposer, range hood with fan, all good quality.	16 linear feet of average quality hardwood veneer cabinets and ceramic tile or good quality laminated plastic counter. Built-in oven, range, dishwasher, garbage disposer, range hood and fan, all average quality.	12 linear feet of low cost hardwood veneer cabinets and laminated plastic counter. Low cost built-in oven, range, dishwasher, garbage disposer and range hood.	8 linear feet of low cost cabinets with laminated plastic counter. Few built-in appliances.
Plumbing	10 to 12 good quality fixtures, copper tubing supply and waste.	8 good quality fixtures.	7 average quality fixtures.	7 low cost fixtures, plastic supply.

Single Family Residences
4 Corners

Estimating Procedure

1. Establish the structure quality class by applying the information on page 11.
2. Multiply the structure floor area (excluding the garage) by the appropriate square foot cost below.
3. Multiply the total from step 2 by the correct location factor listed on page 7.
4. Add, when appropriate, the cost of a porch, garage, heating and cooling equipment, basement, fireplace, carport, appliances and plumbing fixtures beyond that listed in the quality classification. See the cost of these items on pages 24 to 26.

Note

Tract work and highly repetitive jobs may reduce the cost 8 to 12 percent. Add 4 percent to the square foot cost of floors above the second floor level. Work outside metropolitan areas may cost 2 to 6 percent less. When the exterior walls are masonry, add 8 to 9 percent for class 2 and 1 structures and 6 to 7 percent for class 4 and 3 structures.

The building area includes all full story (7' 6'' to 8' high) areas within and including the exterior walls of all floor areas of the building, including small inset areas such as entrances outside the exterior wall but under the main roof. For areas with a ceiling height of less than 80'', see the section on half-story areas on page 27.

Single Family Residence, Class 3

Single Family Residence, Class 4

Square Foot Area

Quality Class	700	800	900	1,000	1,100	1,200	1,300	1,400	1,500	1,600	1,700
1, Best	51.31	48.99	47.11	45.57	44.27	43.15	42.18	41.33	40.58	39.90	39.29
1 & 2	47.95	45.78	44.03	42.58	41.37	40.32	39.42	38.62	37.92	37.29	36.72
2, Good	44.01	42.02	40.41	39.09	37.97	37.01	36.18	35.45	34.80	34.22	33.70
2 & 3	40.34	38.51	37.04	35.82	34.80	33.92	33.16	32.49	31.90	31.37	30.89
3, Average	36.73	35.07	33.73	32.62	31.69	30.89	30.20	29.59	29.05	28.57	28.13
3 & 4	33.16	31.66	30.45	29.45	28.61	27.89	27.26	26.71	26.22	25.79	25.39
4, Low	29.89	28.54	27.45	26.55	25.79	25.14	24.58	24.08	23.64	23.25	22.89

Square Foot Area

Quality Class	1,800	2,000	2,200	2,400	2,600	2,800	3,000	3,200	3,400	3,600	4,000
1, Best	40.82	39.89	39.11	38.46	37.91	37.43	37.01	36.64	36.31	36.01	35.50
1 & 2	36.28	35.45	34.76	34.18	33.69	33.26	32.89	32.56	32.27	32.01	31.55
2, Good	32.49	31.75	31.13	30.61	30.17	29.79	29.46	29.16	28.90	28.66	28.26
2 & 3	29.88	29.20	28.63	28.16	27.75	27.40	27.09	26.82	26.58	26.36	25.99
3, Average	27.42	26.80	26.28	25.84	25.47	25.14	24.86	24.61	24.39	24.19	23.85
3 & 4	25.01	24.43	23.96	23.56	23.22	22.93	22.67	22.44	22.24	22.06	21.75
4, Low	22.90	22.37	21.94	21.57	21.26	20.99	20.76	20.55	20.36	20.20	19.91

Single Family Residences
6 Corners

Estimating Procedure

1. Establish the structure quality class by applying the information on page 11.
2. Multiply the structure floor area (excluding the garage) by the appropriate square foot cost below.
3. Multiply the total from step 2 by the correct location factor listed on page 7.
4. Add, when appropriate, the cost of a porch, garage, heating and cooling equipment, basement, fireplace, carport, appliances and plumbing fixtures beyond that listed in the quality classification. See the cost of these items on pages 24 to 26.

Note

Tract work and highly repetitive jobs may reduce the cost 8 to 12 percent. Add 4 percent to the square foot cost of floors above the second floor level. Work outside metropolitan areas may cost 2 to 6 percent less. When the exterior walls are masonry, add 8 to 9 percent for class 2 and 1 structures and 6 to 7 percent for class 4 and 3 structures.

The building area includes all full story (7' 6'' to 8' high) areas within and including the exterior walls of all floor areas of the building, including small inset areas such as entrances outside the exterior wall but under the main roof. For areas with a ceiling height of less than 80'', see the section on half-story areas on page 27.

Single Family Residence, Class 4

Single Family Residence, Class 2

Square Foot Area

Quality Class	700	800	900	1,000	1,100	1,200	1,300	1,400	1,500	1,600	1,700
1, Best	52.88	50.41	48.44	46.83	45.48	44.34	43.35	42.49	41.74	41.07	40.47
1 & 2	48.99	46.70	44.87	43.38	42.13	41.08	40.16	39.37	38.67	38.05	37.49
2, Good	45.05	42.95	41.27	39.89	38.75	37.77	36.94	36.20	35.56	34.99	34.48
2 & 3	41.24	39.31	37.77	36.52	35.47	34.58	33.81	33.14	32.55	32.03	31.56
3, Average	37.62	35.86	34.46	33.31	32.36	31.54	30.84	30.23	29.70	29.22	28.79
3 & 4	33.92	32.33	31.07	30.04	29.17	28.44	27.81	27.26	26.77	26.34	25.96
4, Low	30.58	29.15	28.01	27.08	26.30	25.64	25.07	24.57	24.13	23.75	23.40

Square Foot Area

Quality Class	1,800	2,000	2,200	2,400	2,600	2,800	3,000	3,200	3,400	3,600	4,000
1, Best	41.77	40.82	40.04	39.38	38.81	38.33	37.91	37.54	37.21	36.91	36.41
1 & 2	37.12	36.27	35.57	34.99	34.49	34.06	33.68	33.35	33.06	32.80	32.35
2, Good	33.32	32.56	31.93	31.41	30.96	30.57	30.24	29.94	29.68	29.44	29.04
2 & 3	30.69	29.99	29.41	28.93	28.52	28.16	27.85	27.58	27.34	27.12	26.75
3, Average	28.11	27.47	26.94	26.50	26.12	25.79	25.51	25.26	25.04	24.84	24.50
3 & 4	25.62	25.04	24.56	24.15	23.81	23.51	23.25	23.03	22.82	22.64	22.33
4, Low	23.46	22.92	22.48	22.11	21.79	21.52	21.28	21.08	20.89	20.73	20.44

Single Family Residences
8 Corners

Estimating Procedure

1. Establish the structure quality class by applying the information on page 11.
2. Multiply the structure floor area (excluding the garage) by the appropriate square foot cost below.
3. Multiply the total from step 2 by the correct location factor listed on page 7.
4. Add, when appropriate, the cost of a porch, garage, heating and cooling equipment, basement, fireplace, carport, appliances and plumbing fixtures beyond that listed in the quality classification. See the cost of these items on pages 24 to 26.

Note

Tract work and highly repetitive jobs may reduce the cost 8 to 12 percent. Add 4 percent to the square foot cost of floors above the second floor level. Work outside metropolitan areas may cost 2 to 6 percent less. When the exterior walls are masonry, add 8 to 9 percent for class 2 and 1 structures and 6 to 7 percent for class 4 and 3 structures.

The building area includes all full story (7' 6'' to 8' high) areas within and including the exterior walls of all floor areas of the building, including small inset areas such as entrances outside the exterior wall but under the main roof. For areas with a ceiling height of less than 80'', see the section on half-story areas on page 27.

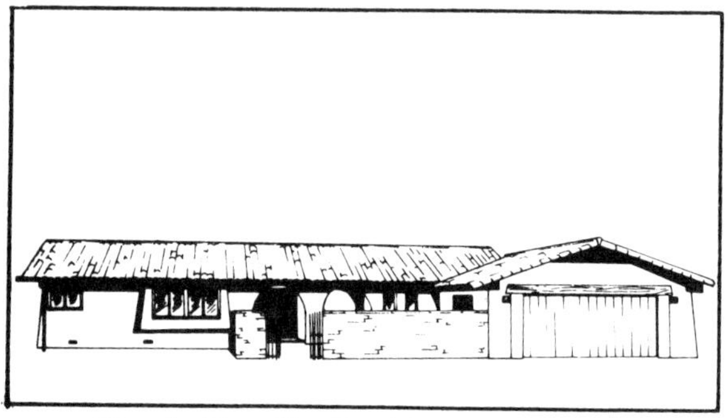

Single Family Residence, Class 1

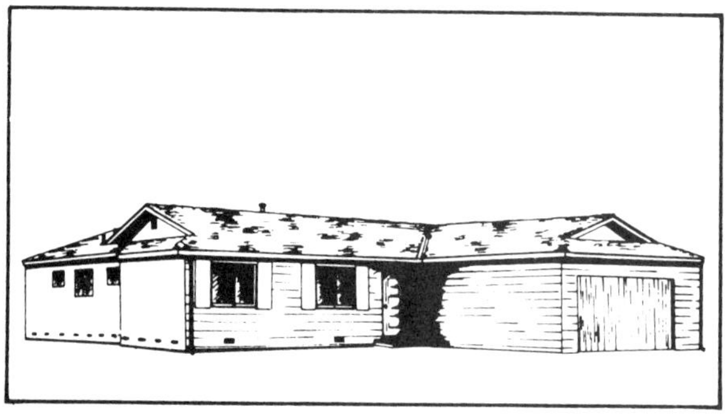

Single Family Residence, Class 3

Square Foot Area

Quality Class	700	800	900	1,000	1,100	1,200	1,300	1,400	1,500	1,600	1,700
1, Best	52.92	50.62	48.76	47.23	45.93	44.82	43.86	43.02	42.27	41.60	40.99
1 & 2	49.41	47.26	45.53	44.10	42.89	41.85	40.95	40.16	39.47	38.84	38.28
2, Good	45.47	43.49	41.89	40.57	39.46	38.51	37.68	36.96	36.31	35.74	35.22
2 & 3	41.67	39.86	38.39	37.18	36.17	35.29	34.54	33.87	33.28	32.75	32.28
3, Average	37.99	36.34	35.00	33.90	32.97	32.18	31.49	30.88	30.34	29.86	29.43
3 & 4	34.26	32.77	31.56	30.57	29.73	29.02	28.39	27.85	27.36	26.93	26.54
4, Low	30.87	29.53	28.44	27.55	26.79	26.15	25.59	25.09	24.66	24.27	23.91

Square Foot Area

Quality Class	1,800	2,000	2,200	2,400	2,600	2,800	3,000	3,200	3,400	3,600	4,000
1, Best	42.69	41.73	40.95	40.29	39.73	39.25	38.83	38.46	38.14	37.85	37.35
1 & 2	37.93	37.09	36.39	35.81	35.31	34.88	34.51	34.18	33.89	33.63	33.19
2, Good	34.10	33.34	32.72	32.19	31.75	31.36	31.03	30.73	30.47	30.24	29.84
2 & 3	31.37	30.67	30.10	29.62	29.20	28.85	28.54	28.27	28.03	27.82	27.45
3, Average	28.75	28.11	27.58	27.14	26.76	26.44	26.16	25.91	25.69	25.49	25.16
3 & 4	26.21	25.63	25.14	24.74	24.40	24.10	23.85	23.62	23.42	23.24	22.93
4, Low	23.96	23.42	22.98	22.61	22.30	22.03	21.80	21.59	21.40	21.24	20.96

Single Family Residences
10 Corners

Estimating Procedure

1. Establish the structure quality class by applying the information on page 11.
2. Multiply the structure floor area (excluding the garage) by the appropriate square foot cost below.
3. Multiply the total from step 2 by the correct location factor listed on page 7.
4. Add, when appropriate, the cost of a porch, garage, heating and cooling equipment, basement, fireplace, carport, appliances and plumbing fixtures beyond that listed in the quality classification. See the cost of these items on pages 24 to 26.

Note

Tract work and highly repetitive jobs may reduce the cost 8 to 12 percent. Add 4 percent to the square foot cost of floors above the second floor level. Work outside metropolitan areas may cost 2 to 6 percent less. When the exterior walls are masonry, add 8 to 9 percent for class 2 and 1 structures and 6 to 7 percent for class 4 and 3 structures.

The building area includes all full story (7' 6'' to 8' high) areas within and including the exterior walls of all floor areas of the building, including small inset areas such as entrances outside the exterior wall but under the main roof. For areas with a ceiling height of less than 80'', see the section on half-story areas on page 27.

Single Family Residence, Class 1

Single Family Residence, Class 2

Square Foot Area

Quality Class	700	800	900	1,000	1,100	1,200	1,300	1,400	1,500	1,600	1,700
1, Best	54.64	52.16	50.19	48.58	47.24	46.10	45.12	44.26	43.50	42.83	42.23
1 & 2	50.65	48.36	46.53	45.04	43.79	42.74	41.83	41.03	40.33	39.71	39.15
2, Good	46.63	44.52	42.84	41.46	40.31	39.34	38.50	37.77	37.13	36.55	36.04
2 & 3	42.69	40.76	39.22	37.96	36.91	36.02	35.25	34.58	33.99	33.47	33.00
3, Average	38.90	37.14	35.74	34.59	33.64	32.82	32.12	31.51	30.98	30.50	30.07
3 & 4	35.09	33.51	32.24	31.21	30.34	29.61	28.98	28.43	27.94	27.51	27.13
4, Low	31.62	30.19	29.05	28.12	27.34	26.68	26.11	25.61	25.18	24.79	24.44

Square Foot Area

Quality Class	1,800	2,000	2,200	2,400	2,600	2,800	3,000	3,200	3,400	3,600	4,000
1, Best	43.62	42.67	41.90	41.24	40.68	40.19	39.76	39.39	39.05	38.75	38.23
1 & 2	38.75	37.92	37.23	36.64	36.14	35.71	35.33	35.00	34.70	34.43	33.97
2, Good	34.83	34.08	33.45	32.93	32.48	32.09	31.75	31.45	31.18	30.94	30.53
2 & 3	32.10	31.40	30.83	30.35	29.93	29.58	29.26	28.98	28.74	28.52	28.14
3, Average	29.40	28.77	28.24	27.80	27.42	27.09	26.81	26.55	26.33	26.12	25.78
3 & 4	26.79	26.22	25.74	25.33	24.99	24.69	24.43	24.20	23.99	23.81	23.49
4, Low	24.50	23.97	23.53	23.16	22.85	22.58	22.34	22.12	21.94	21.77	21.48

Multi-Family Residences Quality Classification

	Class 1 Best Quality	Class 2 Good Quality	Class 3 Average Quality	Class 4 Low Quality
Framing*	Wood frame.	Wood frame.	Wood frame.	Wood frame.
Exterior Walls	Good wood siding or stucco. Masonry veneer on front.	Good quality stucco, good hardboard, plywood or board, masonry veneer on front.	Stucco, hardboard, low cost plywood or low cost shingle.	Light stucco or hardboard.
Windows	Large top quality wood or aluminum.	Average quality wood or aluminum.	Average quality wood or aluminum.	Few and smaller low cost aluminum or wood.
Roofing	Heavy shake or tile.	Good shingle or shake or built-up with heavy rock dress.	Wood or composition shingle or built-up with rock.	Asphalt shingles or built-up, light weight.
Overhang	36" open or 24" closed.	30", open.	16", open.	12" to 16" open.
Gutters	Good quality at all eaves.	6", at all eaves.	Over entrances.	Minimum or none.
Flooring	Terrazzo or tile entry. Good hardwood or carpet. Good sheet vinyl in kitchen.	Hardwood or average quality carpet. Sheet vinyl in kitchen.	Average quality hardwood or low cost carpet. Vinyl asbestos in kitchen.	Vinyl asbestos tile or linoleum.
Interior Walls	Heavy textured wallboard or plaster with putty finish. Good wallpaper or vinyl. Veneer hardwood paneling.	Gypsum board or plaster with good finish. Some wallpaper or paneling.	Gypsum board taped and textured. Some acoustical treatment or wallpaper.	Gypsum board taped and textured.
Trim	Good quality trim with some hardwood.	Painted softwood with some hardwood.	Painted softwood.	Painted softwood.
Closets	Ample, good doors.	Average space and doors.	Average space, low cost doors.	Minimum.
Kitchens	10' good cabinets 10' tile drainboard.	8' average quality cabinets, 8' ceramic tile or good plastic drainboard.	6' low cost cabinets and plastic or ceramic drainboard.	5' low cost cabinets and plastic drainboard.
Bathrooms	Sheet vinyl floor, tile over tub with glass shower door. 6' marble top vanity.	Sheet vinyl or inlaid tile, tile over tub with glass shower door. 4' plastic vanity.	Tile floor, plastic coated hardboard over tub. Glass or plastic shower door. 3' plastic vanity.	Tile floor, plastic coated hardboard over tub.
Plumbing (Costs assume 1 bathroom per unit. See page 27 for the cost of additional bathrooms).	Copper tubing, good quality fixtures.	Galvanized pipe, good quality fixtures.	Average fixtures.	Plastic pipe, low cost fixtures.
Special Features	8' sliding glass door, good quality built-in oven, range with hood, dishwasher.	Average quality disposer and range hood. Low cost dishwasher.	One or two low cost built-in appliances.	None.
***For Masonry Walls** Add 5% to the appropriate square foot cost.	Textured block, tile or brick with masonry facing.	Colored or detailed concrete block, brick or tile with facing.	Colored reinforced concrete block, clay tile or common brick.	Reinforced concrete block or clay tile.

Multi-Family Residences
2 or 3 Units

Estimating Procedure

1. Establish the structure quality class by applying the information on page 16.
2. Multiply the average unit area by the appropriate square foot cost listed below. The average unit area is found by dividing the building area on all floors by the number of units in the building. The building area should include office and utility rooms, interior hallways and interior stairways.
3. Multiply the total from step 2 by the correct location factor listed on page 7.
4. Add, when appropriate, the cost of balconies, porches, garages, heating and cooling equipment, basements, fireplaces, carports, appliances and plumbing fixtures beyond that listed in the quality classification. See the cost of these items on pages 24 to 28.
5. Costs assume one bathroom per unit. Add the cost of additional bathrooms from page 27.

Note

Work outside metropolitan areas may cost 2 to 6 percent less. Add 2 percent to the costs below for second floor areas and 4 percent for third floor areas. Add 5 percent when the exterior walls are masonry.

Multi-Family, Class 1

Multi-Family, Class 3 & 4

Average Unit Area in Square Feet

Quality Class	400	450	500	550	600	650	700	750	800	900	1,000
1, Best	49.50	47.74	46.37	45.28	44.40	43.67	43.06	42.54	42.10	41.40	40.86
1 & 2	45.83	44.20	42.93	41.92	41.11	40.43	39.87	39.39	38.98	38.33	37.83
2, Good	42.42	40.92	39.74	38.81	38.05	37.43	36.90	36.46	36.09	35.48	35.02
2 & 3	39.29	37.89	36.81	35.94	35.24	34.66	34.18	33.77	33.42	32.86	32.44
3, Average	36.38	35.09	34.08	33.28	32.63	32.09	31.65	31.27	30.94	30.43	30.03
3 & 4	33.46	32.27	31.35	30.61	30.01	29.52	29.11	28.76	28.46	27.99	27.62
4, Low	30.79	29.69	28.84	28.16	27.61	27.16	26.78	26.46	26.19	25.75	25.41

Average Unit Area in Square Feet

Quality Class	1,100	1,200	1,300	1,400	1,500	1,600	1,700	1,800	1,900	2,000	2,200
1, Best	40.29	39.85	39.48	39.17	38.92	38.69	38.50	38.33	38.19	38.06	37.84
1 & 2	37.31	36.90	36.57	36.28	36.04	35.84	35.66	35.50	35.37	35.25	35.05
2, Good	34.54	34.16	33.85	33.59	33.37	33.17	33.01	32.87	32.74	32.63	32.45
2 & 3	31.98	31.63	31.34	31.10	30.89	30.71	30.56	30.43	30.31	30.21	30.04
3, Average	29.61	29.29	29.02	28.79	28.60	28.44	28.30	28.18	28.07	27.97	27.82
3 & 4	27.24	26.94	26.69	26.49	26.31	26.16	26.03	25.92	25.82	25.73	25.59
4, Low	25.03	24.76	24.53	24.34	24.18	24.04	23.92	23.82	23.73	23.65	23.51

Multi-Family Residences
4 to 9 Units

Estimating Procedure

1. Establish the structure quality class by applying the information on page 16.
2. Multiply the average unit area by the appropriate square foot cost listed below. The average unit area is found by dividing the building area on all floors by the number of units in the building. The building area should include office and utility rooms, interior hallways and interior stairways.
3. Multiply the total from step 2 by the correct location factor listed on page 7.
4. Add, when appropriate, the cost of balconies, porches, garages, heating and cooling equipment, basements, fireplaces, carports, appliances and plumbing fixtures beyond that listed in the quality classification. See the cost of these items on pages 24 to 28.
5. Costs assume one bathroom per unit. Add the cost of additional bathrooms from page 27.

Note

Work outside metropolitan areas may cost 2 to 6 percent less. Add 2 percent to the costs below for second floor areas and 4 percent for third floor areas. Add 5 percent when the exterior walls are masonry.

Multi-Family, Class 2 & 3

Multi-Family, Class 3

Average Unit Area In Square Feet

Quality Class	400	450	500	550	600	650	700	750	800	900	1,000
1, Best	46.63	44.94	43.64	42.60	41.77	41.09	40.52	40.04	39.64	38.99	38.51
1 & 2	43.26	41.70	40.49	39.53	38.76	38.12	37.60	37.15	36.78	36.18	35.73
2, Good	40.06	38.61	37.49	36.60	35.89	35.30	34.81	34.40	34.05	33.50	33.08
2 & 3	37.09	35.75	34.71	33.89	33.23	32.68	32.23	31.85	31.53	31.02	30.63
3, Average	34.34	33.09	32.13	31.37	30.76	30.26	29.84	29.49	29.19	28.71	28.36
3 & 4	31.46	30.33	29.45	28.75	28.19	27.73	27.34	27.02	26.75	26.31	25.98
4, Low	29.08	28.03	27.21	26.57	26.05	25.62	25.27	24.97	24.72	24.32	24.01

Average Unit Area In Square Feet

Quality Class	1,100	1,200	1,300	1,400	1,500	1,600	1,700	1,800	1,900	2,000	2,200
1, Best	38.03	37.61	37.26	36.97	36.73	36.52	36.34	36.18	36.05	35.93	35.73
1 & 2	35.22	34.83	34.50	34.23	34.01	33.81	33.65	33.50	33.38	33.27	33.09
2, Good	32.60	32.23	31.94	31.69	31.48	31.30	31.14	31.01	30.89	30.79	30.62
2 & 3	30.18	29.85	29.57	29.34	29.15	28.98	28.84	28.71	28.61	28.51	28.36
3, Average	27.95	27.64	27.39	27.17	26.99	26.84	26.71	26.59	26.49	26.41	26.26
3 & 4	25.71	25.42	25.19	24.99	24.83	24.68	24.56	24.46	24.37	24.29	24.15
4, Low	23.65	23.39	23.17	22.99	22.84	22.71	22.60	22.50	22.42	22.34	22.22

Multi-Family Residences
10 or More Units

Estimating Procedure

1. Establish the structure quality class by applying the information on page 16.
2. Multiply the average unit area by the appropriate square foot cost listed below. The average unit area is found by dividing the building area on all floors by the number of units in the building. The building area should include office and utility rooms, interior hallways and interior stairways.
3. Multiply the total from step 2 by the correct location factor listed on page 7.
4. Add, when appropriate, the cost of balconies, porches, garages, heating and cooling equipment, basements, fireplaces, carports, appliances and plumbing fixtures beyond that listed in the quality classification. See the cost of these items on pages 24 to 28.
5. Costs assume one bathroom per unit. Add the cost of additional bathrooms from page 27.

Note

Work outside metropolitan areas may cost 2 to 6 percent less. Add 2 percent to the costs below for second floor areas and 4 percent for third floor areas. Add 5 percent when the exterior walls are masonry.

Multi-Family, Class 3

Multi-Family, Class 3 & 4

Average Unit Area in Square Feet

Quality Class	400	450	500	550	600	650	700	750	800	900	1,000
1, Best	44.04	42.48	41.26	40.30	39.51	38.86	38.32	37.86	37.47	36.84	36.36
1 & 2	40.78	39.33	38.21	37.31	36.58	35.98	35.48	35.06	34.69	34.11	33.67
2, Good	37.75	36.41	35.37	34.54	33.86	33.31	32.84	32.45	32.12	31.58	31.17
2 & 3	34.95	33.71	32.75	31.98	31.36	30.84	30.41	30.05	29.74	29.24	28.86
3, Average	32.37	31.23	30.33	29.62	29.04	28.57	28.17	27.83	27.54	27.08	26.73
3 & 4	29.79	28.73	27.91	27.26	26.72	26.29	25.92	25.61	25.35	24.92	24.60
4, Low	27.41	26.43	25.68	25.07	24.59	24.18	23.85	23.56	23.32	22.93	22.63

Average Unit Area In Square Feet

Quality Class	1,100	1,200	1,300	1,400	1,500	1,600	1,700	1,800	1,900	2,000	2,200
1, Best	35.87	35.48	35.15	34.88	34.65	34.45	34.28	34.13	34.00	33.88	33.69
1 & 2	33.23	32.87	32.57	32.32	32.10	31.92	31.76	31.62	31.50	31.39	31.21
2, Good	30.75	30.42	30.14	29.91	29.71	29.54	29.39	29.26	29.15	29.05	28.88
2 & 3	28.48	28.17	27.91	27.69	27.51	27.35	27.22	27.10	26.99	26.90	26.75
3, Averag	26.36	26.07	25.84	25.64	25.47	25.32	25.20	25.09	24.99	24.90	24.76
3 & 4	24.25	23.99	23.77	23.59	23.43	23.30	23.18	23.08	22.99	22.91	22.78
4, Low	22.31	22.06	21.86	21.69	21.55	21.43	21.32	21.23	21.15	21.07	20.95

Motels
Quality Classification

	Class 1 Best Quality	Class 2 Good Quality	Class 3 Average Quality	Class 4 Low Quality
Framing*	Wood frame.	Wood frame.	Wood frame.	Wood frame.
Windows	Large, good quality.	Average number and quality.	Average number and quality.	Small, few, low cost.
Roofing	Heavy shake, tile or slate.	Medium shake or good built-up with large rock, inexpensive tile.	Wood or good composition shingle, light shake, or good built-up with rock.	Inexpensive shingles or built-up with rock.
Overhang	36" open or 24" closed.	30" open or small closed.	16" open.	12" to 16" open.
Gutters	Good quality at all eaves.	6" at all eaves.	4" at all eaves.	Over entrances.
Exterior Walls	Good wood or stucco, masonry veneer on front.	Good wood siding or stucco with some veneer.	Hardboard, wood shingle, plywood or stucco.	Low cost stucco, hardboard or plywood.
Flooring	Good carpet, good sheet vinyl.	Good carpet, sheet vinyl or inlaid resilient.	Average carpet, average resilient tile in bath.	Minimum tile or low cost carpet.
Interior Walls	Gypsum board with heavy texture or plaster with putty coat. Some good sheet wall cover or paneling.	Gypsum board, taped, textured and painted or plaster. Some wall paper.	Gypsum board taped and textured or colored interior stucco.	Minimum gypsum board.
Trim	Hardwood with some softwood.	Softwood with some hardwood.	Painted softwood.	Painted softwood.
Closets	Ample space with good wood doors.	Adequate, with good wood doors.	Moderate size, low cost plastic, wood or metal doors.	Small, low cost or no doors.
Baths	Vinyl or foil wall cover, ceramic tile over tub with glass shower door, ample mirrors.	Ceramic tile over tub with glass shower door.	Plastic coated hardboard with low cost glass shower door.	Plastic coated hardboard with one small mirror.
Plumbing	Copper tube, good quality fixtures.	Galvanized pipe, good fixtures.	Average cost fixtures.	Plastic pipe, low cost fixtures.
Special Features	8' sliding glass door, 8' to 10' tile pullman in bath.	8' sliding glass door, good tile or plastic top pullman in bath.	Small tile or plastic pullman in bath.	None.
***For Masonry Walls,** add 8% to the appropriate square foot cost.	8" textured face reinforced masonry.	8" colored or detailed reinforced masonry.	8" colored block or common brick, reinforced.	8" painted concrete block.
		Add The Following Amounts Per Kitchen Unit		
Kitchens	Good sink, 8' to 10' of good cabinets and drainboard. $1,600	Average sink and 6' to 8' average cabinets and drainboard. $1,360	Low cost sink, and 5' of cabinets and drainboard. $1,000	Minimum sink, cabinets and drainboard. $800

Note: Add the cost of built-in kitchen fixtures from the table of costs for built-in appliances on page 26.

Motels
9 Units or Less

Estimating Procedure

1. Establish the structure quality class by applying the information on page 20.
2. Multiply the average unit area by the appropriate cost listed below. The average unit area is found by dividing the total building area on all floors (including office and manager's area, utility room, interior hall and stairway area) by the number of units in the building.
3. Multiply the total from step 2 by the correct location factor listed on page 7.
4. Add, when appropriate, the cost of heating and cooling equipment, porches, balconies, exterior stairs, garages, kitchens, built-in kitchen appliances and fireplaces. See pages 24 to 28 and 20.

Note

Add 2 percent for work above the first floor. Work outside metropolitan areas may cost 2 to 6 percent less. Add 8 percent when exterior walls are masonry.

Motel, Class 3 & 4

Average Unit Area In Square Feet

Quality Class	200	225	250	275	300	330	375	425	500	600	720
1, Best	54.46	52.52	50.97	49.70	48.63	47.56	46.28	45.17	43.92	42.73	41.73
1 & 2	49.49	47.73	46.32	45.16	44.19	43.22	42.06	41.05	39.91	38.83	37.92
2, Good	45.01	43.41	42.13	41.07	40.19	39.31	38.25	37.33	36.30	35.32	34.49
2 & 3	40.93	39.47	38.30	37.35	36.55	35.75	34.78	33.95	33.00	32.11	31.36
3, Average	37.21	35.89	34.82	33.95	33.23	32.50	31.62	30.86	30.01	29.19	28.51
3 & 4	33.80	32.60	31.63	30.84	30.18	29.52	28.72	28.04	27.26	26.52	25.90
4, Low	30.71	29.62	28.74	28.02	27.42	26.82	26.10	25.47	24.77	24.09	23.53

Motels
10 to 24 Units

Estimating Procedure

1. Establish the structure quality class by applying the information on page 20.
2. Multiply the average unit area by the appropriate cost listed below. The average unit area is found by dividing the total building area on all floors (including office and manager's area, utility room, interior hall and stairway area) by the number of units in the building.
3. Multiply the total from step 2 by the correct location factor listed on page 7.
4. Add, when appropriate, the cost of heating and cooling equipment, porches, balconies, exterior stairs, garages, kitchens, built-in kitchen appliances and fireplaces. See pages 24 to 28 and 20.

Note

Add 2 percent for work above the first floor. Work outside metropolitan areas may cost 2 to 6 percent less. Add 8 percent when exterior walls are masonry.

Motels, Class 3

Average Unit Area in Square Feet

Quality Class	200	225	250	275	300	330	375	425	500	600	720
1, Best	52.89	51.00	49.49	48.25	47.21	46.18	44.93	43.85	42.64	41.49	40.52
1 & 2	48.07	46.35	44.97	43.85	42.91	41.97	40.83	39.85	38.75	37.70	36.82
2, Good	43.73	42.17	40.92	39.90	39.04	38.18	37.15	36.26	35.26	34.30	33.51
2 & 3	39.73	38.31	37.18	36.25	35.47	34.69	33.75	32.94	32.03	31.17	30.44
3, Average	36.13	34.84	33.81	32.96	32.26	31.55	30.70	29.96	29.13	28.34	27.68
3 & 4	32.82	31.65	30.71	29.94	29.30	28.65	27.88	27.21	26.46	25.74	25.14
4, Low	29.83	28.76	27.91	27.21	26.63	26.04	25.34	24.73	24.04	23.39	22.85

Motels
Over 24 Units

Estimating Procedure

1. Establish the structure quality class by applying the information on page 20.
2. Multiply the average unit area by the appropriate cost listed below. The average unit area is found by dividing the total building area on all floors (including office and manager's area, utility room, interior hall and stairway area) by the number of units in the building.
3. Multiply the total from step 2 by the correct location factor listed on page 7.
4. Add, when appropriate, the cost of heating and cooling equipment, porches, balconies, exterior stairs, garages, kitchens, built-in kitchen appliances and fireplaces. See pages 24 to 28 and 20.

Note

Add 4 percent for work above the first floor. Work outside metropolitan areas may cost 2 to 6 percent less. Add 8 percent when exterior walls are masonry.

Motel, Class 2 &3

Average Unit Area in Square Feet

Quality Class	200	225	250	275	300	330	375	425	500	600	720
1, Best	51.34	49.51	48.04	46.84	45.83	44.83	43.62	42.58	41.40	40.29	39.35
1 & 2	46.67	45.00	43.66	42.57	41.66	40.74	39.65	38.70	37.63	36.62	35.77
2, Good	42.47	40.95	39.73	38.74	37.91	37.08	36.08	35.22	34.24	33.32	32.55
2 & 3	38.58	37.21	36.10	35.20	34.44	33.69	32.78	32.00	31.11	30.27	29.57
3, Average	35.07	33.82	32.82	32.00	31.31	30.62	29.80	29.09	28.28	27.52	26.88
3 & 4	31.85	30.71	29.80	29.05	28.43	27.81	27.06	26.41	25.68	24.99	24.41
4, Low	28.95	27.91	27.09	26.41	25.84	25.28	24.59	24.01	23.34	22.71	22.19

Additional Costs For Residential Structures

Covered Porches

Estimate covered porches by applying a fraction of the main building square foot cost.

Suggestions For Fraction	Suggested Fraction
Ground level floor (usually concrete) without banister, with no ceiling and shed-type roof.	1/4 to 1/3
High (house floor level) floor (concrete or wood) with light banister, no ceiling and shed-type roof.	1/3 to 1/2
Same as above with a finished ceiling and roof like the residence (most typical).	1/2
Same as above but partially enclosed with screen or glass.	1/2 to 2/3
Enclosed lean-to (sleeping porch, etc.) with lighter foundation, wall structure, interior finish or roof than that of house to which it is attached.	1/2 to 3/4
Roofed, enclosed, recessed porch, under the same roof as the main building and with the same type and quality foundation (includes shape costs).	3/4
Roofed, enclosed, recessed porch with the same type roof and foundation as the main building (includes shape costs).	4/4
Good arbor or pergola with floor.	1/4 to 1/3

Uncovered Concrete Decks, cost per square foot

	8″ High	1′ High	2′ High	3′ High	4′ High
Less than 100 square feet	$2.90	$4.10	$6.50	$9.00	$11.70
100 to 200 square feet	2.45	3.60	5.00	6.75	8.90
200 to 400 square feet	2.30	2.70	4.20	5.90	7.40
Over 400 square feet	2.20	2.60	4.00	5.35	6.55

Uncovered Wood Decks, cost per square foot

1″ board floor on 2″ x 6″ joists 24″ o.c.	$2.90 to $3.20
2″ board floor on 4″ x 6″ girders 48″ o.c.	3.75 to 4.00

Porch Roofs, cost per square foot

Type	Cost Per Square Foot	Alternate Roof Covers	Cost Difference Per S.F.
Unceiled shed roof	$2.80 to $3.05	Roll asphalt or corr. metal	Deduct $ .25 to $ .30
Ceiled shed roof	5.70 to 6.50	Asphalt shingles	Deduct .25 to .50
Unceiled gable roof	3.50 to 4.65	Wood shakes	Add .45 to .80
Ceiled gable roof	6.40 to 7.00	Clay or concrete tile	Add 1.00 to 2.50
(Assumes 3 ply built-up roof & gravel)		Slate	Add 2.50 to 3.50

Residential Basements, cost per square foot

Size	Unfinished Basements	Finished Basements
Less than 400 square feet	$14.80 to $15.90	$19.60 to $21.30
400 - 1,000 square feet	10.90 to 12.00	14.80 to 15.90
Over 1,000 square feet	8.00 to 9.40	11.40 to 12.90

All basement costs assume normal soil conditions, 7'-6'' headroom, no plumbing, partitions, or windows. Unfinished basements are based on reinforced concrete floors and walls, open ceilings and minimum lighting. Finished basement costs are based on reinforced concrete floors with vinyl asbestos tile cover, reinforced concrete walls with plywood panel finish, acoustical tile ceiling and lighting similar to average quality residences.

Additional Costs For Residential Structures

Balconies, Standard Wood Frame, cost per square foot

Supported by 4" x 4" posts, 1" wood floor, open on underside, open 2" x 4" railing.	$2.65 to $2.85
Supported by 4" x 4" posts, 1" wood floor, sealed on underside, solid stucco or wood siding on railing.	3.75 to 4.50
Supported by steel columns, lightweight concrete floors, sealed on underside, solid stucco or open grillwork railing.	6.40 to 7.60

Heating and Cooling Equipment.

Prices include wiring and minimum duct work.
Use the higher figures in more extreme climates where greater heating and cooling density is required.
Costs per square foot of floor.

Type	Perimeter Outlets	Overhead Outlets
Central Ducted Air Systems, Single Family		
Forced air heating.	$1.35 to $1.80	$1.20 to $1.55
Forced air heating and cooling.	2.80 to 3.65	2.75 to 3.25
Gravity heat.	1.15 to 1.60	---
Central Ducted Air Systems, Multi-Family		
Forced air heating.	1.30 to 1.80	1.15 to 1.45
Forced air heating and cooling.	2.80 to 3.50	2.65 to 3.35
Motel Units		
Forced air heating.	1.65 to 2.10	1.55 to 1.90
Forced air heating and cooling.	3.00 to 3.30	2.80 to 3.35
Circulating Hot and Cold Water System	$4.85 to $6.75	

Floor and Wall Furnaces, cost each

Single floor unit	$335.00 to $400.00
Dual floor unit	375.00 to 490.00
Single wall unit	350.00 to 430.00
Dual wall unit	360.00 to 495.00
Thermostat control, add	55.00 to 70.00

Electric Baseboard Units, cost each

500 watts, 3'	$135.00 to $145.00
1,000 watts, 4'	185.00 to 195.00
1,500 watts, 6'	220.00 to 250.00
2,000 watts, 8'	265.00 to 440.00
2,500 watts, 10'	335.00 to 360.00
3,000 watts, 12'	410.00 to 440.00

Outside Stairways, cost per square foot

Standard wood frame, wood steps with open risers, open on underside, open 2" x 4" railing.	$3.70 to $4.40
Standard wood frame with solid wood risers, sealed on underside, solid stucco or wood siding on railing.	4.80 to 5.90
Precast concrete steps with open risers, steel frame, pipe rail with ornamental grillwork	13.60 to 16.00

Electric Wall Heaters, cost each

1,000 watts	$175.00 to $210.00
2,000	210.00 to 230.00
3,000	220.00 to 285.00
3,500	230.00 to 300.00
4,000	315.00 to 335.00
4,500	370.00 to 405.00
Add for circulating fan	30.00 to 50.00
Add for thermostat	32.00 to 55.00

Refrigerated Room Coolers, cost each

1/3 ton	$355.00 to $410.00
1/2	365.00 to 445.00
3/4	420.00 to 490.00
1	490.00 to 560.00
1½	550.00 to 640.00
2	720.00 to 830.00

Ton = 12,000 B.T.U.

Additional Costs For Residential Structures

Built-In Equipment

Add these costs only when the item is not included in the specification for the applicable quality class. These costs include plumbing, wiring and installation.

Appliances, cost each.

Drop-in type, gas or electric	
Range with single oven below	$ 465 to $ 720
Range with single oven above	610 to 810
Range with oven above and below	775 to 920
Surface cooking units, four elements	245 to 355
Oven units, single, gas or electric	440 to 550
Oven units, double, gas or electric	590 to 775
Microwave oven, double, with range	1,300 to 1,800
Range hood and fan	135 to 225
Refrigerators, built-in	
Under counter, 5 cubic feet	250 to 340
Single door, 12 cubic feet	445 to 610
Double door, 17 cubic feet	560 to 890
Double door, 22 cubic feet	1,100 to 1,350
Refrigerator, range, oven and sink	1,150 to 1,390
Dishwashers	425 to 660
Garbage disposers	120 to 225

Fireplaces, cost each

	1 Story	2 Story
5′ base, common brick or concrete block, low cost Douglas fir or common brick mantle.	$1,020	$1,300
6′ base, common brick, used brick, face brick or natural stone on interior face with average wood mantle.	1,350	1,750
6′ base, common brick or face brick, good wood mantle.	1,600	2,100
6′ base, common brick, used brick or natural stone on interior face, raised hearth.	2,160	2,450
8′ base, common brick, used brick or natural stone on interior face.	2,600	3,100

Residential Garages and Carports

Attached and detached garages for single family dwellings are usually classified in the same quality class as the main structure. The attached garage assumes a 20 foot wall in common with the main structure. The costs are per square foot of floor area and include only light excavation for foundations. Multiply the square foot cost below by the correct location factor on page 7 to find the square foot cost for any building site. Costs include no interior finish. Where the interior finish is similar to the main dwelling and the garage walls are in vertical alignment with second floor walls, the square foot cost will be about 80% of the main dwelling square foot cost.

Attached Garages for Single Family Dwellings

Square Feet	220	260	280	320	360	400	440	480	540	600	720
Quality Class											
1, Best	15.64	14.87	14.58	14.09	13.72	13.43	13.19	13.00	12.76	12.57	12.29
1 & 2	14.66	13.94	13.66	13.21	12.86	12.59	12.36	12.18	11.96	11.78	11.52
2, Good	13.67	13.00	12.74	12.32	12.00	11.74	11.53	11.36	11.15	10.99	10.75
2 & 3	12.74	12.12	11.88	11.49	11.18	10.95	10.75	10.59	10.40	10.24	10.02
3, Average	11.83	11.25	11.02	10.66	10.38	10.16	9.98	9.83	9.65	9.51	9.30
3 & 4	10.64	10.12	9.92	9.59	9.34	9.14	8.98	8.84	8.68	8.55	8.37
4, Low	9.50	9.03	8.85	8.56	8.33	8.15	8.01	7.89	7.75	7.63	7.46

Detached Garages for Single Family Dwellings

Square Feet	220	260	280	320	360	400	440	480	540	600	720
Quality Class											
1, Best	20.09	18.63	18.06	17.15	16.45	15.90	15.45	15.09	14.65	14.30	13.79
1 & 2	18.80	17.43	16.90	16.05	15.39	14.88	14.46	14.12	13.70	13.38	12.90
2, Good	17.50	16.23	15.74	14.94	14.33	13.85	13.46	13.14	12.76	12.46	12.02
2 & 3	16.33	15.14	14.68	13.94	13.37	12.92	12.56	12.26	11.90	11.62	11.21
3, Average	15.06	13.97	13.54	12.86	12.33	11.92	11.59	11.31	10.98	10.72	10.34
3 & 4	13.49	12.51	12.13	11.52	11.05	10.68	10.38	10.13	9.84	9.60	9.26
4, Low	11.97	11.10	10.76	10.22	9.80	9.48	9.21	8.99	8.73	8.52	8.22

Carports, including asphalt floor, cost between $3.25 and $5.00 per square foot, depending on the type of roof cover.

Additional Costs For Residential Structures

Costs for Multi-Family Residential Bathrooms beyond 1 per unit

	Class 1 Best Quality	Class 2 Good Quality	Class 3 Average Quality	Class 4 Low Quality
2 or 3 units				
2 fixture bath	$1,790	$1,600	$1,380	$1,100
3 fixture bath	2,750	2,400	2,080	1,870
4 fixture bath	3,480	3,300	2,890	2,340
4 to 9 units				
2 fixture bath	1,750	1,490	1,230	1,040
3 fixture bath	2,570	2,280	1,890	1,540
4 fixture bath	3,360	2,900	2,470	2,100
10 or more units				
2 fixture baths	1,540	1,430	1,100	970
3 fixture bath	2,420	2,080	1,800	1,440
4 fixture bath	3,190	2,800	2,340	1,900

Half Story Areas

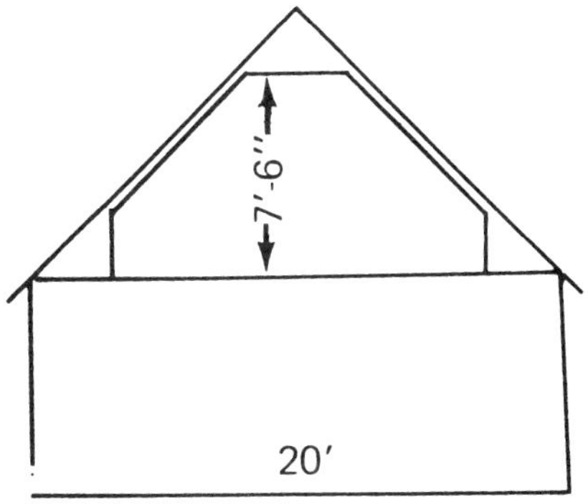

Type A

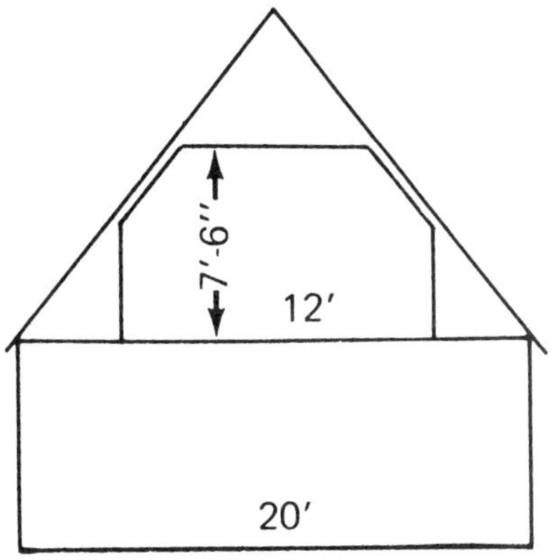

Type B

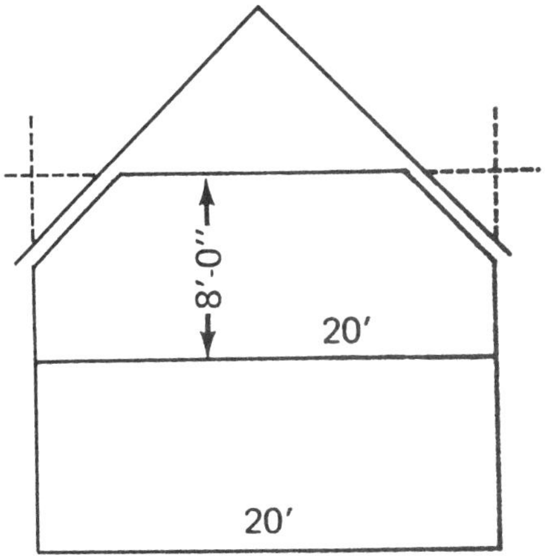

Type C

Use a fraction of the basic square foot cost for figuring the reduced headroom floor area.

Type "C" includes typical dormers.

Type	Same Finish As Main Area	Inferior Finish
A	1/3	1/4
B	1/2	1/3
C	2/3	1/2

Hydraulic Elevators, per shaft cost for car and machinery

Capacity	100 F.P.M.	200 F.P.M.	300 F.P.M.
1,200 lbs.	$17,100	---	---
2,000	18,700	$24,700	$32,800
2,500	23,100	29,200	35,000
3,000	24,700	32,000	37,400
3,500	---	33,600	40,000
4,500	---	35,400	42,400

Add for deluxe car, $3,000.
Add for door frames and door opener per stop over 2: baked enamel $2,500; stainless steel $3,000.

Multi-Family and Motel Garages

Garages built at ground level under a milti-family or motel unit. The costs below include the following components:

1. A reinforced concrete floor in all areas.
2. Exterior walls, on one long side and two short sides, made up of a wood frame and good quality stucco, wood siding or masonry veneer.
3. Steel support columns for the floor above.
4. A finished ceiling in all areas.
5. The difference between the cost of a standard wood frame floor structure at second floor level and one at ground level.
6. An inexpensive light fixture for each 600 square feet.

Where no exterior walls enclose the two short sides, use $5.40 per square foot.

Garages built as separate structures for multi-family or motel units. The costs below include the following components:

1. Foundations.
2. A reinforced concrete floor in all areas.
3. Exterior walls on one long side and two short sides, made up of a wood frame and good quality stucco, wood siding or masonry veneer.
4. Steel support columns supporting the roof.
5. A wood frame roof structure with composition tar and gravel, wood shingle or light shake cover. No interior ceiling finish.
6. An inexpensive light fixture for each 600 square feet.

Use the location modifiers on page 7 to adjust garage costs to any area.

Basement Garages

Costs are listed below for basement garages built on one level, approximately five feet below grade, directly below 2 to 4 story dwelling structures with perimeter walls in vertical alignment. These costs include:

1. Excavation to 5' below ground line.
2. Full wall enclosure.
3. Typical storage facilities.
4. Minimum lighting.
5. Concrete floors.

Access stairways and driveway ramps outside the perimeter walls should be included when calculating the area of the garage.

Use the location modifiers on page 7 to adjust garage costs to any area.

Ground Level Garages

Area	400	800	1,200	2,000	3,000	5,000	10,000	20,000
Cost	10.60	9.50	8.30	8.05	7.60	7.20	6.85	6.65

Separate Structure Garages

Area	400	800	1,200	2,000	3,000	5,000	10,000	20,000
Cost	12.00	10.90	9.80	9.60	9.30	8.60	8.30	8.05

Basement Garages

Type	5,000	7,500	10,000	15,000	20,000	30,000	40,000	60,000
Reinforced concrete exterior walls and columns. Flat con crete roof slab.	16.10	15.20	14.40	13.60	13.40	13.30	13.10	13.00
Concrete block exterior walls, reinforced concrete columns. Flat concrete roof slab.	15.90	14.90	14.10	13.40	13.30	13.10	13.00	12.80
Concrete block exterior walls, steel posts and beams, light concrete/metal roof fireproofed with spray plaster.	15.30	14.10	13.30	13.00	12.40	12.30	12.00	11.70
Concrete block exterior walls, wood posts and beams, light concrete/metal roof fireproofed with spray plaster.	13.60	13.00	12.00	11.30	11.10	10.80	10.60	10.40
Add for each security door.	.75	.50	.35	.23	.19	.14	.10	.09

Cabins And Recreational Dwellings

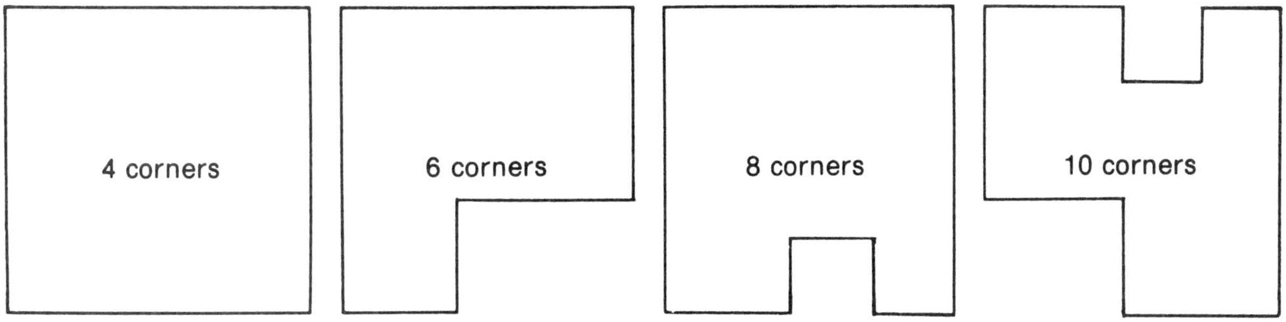

Example of Dwelling Shapes

Cabins and recreational dwellings are designed for single family occupancy, usually on an intermittent basis. These structures are characterized by a more rustic interior and exterior finish and often have construction details which would not meet building requirements in metropolitan areas. Classify these structures into either "conventional type" or "A-frame" construction. Conventional dwellings have an exterior wall which is approximately eight feet high on all sides. A-frame cabins have a sloping roof which reduces the horizontal area 8 feet above the first floor to between 50% and 75% of the first floor area.

The shape of the outside perimeter is an important consideration in estimating the construction cost. Generally, the more complex the shape, the more expensive the structure per square foot of floor area. The shape classification is based on the outline formed by the outer-most exterior walls, including the garage area, regardless of the varying level. Most cabins and recreational dwellings have 4, 6, 8 or 10 corners as illustrated in the example.

Cabins and recreational dwellings are often built under difficult working conditions and in remote sites. Allowance is made here for transportation of materials to typical recreational locations and for the higher labor cost which is usually associated with greater travel distances and adverse conditions. Individual judgments may be necessary in evaluating the cost impact of the dwelling location. The costs assume construction by skilled professional craftsmen. Where non-professional labor or second quality materials are used, use the next lower quality classification than might otherwise apply. If the structure is assembled from prefabricated components, use costs for the next lower half class.

Conventional Recreational Dwellings Quality Classification

	Class 1 Best Quality	Class 2 Good Quality	Class 3 Average Quality	Class 4 Low Quality
Floor Framing	4" x 8" girders 48" o.c. with 2" T&G subfloor, or 2" x 8" to 2" x 10" joists 16" o.c. with 1" plywood subfloor.	4" x 8" girders 48" o.c. with 1¼" plywood or 2" T&G subfloor, or 2" x 8" to 10" joists 16" o.c. with 1" plywood subfloor.	4" x 6" girders 48" o.c. with 1¼" plywood or 2" T&G subfloor, or 2" x 6" to 8" joists 16" o.c. with 1" plywood subfloor.	4" x 6" girders 48" o.c. with 1¼" plywood or 2" x 6" joists 16" o.c. with 1" plywood subfloor.
Roof Framing	Same as floor or heavy glu-lam beams with 2" x 8" or 2" x 10" purlins, 3" T&G flat or low pitch deck and built-up roof.	4" x 8" girders 48" o.c. with 2" or 3" T&G sheathing, or 2" x 6" to 10" rafters 12" to 16" o.c. with 1" sheathing.	4" x 8" girders 48" o.c. with 2" T&G sheathing, or 2" x 6" to 8" rafters 12" to 24" o.c. with 1" sheathing.	4" x 8" girders 48" o.c. with 1¼" plywood or 2" T&G sheathing, or 2" x 6" to 8" rafters up to 24" o.c. with 1" sheathing.
Exterior Finish	Good plywood, lap or board and batt siding. Good trim.	Average to good plywood or board siding or good shingles. Some trim.	Average plywood composition or board siding or shingles. Little or no trim.	Low cost plywood or wood siding or low cost shingle or composition siding.
Windows	Larger area of good insulated wood or metal windows.	Adequate number of good wood or metal windows.	Average quality wood or metal windows.	Few low cost windows.
Roofing	Heavy shakes, usually with steep pitch and several roof planes.	Medium shakes, usually with medium to steep pitch.	Wood or composition shingles with medium to steep pitch.	Low cost composition shingles with medium pitch.
Flooring	Good carpet or hardwood. Sheet vinyl in kitchen and baths.	Average carpet with sheet vinyl or tile in kitchen and baths.	Low cost carpet with tile or linoleum in kitchen and baths.	Linoleum or vinyl asbestos tile.
Interior Finish	Good quality hardwood veneer paneling, with some decorative exposed structural members.	Good textured gypsum wallboard with some plywood or knotty pine paneling.	Textured gypsum wallboard or average quality plywood or hardboard paneling.	Low cost hardboard paneling or gypsum wallboard.
Bathrooms	Two 3-fixture baths and one 2-fixture bath. Good fixtures.	Two 3-fixture baths. Good fixtures.	Two 3-fixture baths.	One 3-fixture bath.
Kitchen	15' to 18' of good hardwood veneer base and wall cabinet. 15' to 18' of good plastic or ceramic drainboard.	12' to 16' of hardwood veneer base cabinet with matching wall cabinets. 12' to 16' of plastic or ceramic drainboard.	8' to 12' of plywood veneer or painted base cabinet with matching wall cabinets. 8' to 12' of plastic drainboard.	8' to 10' of minimum base cabinets with matching wall cabinets. 8' to 10' of minimum plastic drainboard.
Plumbing	Nine good quality fixtures with one large or two 30 gallon water heaters, copper supply piping.	Seven good quality fixtures and one water heater.	Seven average quality fixtures and one water heater.	Four low cost fixtures and one water heater. Plastic supply piping.
Special Features	Built-in oven, range, dishwasher, disposer, range hood with good insulation, good lighting fixtures, insulated sliding glass door and ornate entry door.	Built-in range, oven and range hood, some insulation, 8' sliding glass door, average electric fixtures.	Drop-in range and hood, some insulation, low cost electric fixtures.	Minimum electric fixtures.

Conventional Recreational Dwellings
Four Corners

Estimating Procedure

1. Establish the structure quality class by applying the information on page 30.
2. Multiply the structure floor area by the appropriate cost listed below.
3. Multiply the total from step 2 by the correct location factor listed on page 7.
4. Add, when appropriate, the cost of a deck or porch, paving, fireplace, garage or carport, heating, extra plumbing fixtures, supporting walls, half story areas, construction on hillside lots, and construction in remote areas. See page 39.

Note

Add 4 percent to the square foot cost for floors above the second floor level.

Conventional Recreational Dwelling, Class 3

Square Foot Area

Quality Class	400	500	600	700	800	900	1,000	1,100	1,200	1,300	1,400
1, Best	64.69	58.09	53.55	50.21	47.64	45.59	43.92	42.52	41.33	40.31	39.42
1 & 2	59.15	53.12	48.96	45.91	43.56	41.69	40.16	38.88	37.79	36.86	36.04
2, Good	53.61	48.14	44.38	41.61	39.48	37.78	36.40	35.24	34.25	33.41	32.67
2 & 3	50.32	45.19	41.66	39.06	37.06	35.47	34.17	33.08	32.15	31.36	30.67
3, Average	47.05	42.25	38.95	36.52	34.65	33.16	31.94	30.93	30.06	29.32	28.67
3 & 4	43.41	38.99	35.94	33.70	31.97	30.60	29.47	28.53	27.74	27.05	26.45
4, Low	39.70	35.65	32.86	30.81	29.24	27.98	26.95	26.09	25.37	24.74	24.19

Square Foot Area

Quality Class	1,500	1,600	1,700	1,800	2,000	2,200	2,400	2,600	2,800	3,000	3,200
1, Best	38.37	37.70	37.11	36.57	35.65	34.88	34.23	33.67	33.18	32.75	32.36
1 & 2	35.21	34.60	34.06	33.56	32.72	32.01	31.41	30.90	30.45	30.05	29.70
2, Good	31.93	31.38	30.88	30.44	29.67	29.03	28.49	28.02	27.61	27.25	26.94
2 & 3	29.99	29.47	29.01	28.59	27.87	27.27	26.76	26.32	25.94	25.60	25.30
3, Average	28.11	27.63	27.19	26.80	26.12	25.56	25.08	24.67	24.31	24.00	23.71
3 & 4	26.21	25.75	25.35	24.98	24.35	23.83	23.38	23.00	22.66	22.37	22.11
4, Low	24.32	23.90	23.52	23.18	22.60	22.11	21.70	21.34	21.03	20.76	20.52

Conventional Recreational Dwellings
Six Corners

Estimating Procedure

1. Establish the structure quality class by applying the information on page 30.
2. Multiply the structure floor area by the appropriate cost listed below.
3. Multiply the total from step 2 by the correct location factor listed on page 7.
4. Add, when appropriate, the cost of a deck or porch, paving, fireplace, garage or carport, heating, extra plumbing fixtures, supporting walls, half story areas, construction on hillside lots, and construction in remote areas. See page 39.

Note

Add 4 percent to the square foot cost for floors above the second floor level.

Conventional Recreational Dwelling, Class 4

Square Foot Area

Quality Class	400	500	600	700	800	900	1,000	1,100	1,200	1,300	1,400
1, Best	65.46	58.81	54.25	50.91	48.35	46.31	44.65	43.27	42.10	41.09	40.22
1 & 2	59.88	53.80	49.63	46.58	44.23	42.37	40.85	39.59	38.51	37.59	36.79
2, Good	54.32	48.81	45.02	42.25	40.12	38.43	37.06	35.91	34.94	34.10	33.37
2 & 3	51.03	45.85	42.30	39.69	37.69	36.11	34.81	33.74	32.82	32.04	31.35
3, Average	47.74	42.89	39.57	37.13	35.26	33.78	32.57	31.56	30.71	29.97	29.33
3 & 4	44.07	39.60	36.53	34.28	32.56	31.18	30.07	29.14	28.35	27.67	27.08
4, Low	40.23	36.15	33.35	31.29	29.72	28.47	27.45	26.60	25.88	25.26	24.72

Quality Class	1,500	1,600	1,700	1,800	2,000	2,200	2,400	2,600	2,800	3,000	3,200
1, Best	39.25	38.57	37.96	37.42	36.49	35.71	35.06	34.50	34.02	33.59	33.22
1 & 2	35.92	35.30	34.75	34.25	33.40	32.69	32.09	31.58	31.14	30.75	30.40
2, Good	32.61	32.05	31.55	31.09	30.32	29.68	29.14	28.67	28.27	27.91	27.60
2 & 3	30.69	30.16	29.69	29.26	28.54	27.93	27.42	26.98	26.60	26.27	25.98
3, Average	28.77	28.27	27.83	27.43	26.75	26.18	25.70	25.29	24.94	24.63	24.35
3 & 4	26.86	26.40	25.98	25.61	24.98	24.45	24.00	23.62	23.28	22.99	22.74
4, Low	24.94	24.51	24.13	23.78	23.19	22.70	22.28	21.93	21.62	21.35	21.11

Conventional Recreational Dwellings
Eight Corners

Estimating Procedure

1. Establish the structure quality class by applying the information on page 30.
2. Multiply the structure floor area by the appropriate cost listed below.
3. Multiply the total from step 2 by the correct location factor listed on page 7.
4. Add, when appropriate, the cost of a deck or porch, paving, fireplace, garage or carport, heating, extra plumbing fixtures, supporting walls, half story areas, construction on hillside lots, and construction in remote areas. See page 39.

Note

Add 4 percent to the square foot cost for floors above the second floor level.

Conventional Recreational Dwelling, Class 1 & 2

Square Foot Area

Quality Class	400	500	600	700	800	900	1,000	1,100	1,200	1,300	1,400
1, Best	66.05	59.48	54.96	51.65	49.10	47.07	45.42	44.04	42.87	41.86	40.98
1 & 2	60.47	54.45	50.31	47.28	44.95	43.09	41.58	40.31	39.24	38.32	37.52
2, Good	54.88	49.42	45.66	42.91	40.79	39.11	37.73	36.59	35.61	34.78	34.05
2 & 3	51.59	46.45	42.92	40.33	38.34	36.76	35.47	34.39	33.48	32.69	32.01
3, Average	48.26	43.46	40.16	37.73	35.87	34.39	33.18	32.17	31.32	30.58	29.94
3 & 4	44.41	39.99	36.95	34.72	33.01	31.65	30.53	29.61	28.82	28.14	27.55
4, Low	41.00	36.92	34.11	32.06	30.48	29.22	28.19	27.33	26.61	25.98	25.44

Square Foot Area

Quality Class	1,500	1,600	1,700	1,800	2,000	2,200	2,400	2,600	2,800	3,000	3,200
1, Best	39.91	39.26	38.67	38.15	37.23	36.46	35.79	35.22	34.71	34.27	33.87
1 & 2	36.65	36.06	35.52	35.04	34.19	33.48	32.88	32.35	31.88	31.47	31.11
2, Good	33.31	32.77	32.28	31.84	31.07	30.43	29.88	29.40	28.98	28.60	28.27
2 & 3	31.32	30.81	30.35	29.94	29.22	28.61	28.09	27.64	27.24	26.89	26.58
3, Average	29.39	28.91	28.48	28.09	27.42	26.85	26.36	25.94	25.56	25.24	24.94
3 & 4	27.48	27.03	26.63	26.27	25.64	25.10	24.65	24.25	23.90	23.60	23.32
4, Low	25.55	25.14	24.76	24.43	23.84	23.34	22.92	22.55	22.23	21.94	21.69

Conventional Recreational Dwellings
Ten Corners

Estimating Procedure

1. Establish the structure quality class by applying the information on page 30.
2. Multiply the structure floor area by the appropriate cost listed below.
3. Multiply the total from step 2 by the correct location factor listed on page 7.
4. Add, when appropriate, the cost of a deck or porch, paving, fireplace, garage or carport, heating, extra plumbing fixtures, supporting walls, half story areas, construction on hillside lots, and construction in remote areas. See page 39.

Note

Add 4 percent to the square foot cost for floors above the second floor level.

Conventional Recreational Dwelling, Class 1

Square Foot Area

Quality Class	400	500	600	700	800	900	1,000	1,100	1,200	1,300	1,400
1, Best	66.81	60.25	55.74	52.42	49.86	47.82	46.16	44.77	43.59	42.57	41.68
1 & 2	61.18	55.18	51.04	48.00	45.66	43.79	42.27	41.00	39.92	38.98	38.17
2, Good	55.56	50.11	46.35	43.59	41.46	39.77	38.38	37.23	36.25	35.40	34.66
2 & 3	52.27	47.14	43.61	41.01	39.01	37.42	36.11	35.03	34.10	33.31	32.61
3, Average	48.99	44.18	40.87	38.44	36.56	35.07	33.85	32.83	31.96	31.22	30.57
3 & 4	45.33	40.88	37.82	35.57	33.83	32.45	31.32	30.38	29.57	28.88	28.28
4, Low	41.65	37.56	34.75	32.68	31.08	29.81	28.78	27.91	27.17	26.54	25.99

Square Foot Area

Quality Class	1,500	1,600	1,700	1,800	2,000	2,200	2,400	2,600	2,800	3,000	3,200
1, Best	40.72	40.05	39.44	38.90	37.97	37.20	36.55	35.99	35.50	35.07	34.70
1 & 2	37.33	36.71	36.16	35.66	34.81	34.10	33.51	32.99	32.54	32.15	31.81
2, Good	33.94	33.38	32.88	32.43	31.65	31.01	30.46	30.00	29.59	29.23	28.92
2 & 3	32.00	31.47	30.99	30.57	29.84	29.23	28.72	28.28	27.90	27.56	27.26
3, Average	30.05	29.55	29.10	28.71	28.02	27.45	26.97	26.55	26.20	25.88	25.60
3 & 4	28.14	27.68	27.26	26.89	26.24	25.71	25.26	24.87	24.53	24.24	23.98
4, Low	26.20	25.77	25.38	25.03	24.44	23.94	23.52	23.16	22.85	22.57	22.33

"A-Frame" Cabins
Quality Classification

	Class 1 Best Quality	Class 2 Good Quality	Class 3 Average Quality	Class 4 Low Quality
Floor Framing	4" x 8" girders 48" o.c. with 2" T&G subfloor, or 2" x 6" to 2" x 8" joists 16" o.c. with 1" subfloor.	4" x 8" girders 48" o.c. with 1¼" plywood or 2" T&G subfloor, or 2" x 6" to 2" x 8" joists 16" o.c. with 1" subfloor.	4" x 6" girders 48" o.c. with 1¼" plywood or 2" T&G subfloor, or 2" x 6" joists 16" o.c. with 1" subfloor.	4" x 6" girders 48" o.c. with 1¼" plywood or 2" T&G subfloor, or 2" x 6" joists 16" o.c. with 1" subfloor.
Roof Framing	4" x 8" at 48" o.c. with 2" or 3" T&G sheathing.	4" x 8" at 48" o.c. with 2" or 3" T&G sheathing.	4" x 8" at 48" o.c. with 2" T&G sheathing.	4" x 8" at 48" o.c. with 1¼" plywood or 2" T&G sheathing.
Gable End Finish	Good plywood, lap board or board and batt.	Average to good plywood, or boards.	Average plywood, board or wood shingle.	Low cost plywood, shingle or composition siding.
Windows	Good quality large insulated wood or metal windows.	Average quality insulated wood or metal windows.	Average quality wood or metal windows.	Small glass area of low cost windows.
Roofing	Heavy wood shakes.	Medium wood or aluminum shakes.	Wood or composition shingles.	Low cost composition shingles.
Flooring	Good carpet or hardwood with sheet vinyl in kitchen and baths.	Average to good quality carpet with good tile or sheet vinyl in kitchen and baths.	Average quality carpet with tile in kitchen and baths.	Linoleum or vinyl asbestos tile.
Interior Finish	Good quality hardwood veneer paneling.	Good textured gypsum wallboard, good plywood or knotty pine paneling.	Textured gypsum wallboard or plywood paneling.	Low cost paneling or wallboard.
Bathrooms	Two 3-fixture baths and one 2-fixture bath, good fixtures.	Two 3-fixture baths, good fixtures.	Two 3-fixture baths, average fixtures.	One 3-fixture bath.
Kitchen	15' to 18' of good quality hardwood veneer base cabinet with matching wall cabinets. 15' to 18' of good quality plastic or ceramic tile drainboard.	12' to 16' of hardwood veneer base cabinet with matching wall cabinets. 12' to 16' of plastic or ceramic tile drainboard.	8' to 12' of average quality veneer or painted base cabinets with matching wall cabinets. 8' to 12' of plastic drainboard.	6' to 8' of minimum base cabinets with matching wall cabinets. 6' to 8' of minimum plastic drainboard.
Plumbing	Nine good quality fixtures and one large or two 30 gallon water heaters. Copper supply piping.	Seven good quality fixtures and one water heater.	Seven average quality fixtures and one water heater.	Four low cost fixtures and one water heater. Plastic supply pipe.
Special Features	Built-in oven, range, dishwasher, disposer, range hood with good insulation, good lighting fixtures, insulated sliding glass door and ornate entry door.	Built-in range, oven and range hood, some insulation, 8' sliding glass door, average electric fixtures.	Drop-in range and hood, some insulation, low cost electric fixtures.	Minimum electric fixtures.

"A-Frame" Cabins
Four Corners

Estimating Procedure

1. Establish the structure quality class by applying the information on page 35.
2. Multiply the structure floor area by the appropriate cost listed below.
3. Multiply the total from step 2 by the correct location factor listed on page 7.
4. Add, when appropriate, the cost of a deck or porch, paving, fireplace, garage or carport, heating, extra plumbing fixtures, supporting walls, half story areas, construction on hillside lots, and construction in remote areas. See page 39.

"A-Frame" Cabin, Class 3 & 4

Square Foot Area

Quality Class	400	500	600	700	800	900	1,000	1,100	1,200	1,300	1,400
1, Best	73.39	66.21	61.24	57.57	54.73	52.45	50.59	49.03	47.70	46.55	45.54
1 & 2	66.77	60.25	55.72	52.38	49.79	47.73	46.03	44.61	43.40	42.35	41.44
2, Good	59.99	54.12	50.06	47.06	44.73	42.88	41.35	40.08	38.99	38.05	37.23
2 & 3	56.03	50.56	46.76	43.95	41.78	40.05	38.63	37.43	36.42	35.54	34.77
3, Average	52.01	46.93	43.40	40.80	38.78	37.17	35.85	34.74	33.80	32.99	32.28
3 & 4	46.81	42.24	39.07	36.72	34.91	33.46	32.27	31.27	30.43	29.69	29.05
4, Low	41.67	37.60	34.78	32.69	31.08	29.79	28.73	27.84	27.09	26.43	25.86

Square Foot Area

Quality Class	1,500	1,600	1,700	1,800	2,000	2,200	2,400	2,600	2,800	3,000	3,200
1, Best	43.92	43.20	42.56	41.97	40.97	40.12	39.40	38.78	38.24	37.76	37.33
1 & 2	40.14	39.49	38.90	38.37	37.44	36.67	36.02	35.45	34.95	34.51	34.12
2, Good	36.25	35.66	35.12	34.64	33.81	33.12	32.52	32.01	31.56	31.17	30.81
2 & 3	34.06	33.50	33.00	32.55	31.77	31.11	30.55	30.07	29.65	29.28	28.95
3, Average	31.71	31.19	30.72	30.30	29.57	28.96	28.45	28.00	27.60	27.26	26.95
3 & 4	28.92	28.45	28.02	27.64	26.98	26.42	25.95	25.54	25.18	24.87	24.58
4, Low	25.99	25.57	25.19	24.84	24.25	23.75	23.32	22.95	22.63	22.35	22.09

"A-Frame" Cabins
Six Corners

Estimating Procedure

1. Establish the structure quality class by applying the information on page 35.
2. Multiply the structure floor area by the appropriate cost listed below.
3. Multiply the total from step 2 by the correct location factor listed on page 7.
4. Add, when appropriate, the cost of a deck or porch, paving, fireplace, garage or carport, heating, extra plumbing fixtures, supporting walls, half story areas, construction on hillside lots, and constructiön in remote areas. See page 39.

"A-Frame" Cabin, Class 2 & 3

Square Foot Area

Quality Class	400	500	600	700	800	900	1,000	1,100	1,200	1,300	1,400
1, Best	74.56	67.29	62.28	58.59	55.76	53.50	51.65	50.10	48.79	47.67	46.68
1 & 2	67.73	61.13	56.58	53.23	50.65	48.60	46.92	45.52	44.33	43.30	42.41
2, Good	60.90	54.96	50.87	47.86	45.54	43.70	42.19	40.93	39.86	38.93	38.13
2 & 3	56.90	51.35	47.52	44.71	42.55	40.82	39.41	38.23	37.23	36.37	35.62
3, Average	52.81	47.66	44.11	41.50	39.49	37.89	36.58	35.49	34.56	33.76	33.06
3 & 4	47.56	42.92	39.72	37.37	35.56	34.12	32.94	31.96	31.12	30.40	29.77
4, Low	42.25	38.13	35.29	33.20	31.59	30.31	29.26	28.39	27.65	27.01	26.45

Square Foot Area

Quality Class	1,500	1,600	1,700	1,800	2,000	2,200	2,400	2,600	2,800	3,000	3,200
1, Best	45.14	44.40	43.75	43.16	42.13	41.27	40.54	39.91	39.36	38.87	38.43
1 & 2	41.17	40.50	39.90	39.36	38.42	37.64	36.97	36.40	35.89	35.45	35.05
2, Good	37.19	36.59	36.04	35.56	34.71	34.00	33.40	32.88	32.42	32.02	31.67
2 & 3	34.92	34.35	33.84	33.38	32.59	31.92	31.36	30.87	30.44	30.06	29.73
3, Average	32.61	32.08	31.61	31.18	30.44	29.82	29.29	28.83	28.43	28.08	27.77
3 & 4	29.68	29.20	28.77	28.38	27.70	27.14	26.66	26.24	25.88	25.56	25.27
4, Low	26.65	26.21	25.83	25.48	24.87	24.36	23.93	23.56	23.23	22.95	22.69

"A-Frame" Cabins
Eight Corners

Estimating Procedure

1. Establish the structure quality class by applying the information on page 35.
2. Multiply the structure floor area by the appropriate cost listed below.
3. Multiply the total from step 2 by the correct location factor listed on page 7.
4. Add, when appropriate, the cost of a deck or porch, paving, fireplace, garage or carport, heating, extra plumbing fixtures, supporting walls, half story areas, construction on hillside lots, and construction in remote areas. See page 39.

"A-Frame" Cabin, Class 2

Square Foot Area

Quality Class	400	500	600	700	800	900	1,000	1,100	1,200	1,300	1,400
1, Best	75.78	68.54	63.54	59.84	56.99	54.71	52.84	51.28	49.95	48.80	47.80
1 & 2	68.83	62.26	57.71	54.36	51.76	49.69	48.00	46.58	45.37	44.32	43.41
2, Good	61.77	55.88	51.80	48.78	46.46	44.60	43.08	41.80	40.72	39.78	38.96
2 & 3	57.69	52.18	48.37	45.55	43.38	41.65	40.22	39.03	38.02	37.15	36.38
3, Average	53.47	48.37	44.83	42.23	40.21	38.61	37.29	36.18	35.24	34.43	33.73
3 & 4	48.22	43.62	40.43	38.08	36.26	34.81	33.62	32.63	31.78	31.05	30.41
4, Low	42.86	38.76	35.93	33.84	32.23	30.94	29.88	29.00	28.25	27.60	27.03

Square Foot Area

Quality Class	1,500	1,600	1,700	1,800	2,000	2,200	2,400	2,600	2,800	3,000	3,200
1, Best	46.30	45.55	44.89	44.29	43.26	42.40	41.67	41.04	40.49	40.01	39.58
1 & 2	42.19	41.51	40.90	40.36	39.42	38.64	37.97	37.40	36.90	36.46	36.07
2, Good	38.12	37.50	36.95	36.46	35.61	34.90	34.30	33.79	33.34	32.94	32.59
2 & 3	35.75	35.18	34.66	34.20	33.40	32.74	32.18	31.69	31.27	30.90	30.57
3, Average	33.34	32.80	32.32	31.89	31.15	30.53	30.01	29.55	29.16	28.81	28.51
3 & 4	30.32	29.83	29.40	29.01	28.33	27.77	27.29	26.88	26.52	26.20	25.92
4, Low	27.26	26.82	26.43	26.08	25.47	24.96	24.53	24.16	23.84	23.56	23.31

Additional Costs for Cabins and Recreational Dwellings

Half-Story Costs

For conventional recreational dwellings, use the suggested fractions found on page 27 in the section "Additional Costs for Residential Structures." For "A-Frame" cabins, use one of the following costs: A simple platform with low cost floor cover, minimum partitions, and minimum lighting costs $9.75 to $14.00 per square foot. Average quality half story area with average quality carpet, average number of partitions finished with gypsum wallboard or plywood veneer and average lighting costs $16.90 to $21.00 per square foot. A good quality half story area with good carpet, decorative rustic partitions, ceiling beams and good lighting costs $25.50 to $31.00 per square foot.

Decks and Porches, per square foot

Small (10 to 50 square feet) uncovered wood porch with steps and railing	
1' above ground level	$8.00 to $10.00
4' above ground level	10.00 to 13.60
2" wood deck with steps and railing (300 square foot base)	
1' above ground level	$5.50 to $7.50
4' above ground level	7.50 to 8.75

Fireplaces, cost each

Metal hood with concrete slab	$1,090 to $1,800
Simple concrete block	1,450 to 1,550
Simple block with stone facing	1,800 to 2,200
Simple natural stone	2,200 to 2,650

Flatwork, per square foot

Asphalt paving	$.80 to $1.20
4" concrete	1.70 to 2.50
6" concrete	2.05 to 2.85

Garages and Carports, per square foot

Average carport, no slab (single)	$6.00 to $8.00
Average single garage with slab	14.00 to 15.75
Average double garage with slab	11.50 to 14.25

Heating, cost each

Wall heaters	
35,000 BTU	$490.
65,000 BTU	680.
Central heating, perimeter ducts	$1.60 per S.F.

Extra Plumbing, cost each

Lavatory	$550 to $730
Water closet	680 to 790
Tub	670 to 770
Stall shower	485 to 730
Sink	555 to 740

Supporting Wall Costs

Cabins and recreational dwellings built on sloping lots cost more than if they are built on level lots. The cost of supporting walls of a building that do not enclose any living area should be estimated by using the figures below. These costs include everything above a normal foundation (12" to 18" above ground) up to the bottom of the next floor structure where square foot costs can be applied. In addition to the cost of supporting walls, add the cost of any extra structural members and the higher cost of building on a slope. A good rule of thumb for this is to add $355 for each foot of vertical distance between the highest and the lowest points of intersection of foundation and ground level.

Wood posts, per foot of height

4" x 4"	$1.50 to $1.70
4" x 6"	2.20 to 2.75
6" x 6"	3.50 to 4.00
8" x 8"	5.75 to 6.75
10" x 10"	9.25 to 11.25
12" x 12"	14.00 to 16.75

Brick, per square foot of wall

8" common brick	$6.10 to $6.60
12" common brick	8.90 to 9.40
8" common brick, 1 side face brick	7.30 to 8.00
12" common brick, 1 side face brick	10.00 to 10.40

Reinforced concrete walls,

per cubic foot	$7.40 to $8.90

Reinforced concrete block, per square foot of wall

8" natural	$4.50 to $5.10
8" colored	4.70 to 5.30
8" detailed blocks, natural	4.70 to 4.95
8" detailed blocks, colored	4.90 to 5.20
8" sandblasted	4.75 to 5.35
8" splitface, natural	5.25 to 6.00
8" splitface, colored	5.50 to 6.25
8" slumpstone, natural	6.40 to 7.00
8" slumpstone, colored	7.00 to 7.75
12" natural	6.00 to 6.90

Typical Physical Lives in Years for Residential Structures by Quality Class

	Masonry				Wood Frame			
Building Type	1	2	3	4	1	2	3	4
1 or 2 family	60	60	60	55	60	60	60	55
3 or more units	60	55	55	50	60	55	55	50
Motels	60	55	55	50	60	55	55	50

Raise half classes to the next higher whole class

Normal Percent Good Table for Residential and Agricultural Structures

Average Life In Years

	20 Years		25 Years		30 Years		40 Years	
Age	Rem. Life	% Good	Rem. Life	% Good	Rem. Life	% Good	Rem. Life	% Good
0	20	100	25	100	30	100	40	100
1	19	94	24	95	29	96	39	98
2	18	88	23	90	28	93	38	96
3	17	81	22	86	27	89	37	94
4	16	75	21	81	26	86	36	92
5	15	69	20	77	25	82	35	90
6	14	63	19	72	24	79	34	87
7	13	59	18	68	23	75	33	84
8	12	57	17	63	22	71	32	82
9	11	55	16	60	21	67	31	80
10	11	53	16	58	20	64	30	77
11	10	50	15	56	19	60	29	74
12	9	48	14	54	19	59	28	72
13	8	46	13	53	18	57	27	70
14	7	44	12	51	17	56	27	67
15	7	42	11	49	16	54	26	65
16	6	40	11	48	15	53	25	62
17	5	38	10	46	14	52	24	60
18	5	36	9	44	13	50	23	59
19	4	33	8	43	13	49	22	58
20	4	31	7	41	12	47	21	56
21	3	29	7	39	11	46	21	55
22	3	27	6	37	11	44	20	54
23	3	25	6	35	10	43	19	53
24	3	23	5	34	9	42	18	52
25	2	21	5	32	9	40	17	51
26	2	19	4	30	8	39	17	50
27	2	16	4	29	7	37	16	49
28	2	14	4	27	7	36	15	48
29	2	12	3	25	6	34	14	47
30	1	10	3	24	6	33	14	46
31	--	--	3	22	5	31	13	45
32	--	--	3	20	5	30	12	44
33	--	--	2	18	5	29	12	43
34	--	--	2	17	4	27	11	42
35	--	--	2	15	4	26	11	41
36	--	--	2	13	4	24	10	40
38	--	--	1	10	3	21	9	38
40	--	--	--	--	2	19	7	35
42	--	--	--	--	2	16	6	33
46	--	--	--	--	1	10	5	29
50	--	--	--	--	--	--	4	25
55	--	--	--	--	--	--	3	20
60	--	--	--	--	--	--	2	14
64	--	--	--	--	--	--	1	10

Average Life In Years

	45 Years		50 Years		55 Years		60 Years	
Age	Rem. Life	% Good	Rem. Life	% Good	Rem. Life	% Good	Rem. Life	% Good
0	45	100	50	100	55	100	60	100
2	43	97	48	97	53	98	58	98
4	41	93	46	94	51	96	56	96
6	39	89	44	91	49	94	54	94
8	37	85	42	88	47	91	52	92
10	35	81	39	85	45	88	50	90
12	33	77	38	82	43	85	48	88
14	32	73	36	78	41	82	46	86
16	30	69	35	74	40	79	45	83
18	28	65	33	70	38	76	43	80
20	26	60	31	67	36	73	41	77
22	24	58	29	63	34	69	39	74
24	23	56	28	60	32	65	37	71
26	22	54	26	58	31	62	35	68
28	20	52	24	56	29	60	34	65
30	18	50	23	54	27	58	32	63
32	17	48	21	53	26	56	30	60
34	15	47	20	51	24	55	29	58
36	14	45	18	49	23	53	27	57
38	12	43	17	47	21	51	26	55
40	11	41	16	45	20	50	24	54
42	10	39	14	44	19	48	23	52
44	9	37	13	42	17	46	21	51
46	8	35	12	40	16	45	20	49
48	7	33	11	38	15	43	19	47
50	6	31	10	37	14	41	18	46
52	5	29	9	35	12	40	16	44
54	5	28	8	33	11	38	15	43
56	4	26	7	31	10	36	14	41
58	4	24	6	30	9	35	13	40
60	3	22	5	28	8	33	12	38
62	3	20	4	26	7	31	11	37
64	3	18	4	24	6	30	10	35
66	2	16	3	22	5	28	9	33
68	2	14	3	21	5	27	8	32
70	2	12	3	19	4	25	7	30
72	1	10	2	17	4	23	6	29
76	--	--	2	14	3	20	5	26
80	--	--	1	10	2	17	4	23
84	--	--	--	--	1	10	2	16
96	--	--	--	--	--	--	1	10

COMMERCIAL STRUCTURES SECTION

Section Contents

Structure Type	Page
Urban Stores, Masonry or Concrete	43
Urban Stores, Wood or Wood and Steel	49
Suburban Stores, Masonry or Concrete	56
Suburban Stores, Wood or Wood and Steel	61
Supermarkets, Masonry or Concrete	70
Supermarkets, Wood or Wood and Steel	72
Small Food Stores, Masonry or Concrete	74
Small Food Stores, Wood Frame	76
Discount Houses, Masonry or Concrete	76
Discount Houses, Wood or Wood and Steel	80
Banks and Savings Offices, Masonry or Concrete	82
Banks and Savings Offices, Wood Frame	87
Department Stores, Reinforced Concrete	93
Department Stores, Masonry or Concrete	96
Department Stores, Wood Frame	99
General Office Buildings, Masonry or Concrete	102
General Office Buildings, Wood Frame	110
Medical-Dental Buildings, Masonry or Concrete	118
Medical-Dental Buildings, Wood Frame	126
Convalescent Hospitals, Masonry or Concrete	134
Convalescent Hospitals, Wood Frame	136
Funeral Homes	138
Self Service Restaurants	140
Coffee Shop Restaurants	143
Conventional Restaurants	146
"A-Frame" Restaurants	148
Theaters, Masonry or Concrete	150
Theaters, Wood Frame	156
Mobile Home Parks	161
Service Stations, Wood, Masonry or Painted Steel	163
Service Stations, Porcelain Finished Steel	165
Service Stations, Ranch or Rustic	167
Additional Costs for Service Stations	169
Service Garages, Masonry or Concrete	173
Service Garages, Wood Frame	178
Auto Service Centers, Masonry or Concrete	183
Typical Lives for Commercial Buildings	200
Additional Costs for Commercial Buildings	201

Urban Stores

Urban store buildings are designed for retail sales and are usually found in strip or downtown commercial developments. Square foot costs in this section are representative of a building situation where construction activities are restricted to the immediate site. This restriction tends to make the cost slightly higher than suburban type stores where unlimited use of modern machinery and techniques is possible. Do not use these figures for department stores, discount houses or suburban stores. These building types are evaluated in later sections.

Costs are for shell-type buildings without permanent partitions and include all labor, material and equipment costs for the following:

1. Foundations as required for normal soil conditions.
2. Floor, rear wall, side wall and roof structures.
3. A front wall consisting of vertical support columns or pilasters and horizontal beams spanning the area between these members leaving an open space to receive a display front.
4. Interior floor, wall and ceiling finishes.
5. Exterior wall finish on the side and rear walls.
6. Roof cover.
7. Basic lighting and electrical systems.
8. Rough and finish plumbing.
9. Design and engineering fees.
10. Permits and fees.
11. Utility hook-up.
12. Contractor's contingency, overhead and profit.

The in-place costs of the following components should be added to the basic building cost to arrive at the total structure cost. See the section "Additional Costs for Commercial Structures" on page 201.

1. Heating and air conditioning systems.
2. Elevators and escalators.
3. Fire sprinklers and fire escapes.
4. All display front components.
5. Finish materials on the front wall.
6. Canopies, ramps and docks.
7. Interior partitions.
8. Exterior signs.
9. Mezzanines and basements.
10. Communication systems.

For valuation purposes, urban stores are divided into two building types: 1) masonry or concrete frame and, 2) wood or wood and steel frame. Masonry or concrete urban stores vary widely in cost. Consequently, 6 quality classifications are established. Wood or wood and steel frame urban stores are divided into 4 quality classes.

Urban Stores
Masonry or Concrete
Quality Classification

	Class 1 Best Quality	Class 2 Good Quality	Class 3 High Average Quality	Class 4 Low Average Quality	Class 5 Low Quality	Class 6 Minimum Quality
Foundation	Reinforced concrete.	Reinforced concrete.	Reinforced concrete.	Reinforced concrete.	Reinforced concrete.	Reinforced concrete.
Floor Structure	6″ reinforced concrete on 6″ rock fill or 2″ x 12″ joists 16″ o.c.	4″ to 6″ reinforced concrete on 6″ rock fill or 2″ x 10″ joists 16″ o.c.	4″ reinforced concrete on 6″ rock fill or 2″ x 10″ joists 16″ o.c.	4″ reinforced concrete on 6″ rock fill or 2″ x 8″ joists 16″ o.c.	4″ reinforced concrete on 6″ rock fill or 2″ x 6″ joists 16″ o.c.	4″ reinforced concrete on 4″ rock fill or 2″ x 6″ joists 16″ o.c.
Wall Structure	Reinforced 8″ concrete or 12″ common brick or block.	Reinforced 8″ concrete or 12″ common brick or block.	Reinforced 8″ concrete or 12″ common brick or block.	Reinforced 8″ concrete or 12″ common brick or block.	Reinforced 8″ concrete block or reinforced 6″ concrete.	Reinforced 8″ concrete block or reinforced 6″ concrete or 8″ clay tile or 8″ brick.
Roof Covering	5 ply composition roof on 1″ sheathing with insulation.	5 ply composition roof on 1″ sheathing with insulation.	5 ply composition roof on 1″ sheathing with insulation.	5 ply composition roof on 1″ sheathing with insulation.	4 ply composition roof on 1″ sheathing.	4 ply composition roof on 1″ sheathing.
Floor Finish	Combination solid vinyl tile and terrazzo or very good carpet.	Combination solid vinyl tile and terrazzo or very good carpet.	Solid vinyl tile with some terrazzo or good carpet.	Vinyl asbestos tile with small areas of terrazzo, carpet or solid vinyl tile.	Vinyl asbestos tile.	Minimum tile.
Interior Wall Finish	Plaster on gypsum or metal lath or 2 layers of 5/8″ gypsum wallboard with expensive wall paper or vinyl wall cover.	Plaster on gypsum or metal lath or 2 layers of 5/8″ gypsum wallboard with average wall paper or vinyl wall cover.	Plaster with putty coat finish on gypsum or metal lath, or 5/8″ gypsum wallboard taped, textured and painted, some vinyl wall covering.	Plaster with putty coat finish or gypsum or metal lath, or 5/8″ gypsum wallboard taped, textured and painted or with wall paper.	Lath, 2 coats plaster with putty coat finish or ½″ gypsum wallboard taped, textured and painted.	Interior plaster on masonry. Colored finish.
Ceiling Finish	Acoustical plaster or suspended anodized acoustical metal panels.	Acoustical plaster or suspended anodized acoustical metal panels.	Plaster with putty coat finish and some acoustical plaster or suspended acoustical tile with gypsum wallboard backing.	Plaster with putty coat finish or suspended acoustical tile.	Gypsum wallboard taped and textured or lath, 2 coats of plaster and putty coat finish.	Ceiling tile or gypsum wallboard and paint.
Exterior Wall Finish	Waterproofed and painted finish with face brick on exposed walls.	Waterproofed and painted finish with face brick on exposed walls.	Waterproofed and painted finish, face brick on exposed walls.	Painted finish, face brick on exposed walls.	Painted finish.	Unfinished.
Lighting	Modular encased units and custom designed chandeliers. Many spotlights.	Encased modular units and custom design chandeliers. Many spotlights.	Encased modular units and stock chandeliers. Many spotlights.	Quad open strip fixtures or triple encased louvered strip fixtures. Average number of spotlights.	Triple open strip fixtures or double encased louvered strip fixtures. Some spotlights.	Double open strip flourescent fixtures.
Plumbing (Per each 5,000 S.F.)	6 good fixtures, metal or marble toilet partitions.	6 good fixtures, metal or marble toilet partitions.	6 standard fixtures, metal or marble toilet partitions.	6 standard fixtures, metal toilet partitions.	4 standard commercial fixtures, metal toilet partitions.	4 standard commercial fixtures, wood toilet partitions.
Bath Wall Finish	Ceramic tile or marble or custom mosaic tile.	Ceramic tile or marble or custom mosaic tile.	Ceramic tile or marble or plain mosaic tile.	Gypsum wallboard and paint, some ceramic tile or plastic finish wall board.	Gypsum wallboard and paint.	Gypsum wallboard and paint.

Urban Stores
Masonry or Concrete
First Floor, Length Less Than Twice Width

Estimating Procedure

1. Establish the structure quality class by applying the information on page 43.
2. Compute the building ground floor area. This should include everything within the exterior walls and all insets outside the walls but under the main roof.
3. Add to or subtract from the cost below the appropriate amount from the Wall Height Adjustment Table on page 48 if the first floor wall height is more or less than 16 feet for large stores or 12 feet for small stores.
4. Multiply the adjusted square foot cost by the building area.
5. Deduct, if appropriate, for common walls, using the figures on page 48.
6. Multiply the total cost by the location factor on page 7.
7. Add the cost of heating and cooling equipment, elevators, escalators, fire escapes, fire sprinklers, display fronts, canopies, ramps, docks, interior partitions, mezzanines, basements, and communications systems from pages 201 to 213.
8. Add the cost of second and higher floors from page 47.

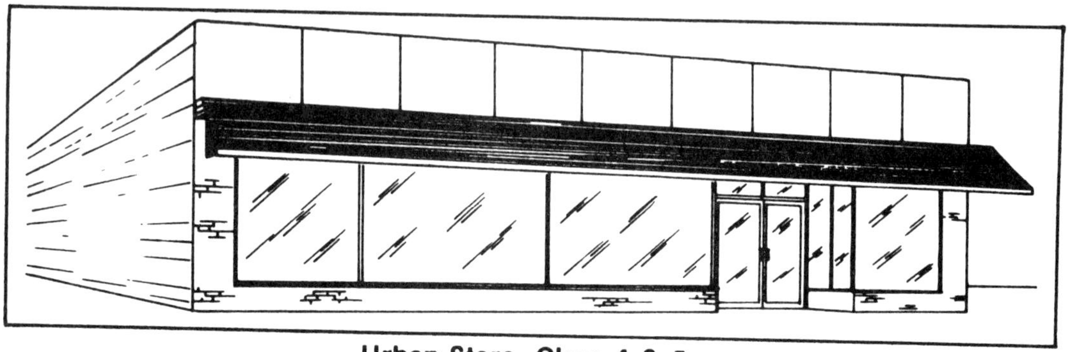

Urban Store, Class 4 & 5

Smaller Stores

Square Foot Area

Quality Class	500	600	700	800	900	1,000	1,200	1,500	1,700	2,000	2,500
4, Low Average	51.91	49.33	47.31	45.69	44.35	43.21	41.38	39.35	38.31	37.05	35.48
4 & 5	48.41	46.00	44.12	42.61	41.36	40.30	38.59	36.70	35.73	34.55	33.09
5, Low	44.92	42.68	40.94	39.53	38.37	37.39	35.80	34.05	33.15	32.06	30.70
5 & 6	42.47	40.35	38.71	37.38	36.28	35.35	33.85	32.19	31.34	30.31	29.02
6, Minimum	40.04	38.04	36.49	35.24	34.20	33.33	31.91	30.35	29.55	28.58	27.36

Larger Stores

Square Foot Area

Quality Class	3.000	3,500	4,000	4,500	5,000	6,000	7,500	10,000	15,000	20,000
1, Best	55.83	54.27	53.03	52.01	51.16	49.81	48.35	46.71	44.82	43.72
1 & 2	51.84	50.38	49.23	48.29	47.50	46.25	44.89	43.37	41.62	40.59
2, Good	48.08	46.73	45.66	44.79	44.06	42.90	41.63	40.23	38.60	37.65
2 & 3	45.39	44.12	43.11	42.29	41.60	40.50	39.31	37.98	36.44	35.55
3, Hi. Avg.	42.72	41.52	40.57	39.80	39.15	38.12	36.99	35.74	34.30	33.45
3 & 4	39.25	38.12	37.22	36.47	35.85	34.85	33.75	32.52	31.08	30.22
4, Low Avg.	35.46	34.44	33.62	32.95	32.39	31.48	30.49	29.38	28.07	27.30
4 & 5	33.36	32.40	31.63	31.00	30.47	29.62	28.69	27.64	26.41	25.69
5, Low	31.28	30.38	29.66	29.07	28.57	27.77	26.90	25.92	24.77	24.09
5 & 6	29.43	28.58	27.90	27.35	26.88	26.13	25.31	24.38	23.30	22.66
6, Minimum	27.55	26.76	26.12	25.60	25.16	24.46	23.69	22.83	21.81	21.21

Urban Stores
Masonry or Concrete
First Floor, Length Between 2 and 4 Times Width

Estimating Procedure

1. Establish the structure quality class by applying the information on page 43.
2. Compute the building ground floor area. This should include everything within the exterior walls and all insets outside the walls but under the main roof.
3. Add to or subtract from the cost below the appropriate amount from the Wall Height Adjustment Table on page 48 if the first floor wall height is more or less than 16 feet for large stores or 12 feet for small stores.
4. Multiply the adjusted square foot cost by the building area.
5. Deduct, if appropriate, for common walls, using the figures on page 48.
6. Multiply the total cost by the location factor on page 7.
7. Add the cost of heating and cooling equipment, elevators, escalators, fire escapes, fire sprinklers, display fronts, canopies, ramps, docks, interior partitions, mezzanines, basements, and communications systems from pages 201 to 213.
8. Add the cost of second and higher floors from page 47.

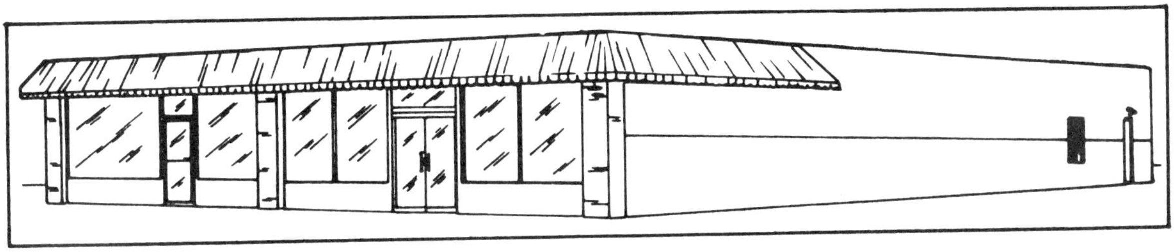

Urban Store, Class 4

Smaller Stores

Square Foot Area

Quality Class	500	600	700	800	900	1,000	1,250	1,500	1,700	2,000	2,500
4, Low Avg.	57.71	54.61	52.21	50.29	48.70	47.37	44.78	42.88	41.67	40.22	38.43
4 & 5	53.88	50.98	48.74	46.95	45.47	44.22	41.80	40.03	38.90	37.55	35.87
5, Low	50.06	47.37	45.29	43.62	42.25	41.09	38.84	37.19	36.15	34.89	33.33
5 & 6	47.49	44.93	42.96	41.38	40.08	38.98	36.84	35.28	34.29	33.10	31.62
6, Minimum	44.96	42.54	40.67	39.18	37.94	36.90	34.88	33.40	32.46	31.34	29.93

Larger Stores

Square Foot Area

Quality Class	3,000	3,500	4,000	4,500	5,000	6,000	7,500	10,000	15,000	20,000
1, Best	58.99	57.27	55.88	54.74	53.78	52.23	50.52	48.60	46.33	44.98
1 & 2	54.95	53.34	52.05	50.99	50.09	48.65	47.06	45.27	43.15	41.89
2, Good	50.87	49.38	48.19	47.20	46.37	45.03	43.56	41.90	39.95	38.78
2 & 3	48.05	46.64	45.51	44.58	43.80	42.54	41.15	39.58	37.73	36.63
3, Hi. Avg.	45.24	43.91	42.85	41.98	41.24	40.05	38.74	37.26	35.52	34.49
3 & 4	41.79	40.46	39.40	38.52	37.79	36.61	35.32	33.87	32.17	31.16
4, Low Avg.	37.75	36.55	35.59	34.80	34.13	33.07	31.90	30.59	29.05	28.14
4 & 5	35.64	34.50	33.60	32.85	32.22	31.22	30.12	28.88	27.43	26.57
5, Low	33.45	32.39	31.54	30.84	30.25	29.31	28.27	27.11	25.75	24.94
5 & 6	31.44	30.44	29.64	28.98	28.43	27.54	26.57	25.48	24.20	23.44
6, Minimum	29.46	28.53	27.78	27.16	26.64	25.81	24.90	23.88	22.68	21.97

Urban Stores
Masonry or Concrete
First Floor, Length More Than 4 Times Width

Estimating Procedure

1. Establish the structure quality class by applying the information on page 43.
2. Compute the building ground floor area. This should include everything within the exterior walls and all insets outside the walls but under the main roof.
3. Add to or subtract from the cost below the appropriate amount from the Wall Height Adjustment Table on page 48 if the first floor wall height is more or less than 16 feet for large stores or 12 feet for small stores.
4. Multiply the adjusted square foot cost by the building area.
5. Deduct, if appropriate, for common walls, using the figures on page 48.
6. Multiply the total cost by the location factor on page 7.
7. Add the cost of heating and cooling equipment, elevators, escalators, fire escapes, fire sprinklers, display fronts, canopies, ramps, docks, interior partitions, mezzanines, basements, and communications systems from pages 201 to 213.
8. Add the cost of second and higher floors from page 47.

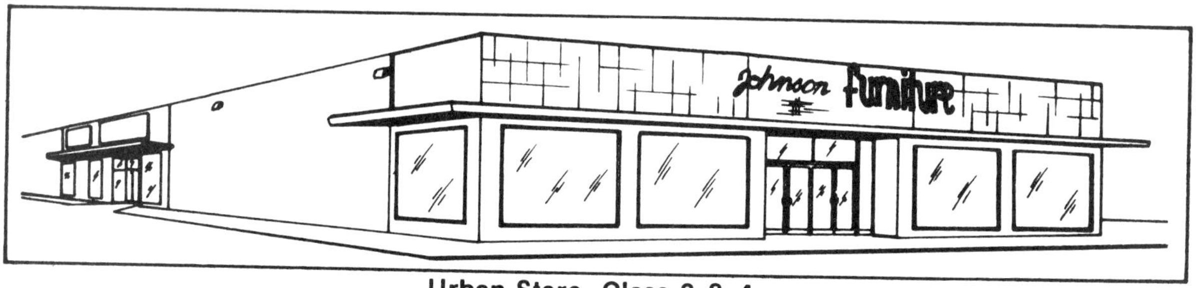

Urban Store, Class 3 & 4

Smaller Stores

Square Foot Area

Quality Class	500	600	700	800	900	1,000	1,200	1,500	1,700	2,000	2,500
4, Low Avg.	64.25	60.48	57.59	55.27	53.37	51.77	49.22	46.42	44.99	43.28	41.16
4 & 5	59.99	56.47	53.76	51.61	49.83	48.34	45.95	43.34	42.01	40.41	38.43
5, Low	55.74	52.47	49.96	47.95	46.30	44.92	42.70	40.27	39.03	37.55	35.71
5 & 6	53.06	49.95	47.56	45.65	44.08	42.76	40.65	38.34	37.16	35.74	33.99
6, Minimum	50.39	47.44	45.17	43.35	41.86	40.61	38.60	36.41	35.29	33.95	32.28

Larger Stores

Square Foot Area

Quality Class	3,000	3,500	4,000	4,500	5,000	6,000	7,500	10,000	15,000	20,000
1, Best	63.49	61.37	59.68	58.29	57.11	55.23	53.17	50.85	48.12	46.51
1 & 2	59.15	57.18	55.60	54.31	53.21	51.46	49.54	47.38	44.84	43.34
2, Good	54.77	52.94	51.48	50.28	49.27	47.65	45.87	43.87	41.51	40.12
2 & 3	49.82	48.16	46.83	45.74	44.82	43.34	41.72	39.90	37.76	36.50
3, Hi. Avg.	48.67	47.05	45.76	44.69	43.79	42.35	40.77	38.99	36.90	35.66
4 & 4	44.96	43.39	42.12	41.09	40.21	38.80	37.26	35.52	33.47	32.26
4, Low Avg.	40.81	39.38	38.24	37.29	36.50	35.22	33.82	32.24	30.38	29.28
4 & 5	38.43	37.09	36.01	35.12	34.37	33.17	31.85	30.36	28.61	27.57
5, Low	36.05	34.79	33.78	32.94	32.24	31.12	29.88	28.48	26.84	25.86
5 & 6	33.94	32.75	31.80	31.01	30.35	29.29	28.12	26.81	25.26	24.35
6, Minimum	31.81	30.70	29.80	29.07	28.45	27.45	26.36	25.13	23.68	22.82

Urban Stores
Masonry or Concrete
Second and Higher Floors

Estimating Procedure

1. Establish the structure quality class. The class for second and higher floors will usually be the same as the first floor.
2. Compute the square foot area of the second floor and each higher floor.
3. Add to or subtract from the square foot cost below the appropriate amount from the Wall Height Adjustment Table on page 48 if the wall height is more or less than 12 feet.
4. Multiply the adjusted square foot cost by the area of each floor.
5. Deduct, if appropriate, for common walls, using the figures on page 48.
6. Add 2% to the cost for each floor above the second floor. For example, the third floor cost would be 102% of the second floor cost and the fourth floor cost would be 104% of the second floor cost.
7. Multiply the total cost for each floor by the location factor on page 7.
8. Add the cost of heating and cooling equipment, escalators, fire escapes, fire sprinklers, and interior partitions from pages 201 to 213.

Length less than twice width

	Square Foot Area										
Quality Class	2,500	3,000	3,500	4,000	4,500	5,000	6,000	7,500	10,000	15,000	20,000
1, Best	49.14	47.75	46.66	45.79	45.07	44.46	43.49	42.41	41.18	39.74	38.88
1 & 2	45.37	44.08	43.08	42.27	41.61	41.05	40.14	39.15	38.02	36.68	35.89
2, Good	41.61	40.43	39.51	38.77	38.16	37.65	36.82	35.91	34.87	33.65	32.92
2 & 3	38.90	37.80	36.94	36.25	35.68	35.20	34.42	33.57	32.60	31.46	30.78
3, Hi. Avg.	36.31	35.28	34.48	33.84	33.31	32.86	32.13	31.34	30.43	29.36	28.73
3 & 4	33.26	32.22	31.42	30.77	30.24	29.78	29.06	28.25	27.34	26.26	25.61
4, Low Avg.	29.76	28.83	28.11	27.53	27.05	26.65	26.00	25.28	24.46	23.49	22.92
4 & 5	27.61	26.75	26.08	25.55	25.10	24.73	24.12	23.45	22.69	21.80	21.26
5, Low	25.48	24.69	24.07	23.58	23.17	22.82	22.26	21.64	20.94	20.12	19.62
5 & 6	23.87	23.13	22.55	22.09	21.70	21.38	20.85	20.28	19.62	18.85	18.38
6, Minimum	22.23	21.54	21.00	20.57	20.21	19.91	19.42	18.88	18.27	17.55	17.12

Length between two and four times width

	Square Foot Area										
Quality Class	2,500	3,000	3,500	4,000	4,500	5,000	6,000	7,500	10,000	15,000	20,000
1. Best	51.05	49.52	48.33	47.37	46.57	45.90	44.81	43.61	42.25	40.63	39.67
1 & 2	47.13	45.71	44.61	43.72	42.99	42.37	41.37	40.26	39.00	37.51	36.62
2, Good	43.20	41.91	40.90	40.08	39.41	38.84	37.92	36.91	35.75	34.39	33.57
2 & 3	40.45	39.23	38.29	37.53	36.90	36.36	35.50	34.55	33.47	32.19	31.43
3, Hi. Avg.	37.74	36.61	35.73	35.02	34.43	33.93	33.13	32.24	31.24	30.04	29.33
3 & 4	34.62	33.47	32.58	31.86	31.26	30.76	29.95	29.06	28.05	26.86	26.15
4, Low Avg.	30.97	29.94	29.14	28.50	27.97	27.52	26.79	26.00	25.09	24.02	23.39
4 & 5	28.80	27.84	27.10	26.50	26.00	25.59	24.91	24.17	23.33	22.34	21.75
5, Low	26.61	25.73	25.04	24.49	24.03	23.64	23.02	22.34	21.56	20.64	20.10
5 & 6	24.93	24.10	23.46	22.94	22.51	22.15	21.57	20.93	20.20	19.34	18.83
6, Minimum	23.26	22.48	21.88	21.40	21.00	20.66	20.12	19.52	18.84	18.04	17.56

Length more than four times width

	Square Foot Area										
Quality Class	2,500	3,000	3,500	4,000	4,500	5,000	6,000	7,500	10,000	15,000	20,000
1, Best	53.09	51.71	50.56	49.60	48.77	48.05	46.86	45.47	43.83	41.78	40.51
1 & 2	49.00	47.72	46.67	45.78	45.01	44.35	43.24	41.97	40.45	38.56	37.39
2, Good	44.91	43.74	42.77	41.95	41.26	40.65	39.63	38.46	37.07	35.34	34.27
2 & 3	42.05	40.95	40.05	39.28	38.63	38.06	37.11	36.01	34.71	33.09	32.08
3, Hi. Avg.	39.22	38.20	37.36	36.65	36.04	35.51	34.62	33.60	32.38	30.87	29.93
3 & 4	36.19	35.11	34.23	33.50	32.87	32.33	31.43	30.40	29.18	27.68	26.75
4, Low Avg.	32.40	31.44	30.65	29.99	29.43	28.94	28.14	27.21	26.13	24.78	23.95
4 & 5	30.11	29.21	28.48	27.87	27.35	26.89	26.14	25.29	24.28	23.03	22.25
5, Low	27.86	27.03	26.36	25.79	25.31	24.89	24.20	23.40	22.47	21.31	20.59
5 & 6	26.11	25.33	24.70	24.17	23.71	23.32	22.67	21.93	21.05	19.97	19.30
6, Minimum	24.39	23.67	23.08	22.58	22.16	21.79	21.18	20.49	19.67	18.66	18.03

Urban Stores Masonry or Concrete Wall Adjustments

Wall Height Adjustment

The square foot costs for urban stores are based on the wall heights listed on each page. The main or first floor height is the distance from the bottom of the floor slab or joists to the top of the roof slab or ceiling joists. Second and higher floors are measured from the top of the floor slab or floor joists to the top of the roof slab or ceiling joists. Add or subtract the amount listed to the square foot cost for each foot more or less than the standard wall height in the tables. For second and higher floors use only 75% of the wall height adjustment cost.

Area	500	600	700	800	900	1,000	1,250	1,500	1,750	2,000	2,500
Adjustment	1.56	1.42	1.31	1.22	1.15	1.10	.98	.90	.84	.78	.71

Area	3,000	3,500	4,000	4,500	5,000	6,000	7,500	10,000	15,000	20,000
Adjustment	.64	.59	.55	.52	.49	.45	.40	.35	.28	.25

Perimeter Wall Adjustment

A common wall exists when two buildings share one wall. Adjust for common walls by deducting the linear foot costs below from the total structure cost. In some structures one or more walls is not owned at all. In this case, deduct the ''No Ownership'' cost per linear foot of wall not owned.

Small Urban Stores (to 2,500 S.F.)
Common Wall, deduct $50.30 per linear foot
No Ownership, deduct $100.00 per linear foot

Large Urban Stores (over 2,500 S.F.)
Common Wall, deduct $95.00 per linear foot
No Ownership, deduct $189.00 per linear foot

Urban Stores
Wood or Wood and Steel Frame
Quality Classification

	Class 1 Best Quality	Class 2 Good Quality	Class 3 Average Quality	Class 4 Low Quality
Foundation	Reinforced concrete.	Reinforced concrete	Reinforced concrete.	Reinforced concrete.
Floor Structure	4" reinforced concrete on 6" rock fill or 2" x 10" joists 16" o.c.	4" reinforced concrete on 6" rock fill or 2" x 8" joists 16" o.c.	4" reinforced concrete on 6" rock fill or 2" x 6" joists 16" o.c.	4" reinforced concrete on 4" rock fill or 2" x 6" joists 16" o.c.
Wall Structure	2" x 6" studs 16" o.c.	2" x 4" or 2" x 6" studs 16" o.c.	2" x 4" studs 16" o.c. up to 14' high, 2" x 6" studs 16" o.c. over 14' high.	2" x 4" studs 16" o.c. up to 14' high, 2" x 6" studs 16" o.c. over 14' high.
Roof Covering	5 ply composition roof on 1" x 6" sheathing with insulation.	5 ply composition roof on 1" x 6" sheathing with insulation.	4 ply composition roof on 1" x 6" sheathing.	4 ply composition roof on 1" x 6" sheathing.
Floor Finish	Sheet vinyl with some terrazzo or good carpet.	Vinyl asbestos tile with small areas of terrazzo, carpet or solid vinyl tile.	Vinyl asbestos tile.	Minimum grade tile.
Interior Wall Finish	Plaster with putty coat finish on gypsum or metal lath, or 5/8" gypsum wallboard taped, textured and painted or some vinyl wall cover.	Plaster with putty coat finish on gypsum or metal lath, or 5/8" gypsum wallboard taped, textured and painted or with wall paper.	Lath, 2 coats plaster with putty coat finish or ½" gypsum wallboard taped, textured and painted.	½" gypsum wallboard taped, textured and painted.
Ceiling Finish	Plaster with putty coat finish and some acoustical plaster or suspended acoustical tile with gypsum wallboard backing.	Plaster with putty coat finish or suspended acoustical tile with exposed grid.	Gypsum wallboard taped and textured or lath, 2 coats of plaster and putty coat finish.	Ceiling tile or gypsum wallboard and paint.
Exterior Wall Finish	Good wood siding.	Average wood siding.	Stucco or average wood siding.	Stucco or inexpensive wood siding.
Lighting	Encased modular units and stock chandeliers. Many spotlights.	Quad open strip fixtures or triple encased louvered strip fixtures. Average number of spotlights.	Triple open strip fixtures or double encased louvered strip fixtures. Some spotlights.	Double open strip fixtures.
Plumbing (Per 5,000 S.F.)	6 standard fixtures, metal or marble toilet partitions.	6 standard fixtures, metal toilet partitions.	4 standard commercial fixtures, metal toilet partitions.	4 standard commercial fixtures, wood toilet partitions.
Bath Wall Finish	Ceramic tile, marble or plain mosaic tile.	Gypsum wallboard and paint, some ceramic tile or plastic finish wallboard.	Gypsum wallboard and paint.	Gypsum wallboard and paint.

Urban Stores
Wood or Wood and Steel Frame
First Floor, Length Less Than Twice Width

Estimating Procedure

1. Establish the structure quality class by applying the information on page 49.
2. Compute the building ground floor area. This should include everything within the exterior walls and all insets outside the walls but under the main roof.
3. Add to or subtract from the cost below the appropriate amount from the Wall Height Adjustment Table on page 54 if the first floor wall height is more or less than 16 feet for large stores or 12 feet for small stores.
4. Multiply the adjusted square foot cost by the building area.
5. Deduct, if appropriate, for common walls, using the figures on page 54.
6. Multiply the total cost by the location factor on page 7.
7. Add the cost of heating and cooling equipment, elevators, escalators, fire escapes, fire sprinklers, display fronts, canopies, ramps, docks, interior partitions, mezzanines, basements, and communications systems from pages 201 to 213.
8. Add the cost of second and higher floors from page 53.

Smaller Stores

Square Foot Area

Quality Class	500	600	700	800	900	1,000	1,250	1,500	1,750	2,000	2,500
2, Good	38.39	37.02	35.94	35.06	34.33	33.72	32.50	31.60	30.90	30.33	29.46
2 & 3	35.59	34.32	33.32	32.51	31.83	31.26	30.13	29.30	28.65	28.12	27.31
3, Average	32.74	31.57	30.65	29.90	29.28	28.75	27.72	26.95	26.35	25.87	25.12
3 & 4	29.74	28.67	27.84	27.16	26.59	26.12	25.18	24.48	23.93	23.49	22.82
4, Low	26.75	25.79	25.04	24.43	23.92	23.49	22.65	22.02	21.53	21.13	20.52

Larger Stores

Square Foot Area

Quality Class	3,000	3,500	4,000	4,500	5,000	6,000	7,500	10,000	15,000	20,000
1, Best	37.69	36.93	36.32	35.81	35.38	34.68	33.92	33.05	32.01	31.40
1 & 2	33.99	33.31	32.75	32.29	31.91	31.28	30.59	29.80	28.87	28.32
2, Good	30.46	29.85	29.35	28.94	28.59	28.03	27.41	26.71	25.87	25.38
2 & 3	28.60	28.02	27.55	27.17	26.84	26.31	25.73	25.07	24.29	23.82
3, Average	26.71	26.17	25.74	25.38	25.07	24.58	24.04	23.42	22.69	22.25
3 & 4	24.85	24.34	23.94	23.60	23.32	22.86	22.36	21.78	21.10	20.70
4, Low	22.99	22.52	22.15	21.84	21.57	21.15	20.68	20.15	19.52	19.15

Urban Stores
Wood or Wood and Steel Frame
First Floor, Length Between 2 and 4 Times Width

Estimating Procedure

1. Establish the structure quality class by applying the information on page 49.
2. Compute the building ground floor area. This should include everything within the exterior walls and all insets outside the walls but under the main roof.
3. Add to or subtract from the cost below the appropriate amount from the Wall Height Adjustment Table on page 54 if the first floor wall height is more or less than 16 feet for large stores or 12 feet for small stores.
4. Multiply the adjusted square foot cost by the building area.
5. Deduct, if appropriate, for common walls, using the figures on page 54.
6. Multiply the total cost by the location factor on page 7.
7. Add the cost of heating and cooling equipment, elevators, escalators, fire escapes, fire sprinklers, display fronts, canopies, ramps, docks, interior partitions, mezzanines, basements, and communications systems from pages 201 to 213.
8. Add the cost of second and higher floors from page 53.

Urban Store, Class 3

Smaller Stores

Square Foot Area

Quality Class	500	600	700	800	900	1,000	1,250	1,500	1,750	2,000	2,500
2, Good	41.71	40.05	38.76	37.72	36.86	36.14	34.72	33.67	32.86	32.21	31.21
2 & 3	38.45	36.92	35.73	34.77	33.98	33.31	32.01	31.04	30.29	29.69	28.77
3, Average	35.41	34.00	32.91	32.02	31.29	30.68	29.47	28.59	27.90	27.34	26.50
3 & 4	32.14	30.86	29.87	29.06	28.40	27.84	26.75	25.95	25.32	24.82	24.05
4, Low	28.84	27.69	26.80	26.08	25.49	24.99	24.01	23.28	22.72	22.27	21.58

Larger Stores

Square Foot Area

Quality Class	3,000	3,500	4,000	4,500	5,000	6,000	7,500	10,000	15,000	20,000
1, Best	39.50	38.61	37.89	37.29	36.78	35.97	35.07	34.05	32.83	32.11
1 & 2	35.72	34.91	34.26	33.72	33.26	32.52	31.71	30.78	29.69	29.03
2, Good	31.96	31.23	30.65	30.16	29.76	29.10	28.37	27.54	26.56	25.97
2 & 3	29.95	29.27	28.72	28.27	27.89	27.27	26.59	25.81	24.89	24.34
3, Average	28.02	27.39	26.88	26.45	26.10	25.52	24.88	24.15	23.29	22.78
3 & 4	26.04	25.45	24.97	24.58	24.25	23.71	23.11	22.44	21.64	21.16
4, Low	24.05	23.50	23.06	22.70	22.39	21.90	21.35	20.72	19.99	19.55

Urban Stores
Wood or Wood and Steel Frame
First Floor, Length More Than 4 Times Width

Estimating Procedure

1. Establish the structure quality class by applying the information on page 49.
2. Compute the building ground floor area. This should include everything within the exterior walls and all insets outside the walls but under the main roof.
3. Add to or subtract from the cost below the appropriate amount from the Wall Height Adjustment Table on page 54 if the first floor wall height is more or less than 16 feet for large stores or 12 feet for small stores.
4. Multiply the adjusted square foot cost by the building area.
5. Deduct, if appropriate, for common walls, using the figures on page 54.
6. Multiply the total cost by the location factor on page 7.
7. Add the cost of heating and cooling equipment, elevators, escalators, fire escapes, fire sprinklers, display fronts, canopies, ramps, docks, interior partitions, mezzanines, basements, and communications systems from pages 201 to 213.
8. Add the cost of second and higher floors from page 53.

Urban Store, Class 3

Smaller Stores

Square Foot Area

Quality Class	500	600	700	800	900	1,000	1,250	1,500	1,750	2,000	2,500
2, Good	45.75	43.50	41.81	40.47	39.39	38.49	36.77	35.54	34.60	33.86	32.75
2 & 3	42.29	40.21	38.65	37.41	36.41	35.58	33.99	32.85	31.99	31.30	30.27
3, Average	38.86	36.95	35.51	34.38	33.46	32.69	31.23	30.19	29.39	28.76	27.82
3 & 4	35.18	33.45	32.15	31.12	30.29	29.60	28.28	27.33	26.61	26.04	25.18
4, Low	31.83	30.27	29.09	28.16	27.41	26.78	25.59	24.73	24.08	23.56	22.78

Larger Stores

Square Foot Area

Quality Class	3,000	3,500	4,000	4,500	5,000	6,000	7,500	10,000	15,000	20,000
1, Best	41.78	40.68	39.80	39.08	38.47	37.50	36.43	35.22	33.81	32.97
1 & 2	37.80	36.81	36.02	35.36	34.81	33.93	32.96	31.87	30.59	29.83
2, Good	33.86	32.98	32.26	31.68	31.18	30.39	29.53	28.55	27.40	26.73
2 & 3	31.76	30.93	30.26	29.71	29.25	28.51	27.70	26.78	25.70	25.07
3, Average	29.68	28.90	28.27	27.76	27.33	26.64	25.88	25.02	24.01	23.42
3 & 4	27.55	26.82	26.24	25.77	25.37	24.72	24.02	23.22	22.29	21.74
4, Low	25.40	24.74	24.20	23.76	23.39	22.80	22.15	21.42	20.56	20.05

Urban Stores
Wood or Wood and Steel Frame
Second and higher floors

Estimating Procedure

1. Establish the structure quality class. The class for second and higher floors will usually be the same as the first floor.
2. Compute the square foot area of the second floor and each higher floor.
3. Add to or subtract from the square foot cost below the appropriate amount from the Wall Height Adjustment Table on page 54 if the wall height is more or less than 12 feet.
4. Multiply the adjusted square foot cost by the area of each floor.
5. Deduct, if appropriate, for common walls, using the figures on page 54.
6. Add 2% to the cost for each floor above the second floor. For example, the third floor cost would be 102% of the second floor cost and the fourth floor cost would be 104% of the second floor cost.
7. Multiply the total cost for each floor by the location factor on page 7.
8. Add the cost of heating and cooling equipment, escalators, fire escapes, fire sprinklers, and interior partitions from pages 201 to 213.

Length less than twice width

Square Foot Area

Quality Class	2,500	3,000	3,500	4,000	4,500	5,000	6,000	7,500	10,000	15,000	20,000
1, Best	32.13	31.45	30.93	30.51	30.15	29.86	29.38	28.85	28.25	27.54	27.11
1 & 2	28.89	28.28	27.81	27.43	27.11	26.85	26.42	25.94	25.40	24.76	24.38
2, Good	25.63	25.09	24.67	24.33	24.05	23.82	23.44	23.01	22.54	21.97	21.63
2 & 3	23.54	23.05	22.66	22.35	22.10	21.88	21.53	21.14	20.70	20.18	19.87
3, Average	21.51	21.06	20.71	20.42	20.19	19.99	19.67	19.32	18.91	18.44	18.15
3 & 4	20.05	19.63	19.30	19.04	18.82	18.63	18.33	18.00	17.63	17.18	16.92
4, Low	18.58	18.19	17.89	17.65	17.44	17.27	16.99	16.69	16.34	15.93	15.68

Length between 2 and 4 times width

Square Foot Area

Quality Class	2,500	3,000	3,500	4,000	4,500	5,000	6,000	7,500	10,000	15,000	20,000
1, Best	33.05	32.31	31.73	31.26	30.87	30.54	30.01	29.42	28.76	27.96	27.49
1 & 2	29.71	29.04	28.52	28.10	27.75	27.45	26.98	26.45	25.85	25.14	24.71
2, Good	26.38	25.78	25.32	24.95	24.64	24.37	23.95	23.48	22.95	22.31	21.94
2 & 3	24.24	23.70	23.27	22.93	22.64	22.40	22.01	21.58	21.09	20.51	20.16
3, Average	22.15	21.65	21.26	20.94	20.68	20.46	20.11	19.72	19.27	18.74	18.42
3 & 4	20.62	20.15	19.79	19.50	19.26	19.05	18.72	18.36	17.94	17.44	17.15
4, Low	19.12	18.69	18.35	18.08	17.86	17.67	17.36	17.02	16.64	16.18	15.90

Length over 4 times width

Square Foot Area

Quality Class	2,500	3,000	3,500	4,000	4,500	5,000	6,000	7,500	10,000	15,000	20,000
1, Best	34.47	33.59	32.91	32.36	31.90	31.52	30.90	30.21	29.43	28.51	27.96
1 & 2	31.00	30.21	29.59	29.10	28.69	28.34	27.78	27.17	26.47	25.64	25.14
2, Good	27.53	26.83	26.28	25.84	25.48	25.17	24.68	24.13	23.51	22.77	22.33
2 & 3	25.32	24.67	24.17	23.77	23.43	23.15	22.69	22.19	21.62	20.94	20.54
3, Average	23.13	22.54	22.08	21.72	21.41	21.15	20.73	20.27	19.75	19.13	18.76
3 & 4	21.54	20.99	20.56	20.22	19.94	19.69	19.31	18.88	18.39	17.82	17.47
4, Low	19.92	19.42	19.02	18.70	18.44	18.22	17.86	17.46	17.01	16.48	16.16

Urban Stores
Wood or Wood and Steel Frame
Wall Adjustments

Wall Height Adjustment

The square foot costs for urban stores are based on the wall heights listed on each page. The main or first floor height is the distance from the bottom of the floor slab or joists to the top of the roof slab or ceiling joists. Second and higher floors are measured from the top of the floor slab or floor joists to the top of the roof slab or ceiling joists. Add or subtract the amount listed to the square foot cost for each foot more or less than the standard wall height in the tables. For second and higher floors use only 75% of the wall height adjustment cost.

Area	500	600	700	800	900	1,000	1,250	1,500	1,750	2,000	2,500
Adjustment	.61	.56	.52	.49	.46	.44	.40	.37	.34	.32	.29

Area	3,000	3,500	4,000	4,500	5,000	6,000	7,500	10,000	15,000	20,000
Adjustment	.25	.23	.22	.21	.20	.19	.18	.16	.15	.14

Perimeter Wall Adjustment

A common wall exists when two buildings share one wall. Adjust for common walls by deducting the linear foot costs below from the total structure cost. In some structures one or more walls is not owned at all. In this case, deduct the "No Ownership" cost per linear foot of wall not owned.

Small Urban Stores (2,500 S.F. or less)
Common Wall, deduct $31.00 per linear foot
No Ownership, deduct $61.80 per linear foot

Large Urban Stores (over 2,500 S.F.)
Common Wall, deduct $44.30 per linear foot
No Onwership, deduct $88.60 per linear foot

Suburban Stores

Suburban stores are usually built as part of shopping centers. They differ from urban stores in that they are built in open areas where modern construction techniques, equipment and more economical designs can be used. They are also subject to greater variations in size and shape than are urban stores. Do not use the figures in this section for department stores, discount houses or urban stores. These building types are evaluated in other sections.

Costs identified "building shell only" do not include permanent partitions, display fronts or finish materials on the front of the building. Costs for "multi-unit buildings" include partitions, display fronts and finish materials on the front of the building. All figures include the following costs:

1. Foundations as required for normal soil conditions.
2. Floor, rear wall, side wall and roof structures.
3. A front wall consisting of vertical support columns or pilasters and horizontal beams spanning the area between these members leaving an open space to receive a display front.
4. Interior floor, wall and ceiling finishes.
5. Exterior wall finish on the side and rear walls.
6. Roof cover.
7. Basic lighting and electrical systems.
8. Rough and finish plumbing.
9. A usual or normal parapet wall.
10. Design and engineering fees.
11. Permits and fees.
12. Utility hook-ups.
13. Contractor's contingency, overhead and mark-up.

The in-place costs of these extra components should be added to the basic building cost to arrive at total structure cost. See the section "Additional Costs for Commercial and Industrial Structures" on page 201.

1. Heating and air conditioning systems.
2. Fire sprinklers.
3. All display front components (shell-type buildings only).
4. Finish materials on the front wall of the building (shell-type building only).
5. Canopies.
6. Interior partitions (shell-type buildings only).
7. Exterior signs.
8. Mezzanines and basements.
9. Loading docks and ramps.
10. Miscellaneous yard improvements.
11. Communications systems.

For valuation purposes suburban stores are divided into two building types: masonry or concrete frame, or wood or wood and steel frame. Each building type is divided into four shape classes:

1) Buildings in which the depth is greater than the front. 2) Buildings in which the front is between **one** and **two** times the depth. 3) Buildings in which the front is between **two** and **four** times the depth. 4) Buildings in which the front is greater than **four** times the depth. Angular buildings should be classed by comparing the sum of the length of all wings to the width of the wings. All areas should be included, but no area should be included as part of two different wings. Note the example at the right.

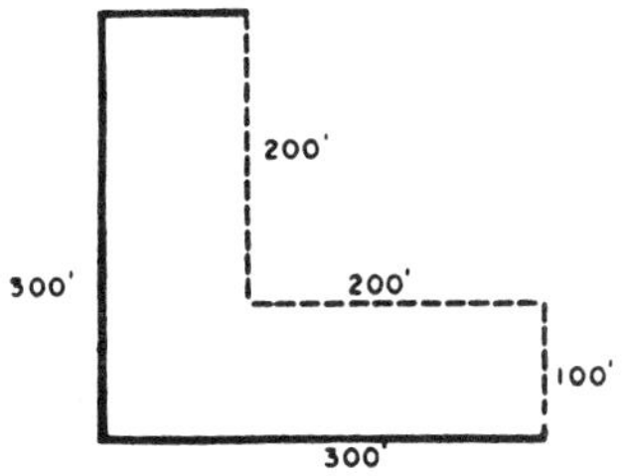

Length = 200 + 300 = 500
Width = 100

Suburban Stores
Masonry or Concrete
Quality Classification

	Class 1 Best Quality	Class 2 Good Quality	Class 3 Average Quality	Class 4 Low Quality
Foundation	Reinforced concrete.	Reinforced concrete.	Reinforced concrete.	Reinforced concrete.
Floor Structure	6" reinforced concrete on 6" rock base.	6" reinforced concrete on 6" rock base.	6" reinforced concrete on 6" rock base.	4" reinforced concrete on 6" rock base.
Wall Structure	8" reinforced decorative concrete block, 6" concrete tilt-up or 8" reinforced brick.	8" reinforced decorative concrete block, 6" concrete tilt-up or 8" reinforced brick.	8" reinforced concrete block, 6" concrete tilt-up or 8" reinforced common brick.	8" reinforced concrete block or 6" concrete tilt-up.
Roof	Glu-lam or steel beams on steel intermediate columns. Panelized roof system, ½" plywood sheathing 5 ply built-up roof with insulation.	Glu-lam or steel beams on steel intermediate columns. Panelized roof system, ½" plywood sheathing 5 ply built-up roof with insulation.	Glu-lam beams on steel intermediate columns. Panelized roof system, ½" plywood sheathing, 4 ply built-up roof.	Glu-lam beams on steel intermediate columns. Panelized roof system, ½" plywood sheathing, 4 ply built-up roof.
Floor Finish	Terrazzo, sheet vinyl or very good carpet.	Vinyl asbestos tile with 50% solid vinyl tile, terrazzo, or good carpet.	Vinyl asbestos tile.	Minimum grade tile.
Interior Wall Finish	Inside of exterior walls furred out with gypsum wallboard or lath and plaster cover. Exterior walls and partitions finished with vinyl wall covers and hardwood veneers.	Interior stucco on inside of exterior walls, gypsum wallboard and texture or paper on partitions, some vinyl wall cover and plywood paneling.	Interior stucco on inside of exterior walls, gypsum wallboard and texture and paint on partitions.	Paint on inside of exterior walls, gypsum wallboard with texture and paint on partitions.
Ceiling Finish	Suspended good grade acoustical tile with gypsum wallboard backing.	Suspended acoustical tile with concealed grid system.	Suspended acoustical tile with exposed grid system.	Exposed beams with ceiling tile or paint.
Lighting	Recessed fluorescent lighting in modular plastic panels.	Continuous recessed 3 tube fluorescent strips with egg crate diffusers, 8' o.c.	Continuous 3 tube fluorescent strips with egg crate diffusers, 8' o.c.	Continuous exposed 2 tube fluorescent strips, 8' o.c.
Exterior	Face brick or stone veneer.	Exposed aggregate, some stone veneer.	Paint on exposed areas, some exposed aggregate.	Paint on exposed areas.
Plumbing	6 good fixtures per 5,000 S.F. of floor area, metal toilet partitions.	6 standard fixtures per 5,000 S.F. of floor area, metal toilet partitions.	4 standard fixtures per 5,000 S.F. of floor area, metal toilet partitions.	4 standard fixtures per 5,000 S.F. of floor area, wood toilet partitions.

The costs on page 59 and 60, and 66 to 69 include display fronts. The quality of the display front will help establish the quality class of the building as a whole. Display fronts are classified as follows:

	Class 1 Best Quality	Class 2 Good Quality	Class 3 Average Quality	Class 4 Low Quality
Bulkhead (0 to 4' high)	Vitrolite, domestic marble or stainless steel.	Black flagstone, terrazzo or good ceramic tile.	Average ceramic tile, Roman brick or imitation flagstone.	Stucco, wood or common brick.
Window Frame	Bronze or stainless steel.	Heavy aluminum.	Aluminum.	Light aluminum with wood stops.
Glass	¼" plate glass with mitered joints.	¼" plate glass, some mitered joints.	¼" plate glass.	Crystal or ¼" plate glass.
Sign Area (4' high)	Vitrolite, domestic marble or stainless steel.	Black flagstone, terrazzo, or good ceramic tile.	Average ceramic tile, Roman brick or imitation flagstone.	Stucco.
Pilasters	Vitrolite, domestic marble.	Black flagstone, terrazzo or good ceramic tile.	Average ceramic tile, Roman brick or imitation flagstone.	Stucco.

Suburban Stores
Masonry or Concrete
Building Shell Only

Estimating Procedure

1. Use these figures to estimate the cost of shell-type buildings without permanent partitions, display fronts or finish material on the front wall of the building.
2. Establish the structure quality class by applying the information on page 56.
3. Compute the building floor area. This should include everything within the exterior walls and all inset areas outside the main walls but under the main roof.
4. Add to or subtract from the cost below the appropriate amount from the Wall Height Adjustment Table (at the bottom of this page) if the wall height is more or less than 16 feet.
5. Multiply the adjusted square foot cost by the building area.
6. Deduct, if appropriate, for common walls, using the figures at the bottom of page 58.
7. Multiply the total cost by the location factor on page 7.
8. Add the cost of the appropriate additional components from page 201. Heating and cooling equipment, fire sprinklers, display fronts, finish materials on the front wall, canopies, interior partitions, exterior signs, mezzanines and basements, loading docks and ramps, yard improvements, and communication systems.

Suburban Store, Class 1 & 2

Depth greater than length of front

Square Foot Area

Quality Class	500	1,000	2,000	2,500	3,000	4,000	5,000	7,500	10,000	12,500	15,000
1, Best	62.15	51.84	44.57	42.71	41.35	39.43	38.12	36.09	34.87	34.05	33.44
1 & 2	56.91	47.47	40.81	39.11	37.86	36.10	34.90	33.04	31.93	31.17	30.61
2, Good	51.70	43.12	37.07	35.53	34.39	32.80	31.71	30.02	29.01	28.32	27.81
2 & 3	48.44	40.41	34.74	33.29	32.23	30.73	29.71	28.13	27.18	26.54	26.06
3, Average	45.18	37.68	32.39	31.05	30.05	28.66	27.71	26.23	25.35	24.75	24.30
3 & 4	42.13	35.14	30.21	28.96	28.03	26.73	25.84	24.46	23.64	23.08	22.67
4, Low	39.70	33.12	28.47	27.29	26.41	25.19	24.35	23.05	22.28	21.75	21.36

Length of front between 1 and 2 times depth

Square Foot Area

Quality Class	500	1,000	2,000	3,000	5,000	7,500	10,000	15,000	20,000	25,000	35,000
1, Best	59.38	49.88	43.20	40.24	37.28	35.42	34.31	32.99	32.21	31.67	30.97
1 & 2	54.28	45.61	39.49	36.79	34.09	32.38	31.37	30.16	29.44	28.95	28.31
2, Good	49.31	41.43	35.87	33.42	30.96	29.41	28.49	27.40	26.75	26.30	25.72
2 & 3	46.17	38.79	33.59	31.29	28.99	27.54	26.67	25.65	25.04	24.62	24.08
3, Average	43.14	36.24	31.38	29.24	27.09	25.73	24.93	23.97	23.40	23.01	22.50
3 & 4	40.44	33.98	29.43	27.41	25.40	24.13	23.37	22.47	21.94	21.57	21.10
4, Low	37.75	31.72	27.46	25.59	23.70	22.52	21.81	20.97	20.48	20.14	19.69

Wall Height Adjustment

Costs above are based on a 16' wall height, measured from the bottom of the floor slab or floor joists to the top of the roof cover. Add or subtract the amount listed to the square foot cost for each foot more or less than 16 feet.

Area	500	1,000	2,000	2,500	3,000	4,000	5,000	7,500	10,000	12,500
Cost	1.02	.75	.54	.48	.44	.38	.34	.27	.23	.20
Area	15,000	20,000	25,000	30,000	35,000	50,000	70,000	75,000	100,000	150,000
Cost	.20	.19	.17	.16	.15	.12	.09	.09	.07	.04

Suburban Stores Masonry or Concrete Building Shell Only

Estimating Procedure

1. Use these figures to estimate the cost of shell-type buildings without permanent partitions, display fronts or finish material on the front wall of the building.
2. Establish the structure quality class by applying the information on page 56.
3. Compute the building floor area. This should include everything within the exterior walls and all inset areas outside the main walls but under the main roof.
4. Add to or subtract from the cost below the appropriate amount from the Wall Height Adjustment Table (on page 57) if the wall height is more or less than 16 feet.
5. Multiply the adjusted square foot cost by the building area.
6. Deduct, if appropriate, for common walls, using the figures at the bottom of page 58.
7. Multiply the total cost by the location factor on page 7.
8. Add the cost of the appropriate additional components from page 201. Heating and cooling equipment, fire sprinklers, display fronts, finish materials on the front wall, canopies, interior partitions, exterior signs, mezzanines and basements, loading docks and ramps, yard improvements, and communication systems.

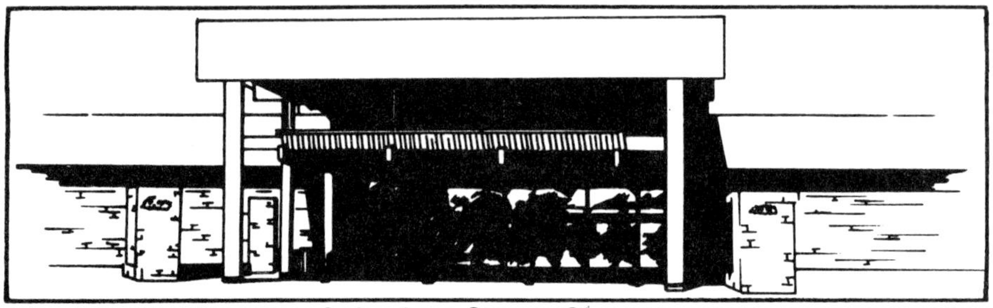

Suburban Store, Class 1

Length between 2 and 4 times depth

Square Foot Area

Quality Class	1,000	2,000	3,000	5,000	7,500	10,000	15,000	20,000	30,000	50,000	70,000
1, Best	51.30	44.25	41.13	38.01	36.04	34.86	33.47	32.64	31.65	30.67	30.14
1 & 2	46.73	40.31	37.47	34.62	32.83	31.76	30.49	29.73	28.84	27.94	27.46
2, Good	42.22	36.42	33.85	31.28	29.66	28.69	27.54	26.86	26.05	25.24	24.81
2 & 3	39.57	34.13	31.73	29.32	27.80	26.89	25.82	25.18	24.42	23.66	23.25
3, Average	36.90	31.83	29.59	27.34	25.92	25.08	24.07	23.48	22.77	22.06	21.68
3 & 4	34.67	29.91	27.80	25.69	24.36	23.56	22.62	22.06	21.40	20.73	20.37
4, Low	32.39	27.94	25.98	24.00	22.76	22.01	21.14	20.61	19.99	19.37	19.03

Length greater than 4 times depth

Square Foot Area

Quality Class	2,000	3,000	5,000	10,000	15,000	20,000	30,000	50,000	75,000	100,000	150,000
1, Best	46.69	43.16	39.63	36.08	34.51	33.58	32.47	31.36	30.66	30.25	29.75
1 & 2	42.06	38.88	35.70	32.50	31.09	30.25	29.25	28.25	27.62	27.25	26.80
2, Good	37.97	35.10	32.23	29.34	28.07	27.31	26.41	25.51	24.94	24.60	24.20
2 & 3	35.57	32.88	30.19	27.48	26.29	25.58	24.74	23.89	23.36	23.04	22.67
3, Average	33.16	30.66	28.15	25.63	24.51	23.85	23.06	22.28	21.78	21.48	21.13
3 & 4	31.06	28.71	26.36	24.00	22.96	22.33	21.60	20.86	20.40	20.12	19.79
4. Low	29.07	26.87	24.67	22.46	21.48	20.90	20.21	19.52	19.09	18.83	18.52

Perimeter Wall Adjustment

A common wall exists when two buildings share one wall. Adjust for common walls by deducting $96.30 per linear foot of common wall from the total structure cost. In some structures one or more walls is not owned at all. In this case, deduct $193.00 per linear foot of wall not owned.

Suburban Stores Masonry or Concrete Multi-Unit Buildings

Estimating Procedure

1. Use these square foot costs to estimate the cost of stores designed for multiple occupancy. These costs include all components of shell buildings plus the cost of display fronts, finish materials on the front of the building and normal interior partitions.
2. Establish the structure quality class by applying the information on page 56. Evaluate the quality of the display front to help establish the correct quality class of the building as a whole. See also pages 207 to 209.
3. Compute the building floor area. This should include everything within the building exterior walls and all inset areas outside the main walls but under the main building roof.
4. Add to or subtract from the cost below the appropriate amount from the Wall Height Adjustment Table (at the bottom of this page) if the wall height is more or less than 16 feet.
5. Multiply the adjusted square foot cost by the building area.
6. Deduct, if appropriate, for common walls, using the figures at the bottom of this page.
7. Multiply the total cost by the location factor on page 7.
8. Add the cost of the appropriate additional components from page 201: Heating and cooling equipment, fire sprinklers, canopies, exterior signs, mezzanines and basements, loading docks and ramps, yard improvements and communications systems.

Depth greater than length of front

Square Foot Area

Quality Class	500	1,000	2,000	2,500	3,000	4,000	5,000	7,500	10,000	12,500	15,000
1, Best	88.45	72.21	60.74	57.81	55.66	52.63	50.57	47.36	45.44	44.14	43.18
1 & 2	78.40	64.00	53.83	51.24	49.33	46.65	44.82	41.97	40.28	39.12	38.27
2, Good	68.35	55.80	46.93	44.67	43.01	40.67	39.07	36.59	35.12	34.11	33.36
2 & 3	62.06	50.66	42.61	40.56	39.05	36.92	35.48	33.22	31.88	30.97	30.29
3, Average	55.72	45.48	38.26	36.42	35.06	33.15	31.85	29.83	28.62	27.80	27.20
3 & 4	51.18	41.78	35.14	33.45	32.20	30.45	29.26	27.40	26.30	25.54	24.98
4, Low	46.65	38.08	32.03	30.49	29.36	27.76	26.67	24.98	23.97	23.28	22.77

Length of front between 1 and 2 times depth

Square Foot Area

Quality Class	500	1,000	2,000	3,000	5,000	7,500	10,000	15,000	20,000	25,000	35,000
1, Best	97.30	78.79	65.59	59.71	53.81	50.07	47.84	45.19	43.61	42.53	41.12
1 & 2	85.55	69.27	57.67	52.50	47.31	44.02	42.06	39.73	38.34	37.40	36.15
2, Good	73.82	59.77	49.76	45.30	40.82	37.99	36.29	34.29	33.09	32.27	31.19
2 & 3	66.66	53.97	44.93	40.91	36.86	34.30	32.77	30.96	29.88	29.14	28.17
3, Average	61.57	49.85	41.50	37.79	34.05	31.68	30.27	28.60	27.60	26.91	26.02
3 & 4	57.66	46.69	38.87	35.38	31.88	29.67	28.35	26.78	25.84	25.20	24.36
4, Low	52.06	42.15	35.09	31.95	28.79	26.79	25.59	24.18	23.33	22.75	22.00

Wall Height Adjustment

Costs above are based on a 16' wall height, measured from the bottom of the floor slab or floor joists to the top of the roof cover. Add or subtract the amount listed to the square foot cost for each foot more or less than 16 feet.

Class	500	1,000	2,000	2,500	3,000	4,000	5,000	7,500	10,000	12,500	15,000	20,000	25,000	30,000	35,000	50,000	70,000	75,000	100,000	150,000
1	5.27	3.71	2.61	2.33	2.13	1.84	1.64	1.33	1.15	1.03	.96	.82	.73	.67	.63	.55	.49	.48	.44	.40
2	3.75	2.64	1.86	1.66	1.51	1.31	1.17	.95	.82	.73	.69	.59	.53	.49	.45	.39	.35	.34	.32	.29
3	2.78	1.96	1.38	1.23	1.12	.97	.86	.70	.61	.54	.52	.44	.39	.36	.34	.29	.26	.26	.24	.22
4	2.12	1.49	1.05	.94	.85	.74	.66	.54	.46	.41	.40	.34	.30	.28	.26	.23	.20	.20	.18	.17

Suburban Stores
Masonry or Concrete
Multi-Unit Buildings

Estimating Procedure

1. Use these square foot costs to estimate the cost of stores designed for multiple occupancy. These costs include all components of shell buildings plus the cost of display fronts, finish materials on the front of the building and normal interior partitions.
2. Establish the structure quality class by applying the information on page 56. Evaluate the quality of the display front to help establish the correct quality class of the building as a whole. See also pages 207 to 209.
3. Compute the building floor area. This should include everything within the building exterior walls and all inset areas outside the main walls but under the main building roof.
4. Add to or subtract from the cost below the appropriate amount from the Wall Height Adjustment Table (at the bottom of page 59) if the wall height is more or less than 16 feet.
5. Multiply the adjusted square foot cost by the building area.
6. Deduct, if appropriate, for common walls, using the figures at the bottom of this page.
7. Multiply the total cost by the location factor on page 7.
8. Add the cost of the appropriate additional components from page 201: Heating and cooling equipment, fire sprinklers, canopies, exterior signs, mezzanines and basements, loading docks and ramps, yard improvements and communications systems.

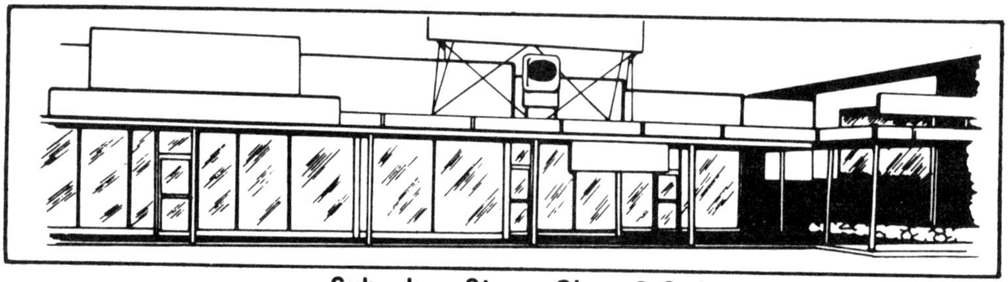

Suburban Store, Class 3 & 4

Length of front between 2 and 4 times depth

Square Foot Area

Quality Class	1,000	2,000	3,000	5,000	7,500	10,000	15,000	20,000	30,000	50,000	70,000
1, Best	88.84	72.88	65.76	58.59	54.06	51.35	48.13	46.20	43.92	41.63	40.41
1 & 2	78.36	64.29	58.01	51.69	47.68	45.29	42.45	40.76	38.74	36.72	35.64
2, Good	67.34	55.25	49.85	44.42	40.98	38.92	36.48	35.02	33.29	31.55	30.63
2 & 3	60.48	49.62	44.77	39.89	36.80	34.96	32.77	31.46	29.90	28.34	27.51
3, Average	53.62	43.99	39.69	35.37	32.63	30.99	29.05	27.89	26.51	25.13	24.39
3 & 4	48.96	40.16	36.24	32.29	29.79	28.30	26.52	25.46	24.20	22.94	22.27
4, Low	44.34	36.37	32.82	29.24	26.98	25.63	24.02	23.06	21.92	20.78	20.17

Length of front greater than 4 times depth

Square Foot Area

Quality Class	2,000	3,000	5,000	10,000	15,000	20,000	30,000	50,000	75,000	100,000	150,000
1, Best	81.46	72.69	63.95	55.22	51.38	49.09	46.39	43.68	41.97	40.96	39.76
1 & 2	71.28	63.60	55.96	48.32	44.96	42.96	40.59	38.22	36.73	35.84	34.79
2, Good	61.17	54.58	48.02	41.47	38.58	36.86	34.83	32.80	31.52	30.76	29.85
2 & 3	54.83	48.93	43.05	37.17	34.58	33.04	31.22	29.40	28.25	27.57	26.76
3, Average	48.70	43.45	38.23	33.01	30.71	29.35	27.73	26.11	25.09	24.49	23.77
3 & 4	44.49	39.70	34.93	30.16	28.06	26.81	25.33	23.86	22.93	22.37	21.71
4, Low	40.18	35.85	31.54	27.24	25.34	24.21	22.88	21.54	20.70	20.20	19.61

Perimeter Wall Adjustment

A common wall exists when two buildings share one wall. Adjust for common walls by deducting $88.00 per linear foot of common wall from the total structure cost. In some structures one or more walls is not owned at all. In this case, deduct $193.00 per linear foot of wall not owned.

Suburban Stores
Wood or Wood and Steel Frame
Quality Classification

	Class 1 Best Quality	Class 2 Good Quality	Class 3 Average Quality	Class 4 Low Quality
Foundation	Reinforced concrete.	Reinforced concrete.	Reinforced concrete.	Reinforced concrete.
Floor Structure	6" reinforced concrete on 6" rock base.	6" reinforced concrete on 6" rock base.	6" reinforced concrete on 6" rock base.	4" reinforced concrete on 6" rock base.
Wall Structure	2" x 6" - 16" o.c.	2" x 6" - 16" o.c.	2" x 6" - 16" o.c.	2" x 4" - 16" o.c.
Roof	Glu-lams or steel beams on steel intermediate columns. Panelized roof system, ½" plywood sheathing, 5 ply built-up roof with insulation.	Glu-lams or steel beams on steel intermediate columns. Panelized roof system, ½" plywood sheathing, 5 ply built-up roof with insulation.	Glu-lams on steel intermediate columns. Panelized roof system, ½" plywood sheathing, 4 ply built-up roof.	Glu-lams on steel intermediate columns. Panelized roof system, ½" plywood sheathing, 4 ply built-up roof.
Floor Finish	Terrazzo, sheet vinyl, or very good carpet.	Vinyl asbestos tile with 50% solid vinyl tile, terrazzo, or good carpet.	Vinyl asbestos tile.	Minimum grade tile.
Interior Wall Finish	Gypsum wallboard or lath and plaster on exterior walls and partitions, finished with vinyl wall covers and hardwood veneers.	Gypsum wallboard, texture and paper on exterior walls and partitions, some vinyl wall cover and plywood paneling.	Gypsum wallboard, texture and paint on interior walls and partitions.	Gypsum wallboard, texture and paint on interior walls and partitions.
Ceiling Finish	Suspended good grade acoustical tile with gypsum board backing.	Suspended acoustical tile with concealed grid system.	Suspended acoustical tile with exposed grid system.	Exposed beams with ceiling tile or painted.
Lighting	Recessed fluorescent lighting in modular plastic panels.	Continuous recessed 3 tube fluorescent strips with egg crate diffusers, 8' o.c.	Continuous 3 tube fluorescent strips with egg crate diffusers, 8' o.c.	Continuous exposed 2 tube fluorescent strips, 8' o.c.
Exterior	Face brick or stone veneer.	Wood siding, some stone veneer.	Stucco on exposed areas, some brick trim.	Stucco on exposed areas.
Plumbing	6 good fixtures per 5,000 S.F. of floor area, metal toilet partitions.	6 standard fixtures per 5,000 S.F. of floor area, metal toilet partitions.	4 standard fixtures per 5,000 S.F. of floor area, metal toilet partitions.	4 standard fixtures per 5,000 S.F. of floor area, wood toilet partitions.

Strip and Island Suburban Stores

For estimating purposes, wood frame suburban stores should be divided into strip type units or island type units. Strip type buildings have a front wall made up of display fronts. The side and rear walls, except for delivery or walk-through doors, are made up of solid, continuous wood frame walls. If there are any display areas in the sides or rear of these buildings, the cost of the display front must be added to the building cost and the cost of the wall that it replaces must be deducted from the building costs.

Island type suburban store buildings have display fronts on the major portion of all four sides. Stores may be arranged so that one store fronts on two sides or they may be partitioned in such a way that there are two separate stores fronting on each side of the building.

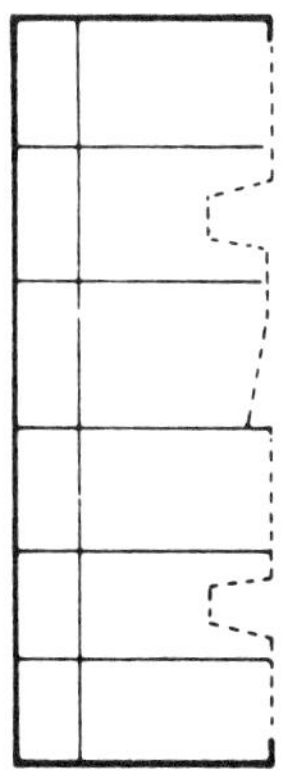

Strip Type

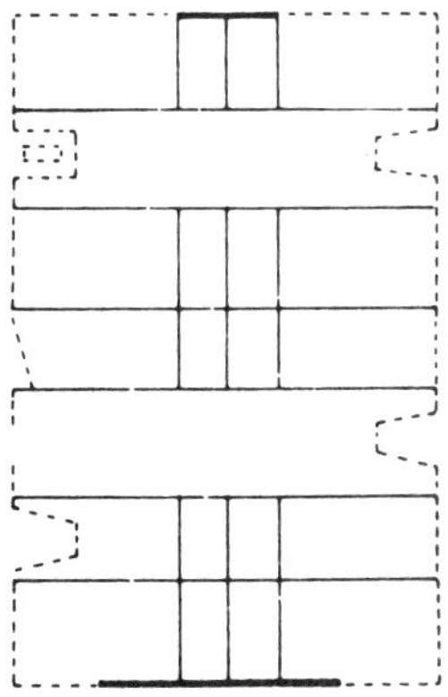

Island Type

Suburban Stores
Wood or Wood and Steel Frame
Building Shell only, Island Type

Estimating Procedure

1. Use these figures to estimate the cost of shell-type buildings without permanent partitions, display fronts or finish material on the front wall of the building.
2. Establish the structure quality class by applying the information on page 61.
3. Compute the building floor area. This should include everything within the exterior walls and all inset areas outside the main walls but under the main roof.
4. Add to or subtract from the square foot cost below the appropriate amount from the Wall Height Adjustment Table (at the bottom of this page) if the wall height is more or less than 16 feet.
5. Multiply the adjusted square foot cost by the building area.
6. Deduct, if appropriate, for common walls, using the figures at the bottom of this page.
7. Multiply the total cost by the location factor on page 7.
8. Add the cost of the appropriate additional components from page 201: Heating and cooling equipment, fire sprinklers, display fronts, finish materials on the front wall, canopies, interior partitions, exterior signs, mezzanines and basements, loading docks and ramps, yard improvements, and communication systems.

Suburban Store, Class 2 & 3

Length less than 1½ times depth

Square Foot Area

Quality Class	3,500	5,000	7,500	10,000	12,500	15,000	20,000	30,000	40,000	50,000	75,000
1, Best	33.33	32.77	32.24	31.93	31.71	31.55	31.32	31.05	30.89	30.78	30.61
1 & 2	29.87	29.37	28.90	28.61	28.42	28.27	28.07	27.83	27.68	27.58	27.43
2, Good	26.43	25.99	25.57	25.32	25.15	25.02	24.84	24.63	24.50	24.41	24.27
2 & 3	24.39	23.99	23.60	23.37	23.21	23.09	22.92	22.73	22.61	22.53	22.40
3, Average	22.39	22.02	21.66	21.45	21.31	21.20	21.04	20.86	20.75	20.68	20.56
3 & 4	20.33	19.99	19.67	19.48	19.34	19.25	19.11	18.94	18.85	18.78	18.67
4, Low	18.29	17.98	17.69	17.52	17.40	17.31	17.19	17.04	16.95	16.89	16.79

Length between 1½ and 2 times depth

Square Foot Area

Quality Class	4,500	5,000	7,500	10,000	15,000	20,000	30,000	40,000	50,000	75,000	100,000
1, Best	33.18	32.99	32.36	32.00	31.60	31.36	31.10	30.94	30.84	30.68	30.59
1 & 2	29.72	29.54	28.98	28.66	28.30	28.09	27.85	27.71	27.62	27.48	27.40
2, Good	26.30	26.15	25.65	25.37	25.05	24.87	24.65	24.53	24.45	24.32	24.25
2 & 3	24.27	24.13	23.67	23.41	23.11	22.94	22.75	22.63	22.56	22.44	22.38
3, Average	22.27	22.14	21.72	21.48	21.21	21.05	20.87	20.77	20.70	20.59	20.53
3 & 4	20.23	20.11	19.73	19.51	19.26	19.12	18.96	18.86	18.80	18.71	18.65
4, Low	18.16	18.05	17.71	17.51	17.29	17.16	17.02	16.93	16.87	16.79	16.74

Wall Height Adjustment

Add or subtract the amount listed to the square foot costs above for each foot of wall height more or less than 16 feet.

Area	3,500	5,000	7,500	10,000	12,500	15,000	20,000	Over 20,000
Cost	.04	.02	.01	.01	.01	.01	.01	.01

Perimeter Wall Adjustment

For a common wall, deduct $17.60 per linear foot of common wall. For no wall ownership, deduct $36.50 per linear foot of wall.

Suburban Stores
Wood or Wood and Steel Frame
Building Shell only, Island Type

Estimating Procedure

1. Use these figures to estimate the cost of shell-type buildings without permanent partitions, display fronts or finish material on the front wall of the building.
2. Establish the structure quality class by applying the information on page 61.
3. Compute the building floor area. This should include everything within the exterior walls and all inset areas outside the main walls but under the main roof.
4. Add to or subtract from the square foot cost below the appropriate amount from the Wall Height Adjustment Table (at the bottom of this page) if the wall height is more or less than 16 feet.
5. Multiply the adjusted square foot cost by the building area.
6. Deduct, if appropriate, for common walls, using the figures at the bottom of this page.
7. Multiply the total cost by the location factor on page 7.
8. Add the cost of the appropriate additional components from page 201: Heating and cooling equipment, fire sprinklers, display fronts, finish materials on the front wall, canopies, interior partitions, exterior signs, mezzanines and basements, loading docks and ramps, yard improvements, and communication systems.

Suburban Store, Class 3 & 4

Length between 2 and 3 times depth

Square Foot Area

Quality Class	10,000	15,000	20,000	30,000	40,000	50,000	75,000	100,000	150,000	200,000	250,000
1, Best	32.49	32.01	31.72	31.39	31.20	31.07	30.86	30.74	30.60	30.52	30.46
1 & 2	29.10	28.66	28.41	28.11	27.94	27.82	27.64	27.53	27.40	27.33	27.28
2, Good	25.74	25.36	25.13	24.87	24.71	24.61	24.45	24.35	24.24	24.18	24.13
2 & 3	23.74	23.38	23.18	22.94	22.79	22.70	22.55	22.46	22.36	22.30	22.25
3, Average	21.76	21.43	21.24	21.02	20.89	20.80	20.67	20.59	20.49	20.44	20.40
3 & 4	19.74	19.45	19.28	19.08	18.96	18.88	18.75	18.68	18.60	18.54	18.51
4, Low	17.73	17.46	17.31	17.13	17.02	16.95	16.84	16.77	16.70	16.65	16.62

Length greater than 3 times depth

Square Foot Area

Quality Class	7,500	10,000	15,000	20,000	30,000	40,000	50,000	75,000	100,000	150,000	200,000
1, Best	32.40	32.19	31.82	31.59	31.29	31.11	30.97	30.76	30.62	30.46	30.36
1 & 2	29.03	28.84	28.50	28.30	28.03	27.87	27.75	27.55	27.43	27.29	27.20
2, Good	25.66	25.49	25.19	25.01	24.78	24.63	24.53	24.35	24.25	24.12	24.04
2 & 3	23.67	23.51	23.24	23.07	22.86	22.72	22.62	22.46	22.37	22.25	22.17
3, Average	21.71	21.58	21.32	21.17	20.97	20.85	20.76	20.61	20.52	20.41	20.34
3 & 4	19.70	19.58	19.35	19.21	19.03	18.92	18.84	18.70	18.62	18.52	18.46
4, Low	17.71	17.60	17.39	17.26	17.10	17.00	16.93	16.81	16.74	16.65	16.59

Wall Height Adjustment

Add or subtract the amount listed to the square foot cost above for each foot of wall height more or less than 16 feet.

Area	3,500	5,000	7,500	10,000	12,500	15,000	20,000	Over 20,000
Cost	.04	.02	.02	.01	.01	.01	.01	.01

Perimeter Wall Adjustment

For a common wall, deduct $17.60 per linear foot of common wall. For no wall ownership, deduct $36.50 per linear foot of wall.

Suburban Stores
Wood or Wood and Steel Frame
Building Shell only, Strip Type

Estimating Procedure

1. Use these figures to estimate the cost of shell-type buildings without permanent partitions, display fronts or finish material on the front wall of the building.
2. Establish the structure quality class by applying the information on page 61.
3. Compute the building floor area. This should include everything within the exterior walls and all inset areas outside the main walls but under the main roof.
4. Add to or subtract from the square foot cost below the appropriate amount from the Wall Height Adjustment Table (at the bottom of this page) if the wall height is more or less than 16 feet.
5. Multiply the adjusted square foot cost by the building area.
6. Deduct, if appropriate, for common walls, using the figures at the bottom of this page.
7. Multiply the total cost by the location factor on page 7.
8. Add the cost of the appropriate additional components from page 201: Heating and cooling equipment, fire sprinklers, display fronts, finish materials on the front wall, canopies, interior partitions, exterior signs, mezzanines and basements, loading docks and ramps, yard improvements, and communication systems.

Length less than 1½ times depth

Square Foot Area

Quality Class	500	1,000	2,000	2,500	3,000	4,000	5,000	7,500	10,000	12,500	15,000
1, Best	52.09	45.00	39.99	38.71	37.77	36.44	35.54	34.14	33.30	32.73	32.31
1 & 2	47.09	40.68	36.15	35.00	34.14	32.95	32.13	30.86	30.10	29.59	29.21
2, Good	42.14	36.41	32.35	31.32	30.55	29.48	28.75	27.62	26.94	26.48	26.14
2 & 3	39.56	34.18	30.37	29.40	28.68	27.68	26.99	25.92	25.29	24.85	24.53
3, Average	36.97	31.94	28.38	27.47	26.81	25.87	25.22	24.23	23.63	23.23	22.93
3 & 4	34.27	29.60	26.31	25.47	24.85	23.98	23.38	22.46	21.91	21.53	21.25
4, Low	31.61	27.31	24.27	23.50	22.92	22.12	21.57	20.72	20.21	19.86	19.61

Length between 1½ and 2 times depth

Square Foot Area

Quality Class	500	1,000	2,000	3,000	5,000	6,000	7,500	10,000	15,000	20,000	35,000
1, Best	50.22	43.57	38.84	36.73	34.61	33.97	33.26	32.46	31.51	30.95	30.05
1 & 2	45.41	39.40	35.11	33.21	31.29	30.71	30.07	29.35	28.49	27.98	27.17
2, Good	40.65	35.27	31.43	29.73	28.01	27.49	26.92	26.28	25.51	25.05	24.32
2 & 3	38.10	33.06	29.46	27.86	26.25	25.77	25.23	24.63	23.91	23.48	22.80
3, Average	35.57	30.86	27.50	26.01	24.51	24.06	23.56	22.99	22.32	21.92	21.28
3 & 4	33.08	28.70	25.58	24.19	22.79	22.37	21.91	21.38	20.75	20.38	19.79
4, Low	30.62	26.57	23.68	22.39	21.10	20.71	20.28	19.79	19.22	18.87	18.32

Wall Height Adjustment

Add or subtract the amount listed to the square foot cost above for each foot of wall height more or less than 16 feet.

Area	500	1,000	2,000	2,500	3,000	4,000	5,000	7,500	10,000	12,500
Cost	.57	.40	.29	.26	.24	.21	.19	.16	.15	.14
Area	15,000	20,000	25,000	30,000	35,000	50,000	70,000	75,000	100,000	150,000
Cost	.08	.06	.05	.04	.04	.03	.02	.02	.02	.02

Perimeter Wall Adjustment

For a common wall, deduct $63.50 per linear foot of common wall. For no wall ownership, deduct $124.00 per linear foot of wall.

Suburban Stores
Wood or Wood and Steel Frame
Building Shell only, Strip Type

Estimating Procedure

1. Use these figures to estimate the cost of shell-type buildings without permanent partitions, display fronts or finish material on the front wall of the building.
2. Establish the structure quality class by applying the information on page 61.
3. Compute the building floor area. This should include everything within the exterior walls and all inset areas outside the main walls but under the main roof.
4. Add to or subtract from the square foot cost below the appropriate amount from the Wall Height Adjustment Table (at the bottom of this page) if the wall height is more or less than 16 feet.
5. Multiply the adjusted square foot cost by the building area.
6. Deduct if appropriate for common walls, using the figures at the bottom of this page.
7. Multiply the total cost by the location factor on page 7.
8. Add the cost of the appropriate additional components from page 201: Heating and cooling equipment, fire sprinklers, display fronts, finish materials on the front wall, canopies, interior partitions, exterior signs, mezzanines, and basements, loading docks and ramps, yard improvements, and communication systems.

Length between 2 and 3 times depth

Square Foot Area

Quality Class	1,000	2,000	3,000	5,000	7,500	10,000	15,000	20,000	30,000	50,000	70,000
1, Best	44.23	39.43	37.30	35.16	33.82	33.02	32.07	31.50	30.83	30.15	29.79
1 & 2	39.99	35.65	33.72	31.79	30.58	29.85	28.99	28.48	27.87	27.26	26.94
2, Good	35.78	31.90	30.17	28.45	27.36	26.71	25.94	25.48	24.94	24.39	24.10
2 & 3	33.62	29.96	28.34	26.72	25.70	25.09	24.37	23.94	23.43	22.92	22.64
3, Average	31.47	28.05	26.53	25.01	24.06	23.49	22.81	22.41	21.93	21.45	21.19
3 & 4	29.27	26.09	24.68	23.27	22.38	21.85	21.22	20.84	20.40	19.95	19.71
4, Low	27.06	24.12	22.82	21.51	20.69	20.20	19.62	19.27	18.86	18.45	18.23

Length greater than 3 times depth

Square Foot Area

Quality Class	2,000	3,000	5,000	10,000	15,000	20,000	30,000	50,000	75,000	100,000	150,000
1, Best	40.73	38.37	36.01	33.65	32.60	31.98	31.24	30.51	30.04	29.77	29.44
1 & 2	36.60	34.47	32.35	30.23	29.29	28.73	28.07	27.41	26.99	26.75	26.45
2, Good	32.75	30.85	28.96	27.06	26.22	25.72	25.13	24.53	24.16	23.94	23.67
2 & 3	30.73	28.95	27.17	25.39	24.60	24.13	23.57	23.02	22.67	22.46	22.21
3, Average	28.74	27.07	25.41	23.74	23.00	22.56	22.04	21.52	21.20	21.00	20.77
3 & 4	26.83	25.27	23.72	22.16	21.47	21.06	20.58	20.09	19.79	19.61	19.39
4, Low	24.89	23.45	22.00	20.56	19.92	19.54	19.09	18.64	18.36	18.19	17.99

Wall Height Adjustment

Add or subtract the amount listed to the square foot cost above for each foot of wall height more or less than 16 feet.

Area	500	1,000	2,000	2,500	3,000	4,000	5,000	7,500	10,000	12,500
Cost	.57	.40	.29	.26	.24	.21	.19	.16	.15	.14
Area	15,000	20,000	25,000	30,000	35,000	50,000	70,000	75,000	100,000	150,000
Cost	.08	.06	.05	.04	.04	.03	.02	.02	.02	.02

Perimeter Wall Adjustment

For a common wall, deduct $63.50 per linear foot of common wall. For no wall ownership, deduct $124.00 per linear foot of wall.

Suburban Stores
Wood or Wood and Steel Frame
Multi-Unit, Island Type

Estimating Procedure

1. Use these square foot costs to estimate the cost of stores designed for multiple occupancy. These costs include all components of shell buildings plus the cost of display fronts, finish materials on the front of the building and normal interior partitions.
2. Establish the structure quality class by applying the information on page 61. Evaluate the quality of the display front to help establish the correct quality class of the building as a whole. The building classes have display fronts as classified on page 56. See also pages 207 to 209.
3. Compute the building floor area. This should include everything within the building exterior walls and all inset areas outside the main walls but under the main building roof.
4. Add to or subtract from the square foot cost below the appropriate amount from the Wall Height Adjustment Table (at the bottom of this page) if the wall height is more or less than 16 feet.
5. Multiply the adjusted square foot cost by the building area.
6. Deduct, if appropriate, for common walls, using the figures at the bottom of this page.
7. Multiply the total cost by the location factor on page 7.
8. Add the cost of the appropriate additional components from page 201: Heating and cooling equipment, fire sprinklers, canopies, exterior signs, mezzanines and basements, loading docks and ramps, yard improvements and communications systems.

Length less than 1½ times depth

Square Foot Area

Quality Class	3,500	5,000	7,500	10,000	12,500	15,000	20,000	30,000	40,000	50,000	75,000
1, Best	72.48	66.48	60.81	57.41	55.08	53.36	50.94	48.05	46.33	45.15	43.32
1 & 2	63.81	58.54	53.54	50.55	48.50	46.98	44.85	42.31	40.79	39.75	38.14
2, Good	54.03	49.57	45.34	42.80	41.07	39.78	37.98	35.83	34.54	33.66	32.29
2 & 3	47.24	43.33	39.63	37.42	35.90	34.78	33.20	31.32	30.19	29.43	28.23
3, Average	40.47	37.13	33.96	32.06	30.76	29.80	28.44	26.83	25.87	25.21	24.19
3 & 4	35.86	32.89	30.09	28.41	27.25	26.40	25.20	23.78	22.92	22.34	21.43
4, Low	31.38	28.79	26.33	24.86	23.85	23.11	22.06	20.81	20.06	19.55	18.76

Length between 1½ and 2 times depth

Square Foot Area

Quality Class	4,500	5,000	7,500	10,000	15,000	20,000	30,000	40,000	50,000	75,000	100,000
1, Best	70.70	68.69	62.07	58.26	53.87	51.33	48.37	46.65	45.48	43.70	42.65
1 & 2	61.65	59.90	54.13	50.80	46.98	44.76	42.18	40.68	39.66	38.11	37.19
2, Good	52.63	51.13	46.20	43.37	40.10	38.21	36.01	34.72	33.86	32.53	31.75
2 & 3	46.05	44.74	40.43	37.95	35.09	33.43	31.51	30.38	29.63	28.46	27.78
3, Average	39.50	38.37	34.67	32.54	30.09	28.67	27.02	26.06	25.41	24.41	23.83
3 & 4	35.29	34.29	30.98	29.08	26.89	25.62	24.14	23.28	22.70	21.81	21.29
4, Low	31.14	30.25	27.33	25.66	23.72	22.60	21.30	20.54	20.03	19.25	18.78

Wall Height Adjustment

Add or subtract the amount listed to the square foot cost above for each foot of wall height more or less than 16 feet.

Square Foot Area

Class	3,500	5,000	7,500	10,000	12,500	15,000	20,000	30,000	40,000	50,000	75,000	100,000	150,000
1	3.65	3.06	2.51	2.19	1.96	1.80	1.56	1.29	1.12	1.01	.83	.73	.61
2	2.46	2.06	1.69	1.47	1.32	1.21	1.05	.87	.76	.68	.56	.49	.41
3	1.65	1.39	1.14	.99	.89	.81	.71	.58	.51	.46	.38	.33	.27
4	1.18	.99	.82	.71	.64	.58	.51	.42	.36	.33	.27	.24	.20

Perimeter Wall Adjustment

For common wall, deduct per linear foot: Class 1, $350; Class 2, $235; Class 3, $160; Class 4, $105. For no wall ownership, deduct per linear foot: Class 1, $685; Class 2, $450; Class 3, $315; Class 4, $210.

Suburban Stores
Wood or Wood and Steel Frame
Multi-Unit, Island Type

Estimating Procedure

1. Use these square foot costs to estimate the cost of stores designed for multiple occupancy. These costs include all components of shell buildings plus the cost of display fronts, finish materials on the front of the building and normal interior partitions.
2. Establish the structure quality class by applying the information on page 61. Evaluate the quality of the display front to help establish the correct quality class of the building as a whole. The building classes have display fronts as classified on page 56. See also pages 207 to 209.
3. Compute the building floor area. This should include everything within the building exterior walls and all inset areas outside the main walls but under the main building roof.
4. Add to or subtract from the square foot cost below the appropriate amount from the Wall Height Adjustment Table (at the bottom of this page) if the wall height is more or less than 16 feet.
5. Multiply the adjusted square foot cost by the building area.
6. Deduct, if appropriate, for common walls, using the figures at the bottom of this page.
7. Multiply the total cost by the location factor on page 7.
8. Add the cost of the appropriate additional components from page 201: Heating and cooling equipment, fire sprinklers, canopies, exterior signs, mezzanines and basements, loading docks and ramps, yard improvements and communications systems.

Length between 2 and 3 times depth

Square Foot Area

Quality Class	7,500	10,000	15,000	20,000	30,000	40,000	50,000	75,000	100,000	150,000	200,000
1, Best	62.53	58.91	54.58	51.99	48.91	47.06	45.80	43.83	42.65	41.26	40.42
1 & 2	54.67	51.50	47.72	45.45	42.75	41.14	40.04	38.32	37.29	36.07	35.34
2, Good	46.84	44.13	40.89	38.95	36.64	35.25	34.31	32.83	31.95	30.90	30.28
2 & 3	41.14	38.76	35.91	34.21	32.18	30.96	30.13	28.84	28.06	27.14	26.60
3, Average	35.52	33.46	31.00	29.53	27.78	26.73	26.01	24.89	24.23	23.43	22.96
3 & 4	31.59	29.76	27.58	26.27	24.71	23.78	23.14	22.14	21.55	20.84	20.42
4, Low	27.71	26.10	24.19	23.04	21.67	20.85	20.29	19.42	18.90	18.28	17.91

Length greater than 3 times depth

Square Foot Area

Quality Class	10,000	15,000	20,000	30,000	40,000	50,000	75,000	100,000	150,000	200,000	250,000
1, Best	62.03	57.04	54.07	50.57	48.49	47.07	44.87	43.56	42.01	41.09	40.46
1 & 2	54.26	49.89	47.30	44.23	42.41	41.17	39.25	38.10	36.75	35.94	35.39
2, Good	46.51	42.77	40.55	37.92	36.36	35.30	33.65	32.66	31.50	30.81	30.34
2 & 3	40.89	37.60	35.65	33.34	31.96	31.03	29.58	28.72	27.69	27.09	26.67
3, Average	35.30	32.46	30.77	28.78	27.59	26.79	25.53	24.79	23.91	23.38	23.02
3 & 4	31.43	28.90	27.40	25.62	24.57	23.85	22.73	22.07	21.29	20.82	20.50
4, Low	27.57	25.35	24.03	22.48	21.55	20.92	19.94	19.36	18.67	18.26	17.98

Wall Height Adjustment

Add or subtract the amount listed to the square foot cost above for each foot of wall height more or less than 16 feet.

Class	7,500	10,000	12,500	15,000	20,000	30,000	40,000	50,000	75,000	100,000	150,000	200,000	250,000
1	2.53	2.20	1.97	1.80	1.57	1.29	1.12	1.01	.83	.73	.60	.53	.47
2	1.71	1.49	1.33	1.22	1.06	.87	.76	.68	.56	.49	.41	.35	.32
3	1.15	1.00	.89	.82	.71	.58	.51	.46	.38	.33	.27	.24	.21
4	.83	.72	.64	.59	.51	.42	.37	.33	.27	.24	.20	.17	.16

Perimeter Wall Adjustment

Class	For a Common Wall, Deduct per Linear Foot.	For No Wall Ownership, Deduct per Linear Foot.
1	$350.00	$685.00
2	235.00	450.00
3	160.00	315.00
4	105.00	210.00

Suburban Stores
Wood or Wood and Steel Frame
Multi-Unit, Strip Type

Estimating Procedure

1. Use these square foot costs to estimate the cost of stores designed for multiple occupancy. These costs include all components of shell buildings plus the cost of display fronts, finish materials on the front of the building and normal interior partitions.
2. Establish the structure quality class by applying the information on page 61. Evaluate the quality of the display front to help establish the correct quality class of the building as a whole. The building classes have display fronts as classified on page 56. See also pages 207 to 209.
3. Compute the building floor area. This should include everything within the building exterior walls and all inset areas outside the main walls but under the main roof.
4. Add to or subtract from the square foot cost below the appropriate amount from the Wall Height Adjustment Table (at the bottom of this page) if the wall height is more or less than 16 feet.
5. Multiply the adjusted square foot cost by the building area.
6. Deduct, if appropriate, for common walls, using the figures at the bottom of this page.
7. Multiply the total cost by the location factor on page 7.
8. Add the cost of the appropriate additional components from page 201: Heating and cooling equipment, fire sprinklers, canopies, exterior signs, mezzanines and basements, loading docks and ramps, yard improvements and communications systems.

Suburban Store, Class 3

Length less than 1½ times depth

Square Foot Area

Quality Class	500	1,000	2,000	2,500	3,000	4,000	5,000	7,500	10,000	12,500	15,000
1, Best	76.38	63.69	54.74	52.46	50.78	48.42	46.81	44.31	42.82	41.81	41.06
1 & 2	66.94	55.82	47.97	45.97	44.50	42.44	41.03	38.84	37.53	36.64	35.98
2, Good	57.54	47.98	41.24	39.52	38.26	36.48	35.27	33.38	32.26	31.50	30.93
2 & 3	51.88	43.26	37.18	35.63	34.49	32.89	31.80	30.10	29.09	28.40	27.89
3, Average	46.35	38.65	33.22	31.84	30.81	29.38	28.41	26.89	25.99	25.37	24.92
3 & 4	42.25	35.23	30.28	29.02	28.09	26.78	25.89	24.51	23.69	23.13	22.71
4, Low	38.14	31.80	27.33	26.20	25.36	24.18	23.38	22.13	21.38	20.88	20.50

Length between 1½ and 2 times depth

Square Foot Area

Quality Class	500	1,000	2,000	3,000	5,000	7,500	10,000	15,000	20,000	25,000	35,000
1, Best	85.23	70.14	59.45	54.72	49.96	46.97	45.18	43.06	41.80	40.93	39.80
1 & 2	74.47	61.28	51.94	47.81	43.66	41.04	39.48	37.62	36.52	35.77	34.78
2, Good	63.77	52.47	44.48	40.94	37.38	35.14	33.80	32.22	31.27	30.63	29.78
2 & 3	57.27	47.12	39.94	36.76	33.57	31.56	30.36	28.93	28.08	27.50	26.74
3, Average	50.87	41.86	35.48	32.65	29.82	28.03	26.96	25.70	24.94	24.43	23.75
3 & 4	46.05	37.90	32.12	29.56	27.00	25.38	24.41	23.27	22.58	22.12	21.51
4, Low	41.38	34.05	28.86	26.56	24.26	22.80	21.94	20.91	20.29	19.87	19.32

Wall Height Adjustment

Add or subtract the amount listed to the square foot cost above for each foot of wall height more or less than 16 feet.

Class	500	1,000	2,000	2,500	3,000	4,000	5,000	7,500	10,000	12,500	15,000	20,000	25,000	30,000
1	5.04	3.52	2.47	2.21	2.01	1.74	1.56	1.28	1.11	1.00	.92	.80	.72	.66
2	3.15	2.20	1.54	1.38	1.26	1.09	.98	.80	.70	.62	.57	.50	.45	.41
3	2.21	1.54	1.08	.97	.88	.76	.68	.56	.49	.44	.40	.35	.31	.29
4	1.56	1.09	.77	.68	.62	.54	.48	.40	.34	.31	.28	.25	.22	.20

Perimeter Wall Adjustment

For a common wall, deduct $63.50 per linear foot. For no wall ownership, deduct $124.00 per linear foot.

Suburban Stores
Wood or Wood and Steel Frame
Multi-Unit, Strip Type

Estimating Procedure

1. Use these square foot costs to estimate the cost of stores designed for multiple occupancy. These costs include all components of shell buildings plus the cost of display fronts, finish materials on the front of the building and normal interior partitions.
2. Establish the structure quality class by applying the information on page 61. Evaluate the quality of the display front to help establish the correct quality class of the building as a whole. The building classes have display fronts as classified on page 56. See also pages 207 to 209.
3. Compute the building floor area. This should include everything within the building exterior walls and all inset areas outside the main walls but under the main roof.
4. Add to or subtract from the square foot cost below the appropriate amount from the Wall Height Adjustment Table (at the bottom of this page) if the wall height is more or less than 16 feet.
5. Multiply the adjusted square foot cost by the building area.
6. Deduct, if appropriate, for common walls, using the figures at the bottom of this page.
7. Multiply the total cost by the location factor on page 7.
8. Add the cost of the appropriate additional components from page 201: Heating and cooling equipment, fire sprinklers, canopies, exterior signs, mezzanines and basements, loading docks and ramps, yard improvements and communications systems.

Suburban Store, Class 4

Length between 2 and 3 times depth

Quality Class	1,000	2,000	3,000	5,000	7,500	10,000	15,000	20,000	30,000	50,000	70,000
1, Best	80.22	66.62	60.60	54.58	50.79	48.53	45.85	44.25	42.36	40.46	39.45
1 & 2	69.96	58.09	52.85	47.60	44.29	42.32	39.98	38.59	36.94	35.28	34.40
2, Good	59.71	49.58	45.11	40.62	37.80	36.12	34.12	32.93	31.53	30.11	29.36
2 & 3	53.50	44.42	40.41	36.40	33.87	32.36	30.57	29.51	28.25	26.98	26.31
3, Average	47.27	39.26	35.71	32.16	29.93	28.60	27.02	26.08	24.96	23.84	23.25
3 & 4	42.80	35.54	32.33	29.12	27.09	25.89	24.46	23.61	22.60	21.59	21.05
4, Low	38.35	31.84	28.97	26.09	24.28	23.20	21.92	21.15	20.25	19.34	18.86

Length greater than 3 times depth

Quality Class	2,000	3,000	5,000	10,000	15,000	20,000	30,000	50,000	75,000	100,000	150,000
1, Best	73.21	65.88	58.63	51.44	48.29	46.42	44.22	42.02	40.63	39.81	38.84
1 & 2	63.78	57.39	51.08	44.81	42.07	40.44	38.52	36.60	35.40	34.68	33.84
2, Good	54.39	48.94	43.55	38.21	35.87	34.49	32.85	31.21	30.19	29.57	28.85
2 & 3	46.98	42.28	37.62	33.01	30.99	29.79	28.38	26.96	26.08	25.55	24.92
3, Average	42.93	38.63	34.38	30.17	28.32	27.22	25.93	24.64	23.83	23.35	22.78
3 & 4	39.10	35.19	31.31	27.47	25.79	24.79	23.62	22.44	21.70	21.26	20.74
4, Low	35.24	31.71	28.22	24.76	23.25	22.35	21.29	20.23	19.56	19.16	18.70

Wall Height Adjustment

Add or subtract the amount listed to the square foot of floor cost for each foot of wall height more or less than 16 feet.

Class	1,000	2,000	3,000	5,000	7,500	10,000	15,000	20,000	30,000	50,000	70,000	75,000	100,000	150,000
1	3.52	2.47	2.01	1.57	1.29	1.12	.93	.81	.68	.54	.47	.45	.40	.34
2	2.17	1.53	1.25	.97	.80	.69	.57	.50	.42	.33	.29	.28	.25	.21
3	1.55	1.09	.89	.69	.57	.50	.41	.36	.30	.24	.21	.20	.18	.15
4	1.12	.79	.64	.50	.41	.36	.30	.26	.22	.17	.15	.15	.13	.11

Perimeter Wall Adjustment

For a common wall, deduct $63.50 per linear foot. For no wall ownership, deduct $124.00 per linear foot.

Supermarkets
Masonry or Concrete
Quality Classification

	Class 1 Best Quality	Class 2 Good Quality	Class 3 Average Quality	Class 4 Low Quality
Foundation	Reinforced concrete.	Reinforced concrete.	Reinforced concrete.	Reinforced concrete.
Floor Structure	4" reinforced concrete on 6" rock fill.	4" reinforced concrete on 6" rock fill.	4" reinforced concrete on 6" rock fill.	4" reinforced concrete on 6" rock fill.
Wall Structure	6" concrete tilt-up or ornamental block or brick.	6" concrete tilt-up, colored concrete block or brick.	6" concrete tilt-up or 8" concrete block.	6" concrete tilt-up or 8" concrete block.
Roof Structure	Glu-lams or steel "I" beams on steel intermediate columns, 2" x 12" purlins 16" o.c., ½" plywood sheathing.	Glu-lams or steel "I" beams on steel intermediate columns, 2" x 12" purlins 16" o.c., ½" plywood sheathing.	Glu-lams or steel "I" beams on steel intermediate columns, 2" x 12" purlins 16" o.c., ½" plywood sheathing.	Glu-lams or steel "I" beams on steel intermediate columns, 3" x 12" purlins 3' o.c., ½" plywood sheathing.
Floor Finish	Terrazzo in sales area. Sheet vinyl or carpet in cashiers' area.	Vinyl asbestos tile in sales area. Terrazzo, solid vinyl tile or carpet in cashiers' area.	Vinyl asbestos tile in sales area.	Minimum grade tile in sales area.
Interior Wall Finish	Inside of exterior walls furred out with gypsum wallboard and paint or interior stucco, interior stucco or gypsum wallboard and vinyl wall cover on partitions.	Paint on inside of exterior walls, gypsum wallboard and paint or vinyl wall cover on partitions.	Paint on inside of exterior walls, wallboard and paint on partitions.	Paint on inside of exterior walls, wallboard and paint on partitions.
Ceiling Finish	Suspended acoustical tile, dropped ceiling over meat and produce departments.	Suspended acoustical tile or gypsum board and acoustical texture, dropped ceiling over meat and produce departments.	Ceiling tile on roof purlins, dropped ceiling over meat department.	Open.
Front	A large amount of plate glass in good aluminum frames, (18'-22' high for ¾ of width), brick or stone veneer on remainder, 1 pair of good automatic doors per 7,000 S.F. of floor area, anodized aluminum sunshade over glass area, 8' canopy across front, 10'-12' raised walk across front.	A large amount of plate glass in good aluminum frames, (16'-18' high for 2/3 of width), brick or stone veneer on remainder, 1 pair of good automatic doors per 10,000 S.F. of floor area, 8' canopy across front, 10' raised walk across front.	A moderate amount of plate glass in average quality aluminum frames, (12'-16' high for 2/3 of width), exposed aggregate on remainder, 1 pair average automatic doors per 10,000 S.F. of floor area, 6' canopy across front, 8' raised walk across front.	Stucco or exposed aggregate with a small amount of plate glass in an inexpensive aluminum frame, (6'-10' high for ½ of width) 6' canopy across front, 6' ground level walk across front.
Exterior Wall Finish	Large ornamental rock or brick veneer.	Large ornamental rock or brick veneer.	Paint, some exposed aggregate.	Paint.
Roof Cover	5 ply built-up roofing with large rock.	5 ply built-up roofing with tar and rock.	4 ply built-up roofing.	4 ply built-up roofing.
Plumbing	2 rest rooms with 3 fixtures, floor piping and drains to refrigerated cases, 2 double sinks with drain board.	2 rest rooms with 3 fixtures each, floor piping and drains to refrigerated cases, 2 double sinks with drain board.	2 rest rooms with 2 fixtures each, floor piping and drains to refrigerated cases, 2 double sinks.	1 rest room with 2 fixtures, floor piping and drains to refrigerated cases.
Electrical	Conduit wiring, recessed 4 tube fluorescent fixtures 8' o.c., 30-40 spotlights.	Conduit wiring, 4 tube fluorescent fixtures with diffusers 8' o.c., 30-40 spotlights.	Conduit wiring, 3 tube fluorescent fixtures, 8' o.c., 5 or 10 spotlights.	Conduit wiring, double tube fluorescent fixtures, 8' o.c.

Square foot costs include the following components: Foundations as required for normal soil conditions. Floor, wall, and roof structures. Interior floor, wall and ceiling finishes. Exterior wall finish and roof cover. Display fronts. Interior partitions. Entry and delivery doors. A canopy and walk across the front of the building as described in the applicable building specifications. Basic lighting and electrical systems. Rough and finish plumbing. All plumbing, piping and wiring necessary to operate the usual refrigerated cases and vegetable cases. Design and engineering fees. Permits and hook-up fees. Contractor's mark-up.

Supermarkets
Masonry or Concrete

Estimating Procedure

1. Establish the structure quality class by using the information on page 70.
2. Compute the building floor area. This should include everything within the building exterior walls and all insets outside the main walls but under the main building roof.
3. Add to or subtract from the square foot cost below the appropriate amount from the Wall Height Adjustment Column (at the bottom of this page) if the wall height is more or less than 20 feet.
4. Multiply the adjusted square foot cost by the building area.
5. Deduct, if appropriate, for common walls, using the figures at the bottom of this page.
6. Multiply the total cost by the locaton factor listed on page 7.
7. Add the cost of heating and cooling equipment, fire sprinklers, exterior signs, yard improvements, loading docks, ramps and walk-in boxes if they are an integral part of the building. See pages 201 to 213.

Supermarket, Class 2

Square Foot Area

Quality Class	5,000	7,500	10,000	12,500	15,000	20,000	25,000	30,000	35,000	40,000	50,000
Exceptional	57.21	53.20	50.80	49.15	47.93	46.22	45.04	44.18	43.50	42.96	42.12
1, Best	53.72	49.96	47.70	46.15	45.01	43.40	42.30	41.48	40.85	40.34	39.55
1 & 2	49.26	45.81	43.74	42.33	41.28	39.80	38.79	38.04	37.46	36.99	36.27
2, Good	44.27	41.17	39.31	38.04	37.09	35.77	34.86	34.19	33.67	33.24	32.60
2 & 3	41.06	38.18	36.46	35.27	34.40	33.17	32.33	31.71	31.22	30.83	30.23
3, Average	37.45	34.83	33.25	32.18	31.38	30.26	29.49	28.92	28.48	28.12	27.58
3 & 4	33.61	31.26	29.84	28.88	28.16	27.15	26.47	25.96	25.56	25.24	24.75
4, Low	29.56	27.49	26.25	25.40	24.77	23.88	23.27	22.83	22.48	22.20	21.76
Wall Height Adjustment *	.61	.50	.43	.39	.35	.31	.27	.25	.23	.22	.19

*Wall Height Adjustment

Add or subtract the amount listed in this column to the square foot of floor cost for each foot of wall height more or less than 20 feet.

Perimeter Wall Adjustment

A common wall exists when two buildings share one wall. Adjust for common walls by deducting the linear foot costs below from the total structure cost. In some structures one or more walls is not owned at all. In this case, deduct the ''No Ownership'' cost per linear foot of wall not owned.

For Common Wall, deduct $102.00 per linear foot.
For No Wall Ownership, deduct $204.00 per linear foot.

Supermarkets
Wood or Wood and Steel Frame
Quality Classification

	Class 1 Best Quality	Class 2 Good Quality	Class 3 Average Quality	Class 4 Low Quality
Foundation	Reinforced concrete.	Reinforced concrete.	Reinforced concrete.	Reinforced concrete.
Floor Structure	4" reinforced concrete on 6" rock fill.	4" reinforced concrete on 6" rock fill.	4" reinforced concrete on 6" rock fill.	4" reinforced concrete on 6" rock fill.
Wall Structure	2" x 6" - 16" o.c.	2" x 6" - 16" o.c.	2" x 6" - 16" o.c.	2" x 4" - 16" o.c.
Roof Structure	Glu-lams or steel "I" beams on steel intermediate columns, 2" x 12" purlins 16" o.c., ½" plywood sheathing.	Glu-lams or steel, "I" beams on steel intermediate columns, 2" x 12" purlins 16" o.c., ½" plywood sheathing.	Glu-lams or steel "I" beams on steel intermediate columns, 2" x 12" purlins 16" o.c., ½" plywood sheathing.	Glu-lams or steel "I" beams on steel intermediate columns, 3" x 12" purlins 3' o.c., ½" plywood sheathing.
Floor Finish	Terrazzo in sales area. Sheet vinyl or carpet in cashiers' area.	Vinyl asbestos tile in sales area. Terrazzo, solid vinyl tile or carpet in cashiers' area.	Vinyl asbestos tile in sales area.	Minimum tile or inexpensive vinyl asbestos tile in sales area.
Interior Wall Finish	Gypsum wallboard and vinyl wall cover or interior stucco on inside of exterior walls and on partitions.	Gypsum wallboard, texture and paint or vinyl wall cover on inside of exterior walls, and on partitions.	Gypsum wallboard, texture and paint on inside of exterior walls, and on partitions.	Gypsum wallboard and paint on inside of exterior walls, and on partitions.
Ceiling Finish	Suspended acoustical tile, dropped ceiling over meat and produce departments.	Suspended acoustical tile or gypsum board and acoustical texture, dropped ceiling over meat and produce departments.	Ceiling tile on roof purlins, dropped ceiling over meat department.	Open.
Front	A large amount of plate glass in good aluminum frames, (18'-22' high for ¾ of width), brick or stone veneer on remainder, 1 pair good automatic doors per 7,000 S.F. of floor area, anodized aluminum sunshade over glass area, 8' canopy across front, 10'-12' raised walk across front.	A large amount of plate glass in good aluminum frames, (16'-18' high for 2/3 of width), brick or stone veneer on remainder, 1 pair good automatic doors per 10,000 S.F. of floor area, 9' canopy across front, 10' raised walk across front.	Moderate amount of plate glass in average quality aluminum frames, (12'-16' high for 2/3 of width), wood siding on remainder, 1 pair average automatic doors per 10,000 S.F. of floor area, 6' canopy across front, 8' raised walk across front.	Stucco with small amount of plate glass in an inexpensive aluminum frame, (6'-10' high for ½ of width), 6' canopy across front, 6' ground level walk across front.
Exterior Wall Finish	Large ornamental rock or brick veneer.	Good wood siding, some masonry veneer.	Stucco or wood siding.	Stucco.
Roof Cover	5 ply built-up roofing with large rock.	5 ply built-up roofing with tar and rock.	4 ply built-up roofing.	4 ply built-up roofing.
Plumbing	2 rest rooms with 3 fixtures each, floor piping and drains to refrigerated cases, 2 double sinks with drain board.	2 rest rooms with 3 fixtures each, floor piping and drains to refrigerated cases, 2 double sinks with drain board.	2 rest rooms with 2 fixtures each, floor piping and drains to refrigerated cases, 2 double sinks.	1 rest room with 2 fixtures, floor piping and drains to refrigerated cases.
Electrical	Conduit wiring, recessed 4 tube fluorescent fixtures, 8' o.c., 30 to 40 spotlights.	Conduit wiring, 4 tube fluorescent fixtures with diffusers, 8' o.c., 30 to 40 spotlights.	Conduit wiring, 3 tube fluorescent fixtures, 8' o.c., 5 or 10 spotlights.	Conduit wiring, double tube fluorescent fixtures, 8' o.c.

Square foot costs include the following components: Foundations as required for normal soil conditions. Floor, wall, and roof structures. Interior floor, wall and ceiling finishes. Exterior wall finish and roof cover. Display fronts. Interior partitions. Entry and delivery doors. A canopy and walk across the front of the building as described in the applicable building specifications. Basic lighting and electrical systems. Rough and finish plumbing. All plumbing, piping and wiring necessary to operate the usual refrigerated cases and vegetable cases. Design and engineering fees. Permits and hook-up fees. Contractor's mark-up.

Supermarkets
Wood or Wood and Steel Frame

Estimating Procedure

1. Establish the structure quality class by using the information on page 72.
2. Compute the building floor area. This should include everything within the building exterior walls and all insets outside the main walls but under the main building roof.
3. Add to or subtract from the square foot cost below the appropriate amount from the Wall Height Adjustment Column (at the bottom of this page) if the wall height is more or less than 20 feet.
4. Multiply the adjusted square foot cost by the building area.
5. Deduct, if appropriate, for common walls, using the figures at the bottom of this page.
6. Multiply the total cost by the location factor listed on page 7.
7. Add the cost of heating and cooling equipment, fire sprinklers, exterior signs, yard improvements, loading docks, ramps and walk-in boxes if they are an integral part of the building. See pages 201 to 213.

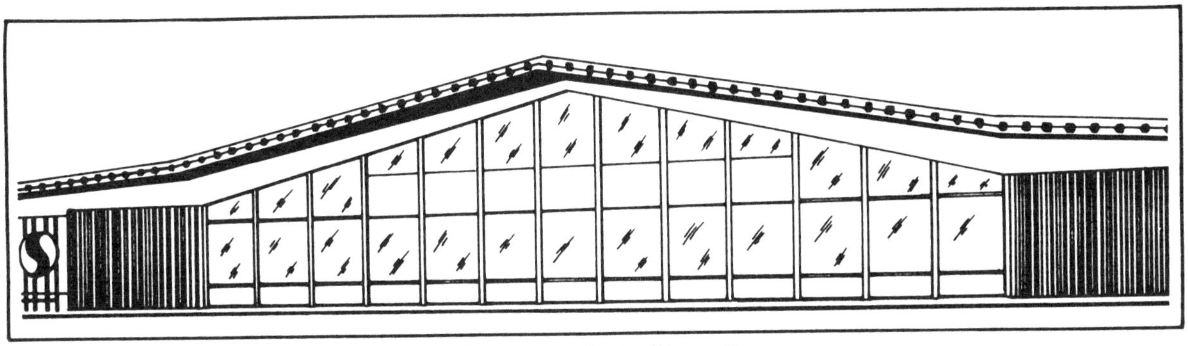

Supermarket, Class 3

Square Foot Area

Quality Class	5,000	7,500	10,000	12,500	15,000	20,000	25,000	30,000	35,000	40,000	50,000
Exceptional	51.08	48.13	46.36	45.15	44.25	42.99	42.13	41.49	40.99	40.59	39.98
1, Best	47.97	45.20	43.54	42.40	41.56	40.37	39.56	38.96	38.49	38.12	37.54
1 & 2	43.61	41.09	39.58	38.54	37.78	36.70	35.96	35.42	34.99	34.65	34.13
2, Good	38.81	36.57	35.22	34.30	33.62	32.66	32.00	31.52	31.14	30.84	30.37
2 & 3	35.90	33.82	32.58	31.73	31.10	30.21	29.60	29.15	28.81	28.52	28.09
3, Average	32.79	30.90	29.76	28.98	28.41	27.60	27.04	26.63	26.31	26.05	25.66
3 & 4	29.70	27.98	26.95	26.25	25.72	24.99	24.49	24.12	23.83	23.60	23.24
4, Low	26.47	24.94	24.02	23.39	22.93	22.27	21.83	21.50	21.24	21.03	20.71
Wall Height Adjustment *	.32	.27	.24	.22	.21	.19	.17	.16	.15	.14	.13

*Wall Height Adjustment

Add or subtract the amount listed in this column to the square foot of floor cost for each foot of wall height more or less than 20 feet.

Perimeter Wall Adjustment

A common wall exists when two buildings share one wall. Adjust for common walls by deducting the linear foot costs below from the total structure cost. In some structures one or more walls is not owned at all. In this case, deduct the "No Ownership" cost per linear foot of wall not owned.

For Common Wall, deduct $70.00 per linear foot.
For No Wall Ownership, deduct $140.00 per linear foot.

Small Food Stores Masonry Construction Quality Classification

	Class 1 Best Quality	Class 2 Good Quality	Class 3 Average Quality	Class 4 Low Quality
Foundation	Reinforced concrete.	Reinforced concrete.	Reinforced concrete.	Reinforced concrete.
Floor Structure	4" reinforced concrete on 6" rock fill.	4" reinforced concrete on 6" rock fill.	4" reinforced concrete on 6" rock fill.	4" reinforced concrete on 6" rock fill.
Wall Structure	8" colored concrete block.	8" concrete block.	8" concrete block.	8" concrete block.
Roof Structure	Glu-lams or steel "I" beams on steel intermediate columns, 2" x 12" purlins 16" o.c., ½" plywood sheathing.	Glu-lams or steel "I" beams, 2" x 12" purlins 16" o.c., ½" plywood sheathing.	Glu-lams, 3" x 12" purlins 3' o.c., ½" plywood sheathing.	Glu-lams, 3" x 12" purlins 3' o.c., ½" plywood sheathing.
Floor Finish	Vinyl asbestos tile in sales area.	Vinyl asbestos tile in sales area.	Minimum grade tile in sales area.	Concrete.
Interior Wall Finish	Paint on inside of exterior walls, gypsum wallboard, texture and paint or vinyl wall cover on partitions.	Paint on inside of exterior walls, gypsum wallboard, texture and paint on partitions.	Paint on inside of exterior walls, gypsum wallboard and paint on partitions.	Paint on inside of exterior walls, gypsum wallboard and paint on partitions.
Ceiling Finish	Suspended acoustical tile or gypsum board and acoustical texture.	Ceiling tile on roof purlins.	Open.	Open.
Front	A large amount of plate glass in good aluminum frames (10' to 12' high for 2/3 of width), brick or stone veneer on remainder, 1 pair good aluminum and glass doors per 3,000 S.F. of floor area, 8' canopy across front, 10' raised walk across front.	A moderate amount of plate glass in average aluminum frames (8' to 10' high for 2/3 of width), exposed aggregate on remainder, 1 pair average aluminum and glass doors per 3,000 S.F. of floor area, 6' canopy across front, 8' raised walk across front.	Painted concrete block with a small amount of plate glass in an inexpensive aluminum frame (6' to 8' high for ½ of width), 6' canopy across front, 6' ground level walk across front.	Painted concrete block with small amount of crystal glass in wood frames, wood and glass doors, small canopy over entrance, 6' ground level walk at entrances.
Exterior Wall Finish	Colored block.	Paint.	Paint.	Paint.
Roof Cover	5 ply built-up roofing with tar and rock.	4 ply built-up roofing.	4 ply built-up roofing.	4 ply built-up roofing.
Plumbing	2 rest rooms with 3 fixtures each, floor piping and drains to refrigerated cases.	1 rest room with 3 fixtures, floor piping and drains to refrigerated cases.	1 rest room with 2 fixtures, floor piping and drains to refrigerated cases.	1 rest room with 2 fixtures, floor piping and drains to refrigerated cases.
Electrical	Conduit wiring, 4 tube fluorescent fixtures with diffusers, 8' o.c., 5 spotlights.	Conduit wiring, 3 tube fluorescent fixtures, 8' o.c.	Conduit wiring, double tube fluorescent fixtures, 8' o.c.	Conduit wiring, incandescent fixtures, 10' o.c. or single tube fluorescent fixtures, 8' o.c.

Square foot costs include the following components: Foundations as required for normal soil conditions. Floor, wall, and roof structures. Interior floor, wall and ceiling finishes. Exterior wall finish and roof cover. Display fronts. Interior partitions. Entry and delivery doors. A canopy and walk across the front of the building as described in the applicable building specifications. Basic lighting and electrical systems. Rough and finish plumbing. All plumbing, piping and wiring necessary to operate the usual refrigerated cases and vegetable cases. Design and engineering fees. Permits and hook-up fees. Contractor's mark-up.

Small Food Stores
Masonry Construction

Estimating Procedure

1. Establish the structure quality class by using the information on page 74.
2. Compute the building floor area. This should include everything within the building exterior walls and all insets outside the main walls but under the main building roof.
3. Add to or subtract from the square foot cost below the appropriate amount from the Wall Height Adjustment Column (at the bottom of this page) if the wall height is more or less than 12 feet.
4. Multiply the adjusted square foot cost by the building area.
5. Deduct, if appropriate, for common walls, using the figures at the bottom of this page.
6. Multiply the total cost by the location factor listed on page 7.
7. Add the cost of heating and cooling equipment, fire sprinklers, exterior signs, yard improvements, loading docks, ramps and walk-in boxes if they are an integral part of the building. See pages 201 to 213.

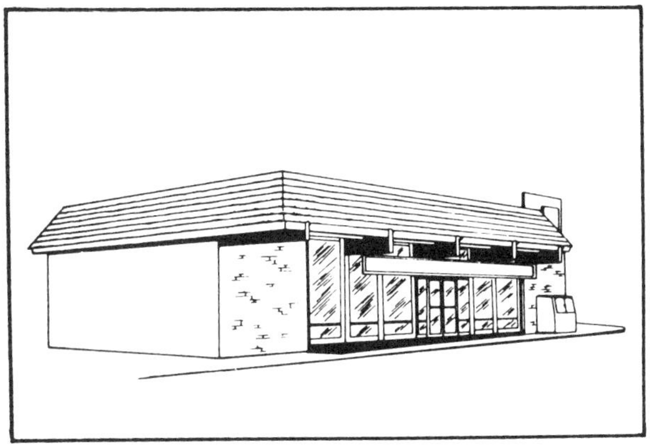

Small Food Store, Class 1 & 2

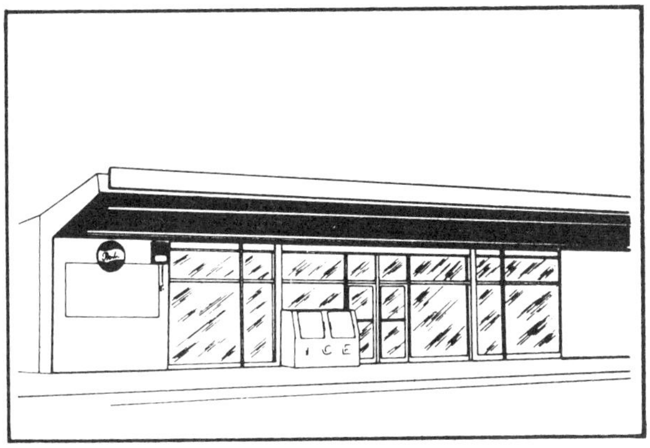

Small Food Store, Class 2

Square Foot Area

Quality Class	500	1,000	1,500	2,000	2,500	3,000	3,500	4,000	4,500	5,000	6,000
1, Best	83.65	63.37	55.40	50.95	48.04	45.96	44.38	43.13	42.11	41.26	39.92
1 & 2	76.23	57.89	50.67	46.64	44.00	42.11	40.67	39.54	38.62	37.85	36.62
2, Good	68.85	52.45	45.98	42.35	39.98	38.28	36.99	35.96	35.13	34.43	33.33
2 & 3	63.74	48.88	42.84	39.39	37.11	35.46	34.20	33.20	32.38	31.69	30.59
3, Average	58.42	44.71	39.19	36.06	33.99	32.51	31.37	30.47	29.73	29.12	28.13
3 & 4	47.89	40.34	36.22	33.54	31.62	30.15	28.97	28.01	27.20	26.50	25.37
4, Low	44.14	36.94	33.10	30.61	28.84	27.48	26.41	25.52	24.78	24.14	23.10
Wall Height Adjustment *	1.95	1.37	1.12	.97	.86	.79	.73	.68	.65	.61	.56

*Wall Height Adjustment

Add or subtract the amount listed in this column to the square foot of floor cost for each foot of wall height more or less than 12 feet.

Perimeter Wall Adjustment

For Common Wall, deduct $54.00 per linear foot.
For No Wall Ownership, deduct $108.00 per linear foot.

Small Food Stores
Wood Frame Construction
Quality Classification

	Class 1 Best Quality	Class 2 Good Quality	Class 3 Average Quality	Class 4 Low Quality
Foundation	Reinforced concrete.	Reinforced concrete.	Reinforced concrete.	Reinforced concrete.
Floor Structure	4" reinforced concrete on 6" rock fill.	4" reinforced concrete on 6" rock fill.	4" reinforced concrete on 6" rock fill.	4" reinforced concrete on 6" rock fill.
Wall Structure	2" x 6" - 16" o.c.	2" x 6" - 16" o.c.	2" x 4" - 16" o.c.	2" x 4" - 16" o.c.
Roof Structure	Glu-lams or steel "I" beams on steel intermediate columns, 2" x 12" purlins 16" o.c., ½" plywood sheathing.	Glu-lams or steel "I" beams on steel intermediate columns, 2" x 12" purlins 16" o.c., ½" plywood sheathing.	Glu-lams or steel "I" beams, 3" x 12" purlins 3' o.c., ½" plywood sheathing.	Glu-lams, 3" x 12" purlins 3' o.c., ½" plywood sheathing.
Floor Finish	Vinyl asbestos tile in sales area.	Vinyl asbestos tile in sales area.	Inexpensive vinyl asbestos tile in sales area.	Concrete.
Interior Wall Finish	Gypsum wallboard, texture and paint or vinyl wall cover on inside of exterior walls, and on partitions.	Gypsum wallboard, texture and paint on inside of exterior walls, and on partitions.	Gypsum wallboard and paint on inside of exterior walls, and on partitions.	Gypsum wallboard and paint on inside of exterior walls, and on partitions.
Ceiling Finish	Suspended acoustical tile or gypsum board and acoustical texture.	Ceiling tile on roof purlins.	Open.	Open.
Front	A large amount of plate glass in good aluminum frames (10' to 12' high for 2/3 of width), brick or stone veneer on remainder, 1 pair good aluminum and glass doors per 3,000 S.F.of floor area, 8' canopy across front, 10' raised walk across front.	A moderate amount of plate glass in average quality aluminum frames (8' to 10' high for 2/3 of width), ornamental concrete block on remainder, 1 pair average aluminum and glass doors per 3,000 S.F.of floor area, 6' canopy across front, 8' raised walk across front.	Painted stucco with a small amount of plate glass in an inexpensive aluminum frame (6' to 8' high for ½ of width) 6' canopy across front, 6' ground level walk across front.	Stucco with small amount of crystal glass in wood frames, wood and glass doors, small canopy over entrance, 6' ground level walk at entrances.
Exterior Wall Finish	Stucco and paint or wood siding.	Stucco and paint.	Stucco.	Stucco.
Roof Cover	5 ply built-up roofing with tar and rock.	4 ply built-up roofing.	4 ply built-up roofing.	4 ply built-up roofing.
Plumbing	2 rest rooms with 3 fixtures each, floor piping and drains to refrigerated cases.	1 rest room with 3 fixtures, floor piping and drains to refrigerated cases.	1 rest room with 2 fixtures, floor piping and drains to refrigerated cases.	1 rest room with 2 fixtures, floor piping and drains to refrigerated cases.
Electrical	Conduit wiring, 4 tube fluorescent fixtures with diffusers, 8' o.c., 5 spotlights.	Conduit wiring, 3 tube fluorescent fixtures, 8' o.c.	Conduit wiring, double tube fluorescent fixtures, 8' o.c.	Conduit wiring, incandescent fixtures, 10' o.c. or single tube fluorescent fixtures, 8' o.c.

Square foot costs include the following components: Foundations as required for normal soil conditions. Floor, wall, and roof structures. Interior floor, wall and ceiling finishes. Exterior wall finish and roof cover. Display fronts. Interior partitions. Entry and delivery doors. A canopy and walk across the front of the building as described in the applicable building specifications. Basic lighting and electrical systems. Rough and finish plumbing. All plumbing, piping and wiring necessary to operate the usual refrigerated cases and vegetable cases. Design and engineering fees. Permits and hook-up fees. Contractor's mark-up.

Small Food Stores
Wood Frame Construction

Estimating Procedure

1. Establish the structure quality class by using the information on page 76.
2. Compute the building floor area. This should include everything within the building exterior walls and all insets outside the main walls but under the main building roof.
3. Add to or subtract from the square foot cost below the appropriate amount from the Wall Height Adjustment Column (at the bottom of this page) if the wall height is more or less than 12 feet.
4. Multiply the adjusted square foot cost by the building area.
5. Deduct, if appropriate, for common walls, using the figures at the bottom of this page.
6. Multiply the total cost by the location factor listed on page 7.
7. Add the cost of heating and cooling equipment, fire sprinklers, exterior signs, yard improvements, loading docks, ramps and walk-in boxes if they are an integral part of the building. See pages 201 to 213.

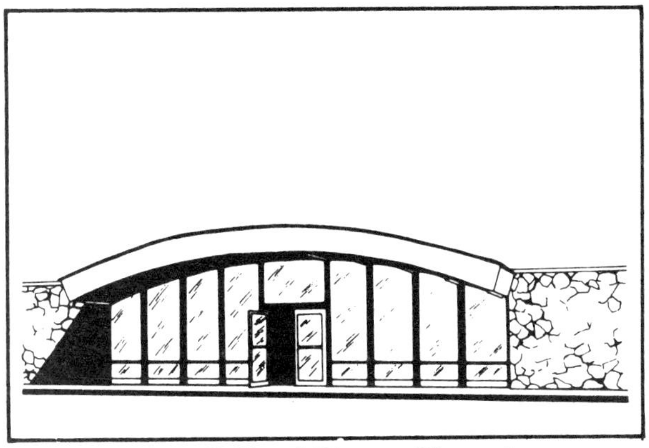

Small Food Store, Class 1

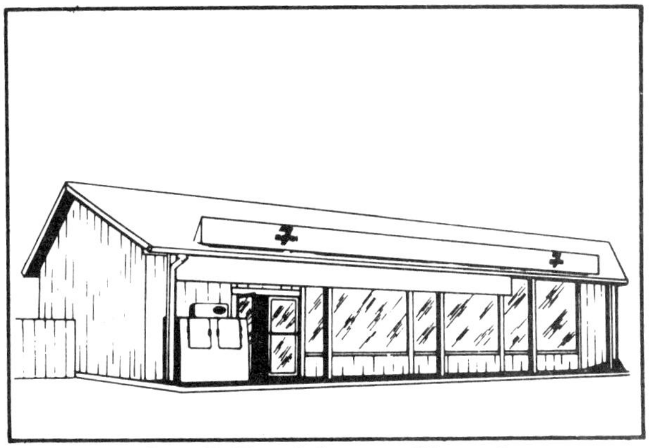

Small Food Store, Class 3

Square Foot Area

Quality Class	500	1,000	1,500	2,000	2,500	3,000	3,500	4,000	4,500	5,000	6,000
1, Best	72.45	55.22	48.70	45.13	42.83	41.21	39.99	39.04	38.27	37.62	36.62
1 & 2	66.27	50.61	44.68	41.44	39.35	37.88	36.78	35.91	35.21	34.63	33.71
2, Good	60.12	46.04	40.69	37.77	35.89	34.56	33.56	32.77	32.14	31.61	30.78
2 & 3	54.18	41.61	36.84	34.23	32.55	31.36	30.47	29.77	29.21	28.74	28.00
3, Average	48.30	37.21	33.01	30.72	29.24	28.19	27.41	26.79	26.30	25.88	25.23
3 & 4	43.33	33.43	29.69	27.65	26.33	25.40	24.70	24.16	23.72	23.35	22.77
4, Low	38.55	29.52	26.21	24.45	23.33	22.55	21.97	21.52	21.16	20.86	20.40
Wall Height Adjustment *	1.15	.84	.69	.60	.54	.49	.45	.42	.40	.38	.34

***Wall Height Adjustment**

Add or subtract the amount listed in this column to the square foot of floor cost for each foot of wall height more or less than 12 feet.

Perimeter Wall Adjustment

For Common Wall, deduct $34.00 per linear foot.
For No Wall Ownership, deduct $68.00 per linear foot.

Discount Houses
Masonry or Concrete
Quality Classification

	Class 1 Best Quality	Class 2 Good Quality	Class 3 Average Quality	Class 4 Low Quality
Foundation	Reinforced concrete.	Reinforced concrete.	Reinforced concrete.	Reinforced concrete.
Floor Structure	4" reinforced concrete on 6" rock fill.	4" reinforced concrete on 6" rock fill.	4" reinforced concrete on 6" rock fill.	4" reinforced concrete on 6" rock fill.
Wall Structure	6" concrete tilt-up or ornamental block or brick.	6" concrete tilt-up, colored concrete block or brick.	6" concrete tilt-up or 8" concrete block.	6" concrete tilt-up or 8" concrete block.
Roof Structure	Glu-lams or steel "I" beams on steel intermediate columns, plywood sheathing.	Glu-lams or steel "I" beams on steel intermediate columns, plywood sheathing.	Glu-lams or steel "I" beams on steel intermediate columns, plywood sheathing.	Glu-lams or steel "I" beams on steel intermediate columns, plywood sheathing.
Floor Finish	Terrazzo, sheet vinyl or good carpet in sales area.	Vinyl asbestos tile in sales area with some terrazzo, solid vinyl tile, or carpet.	Vinyl asbestos tile in sales area.	Inexpensive vinyl asbestos tile in sales area.
Interior Wall Finish	Inside of exterior walls furred out with gypsum wallboard and paint or interior stucco or gypsum wallboard and vinyl wall cover on partitions.	Paint on inside of exterior walls, gypsum wallboard and paint or vinyl wall cover on partitions.	Paint on inside of exterior walls, gypsum wallboard and paint on partitions.	Paint on inside of exterior walls, gypsum wallboard and paint on partitions.
Ceiling Finish	Suspended acoustical tlle, dropped ceilings in some areas.	Suspended acoustical tile or gypsum board and acoustical texture, dropped ceilings in some areas.	Ceiling tile on roof purlins, dropped ceiling over some areas.	Open.
Front	A large amount of plate glass in good aluminum frames (18'-22' high for ¾ of width) brick or stone veneer on remainder, 1 pair good automatic doors per 7,000 S.F. of floor area, anodized aluminum sunshade over glass area, 8' canopy across front, 10'-12' raised walk across front.	A large amount of plate glass in good aluminum frames (16'-18' high for 2/3 of width) brick or stone veneer on remainder, 1 pair good automatic doors per 10,000 S.F. of floor area, 8' canopy across front, 10' raised walk across front.	A moderate amount of plate glass in average quality aluminum frames, (12'-16' high extending 20' on each side of entrances), exposed aggregate on remainder, 1 pair average automatic doors per 20,000 S.F. of floor area, 6' canopy over glass area.	Stucco or exposed aggregate with a small amount of plate glass in inexpensive alumimum frames, (6'-10' high for ½ of width), 6' canopy at entrances, 6' ground level walk across front.
Exterior Wall Finish	Large ornamental rock or brick veneer on exposed walls.	Exposed aggregate on exposed walls.	Paint, some exposed aggregate.	Paint
Roof Cover	5 ply built-up roofing with large rock.	5 ply built-up roofing with tar and rock.	4 ply built-up roofing.	4 ply built-up roofing.
Plumbing	2 rest rooms with 8 fixtures each, floor piping and drains to refrigerated cases.	2 rest rooms with 6 fixtures each, floor piping and drains to refrigerated cases.	2 rest rooms with 4 fixtures each, floor piping and drains to refrigerated cases.	2 rest rooms with 2 fixtures each, floor piping and drains to refrigerated cases.
Electrical	Conduit wiring, recessed 4 tube fluorescent fixtures 8' o.c., 60-80 spotlights.	Conduit wiring, 4 tube fluorescent fixtures with diffusers 8' o.c. 60-80 spotlights.	Conduit wiring, 3 tube fluorescent fixtures 8' o.c., 20 to 40 spotlights.	Conduit wiring, double tube fluorescent fixtures, 8' o.c.

Square foot costs include the following components: Foundations as required for normal soil conditions. Floor, wall, and roof structures, interior floor, wall and ceiling finishes. Exterior wall finish and roof cover. Display fronts. Interior partitions. Entry and delivery doors. A canopy and walk across the front of the building as described in the applicable building specifications. Basic lighting and electrical systems. Rough and finish plumbing. Design and engineering fees. Permit and hook-up fees. Contractor's mark-up.

Discount Houses
Masonry or Concrete

Estimating Procedure

1. Establish the structure quality class by using the information on page 78.
2. Compute the building floor area. This should include everything within the building exterior walls and all insets outside the main walls but under the main building roof.
3. Add to or subtract from the square foot cost below the appropriate amount from the Wall Height Adjustment Column (at the bottom of this page) if the wall height is more or less than 20 feet.
4. Multiply the adjusted square foot cost by the building area.
5. Deduct, if appropriate, for common walls, using the figures at the bottom of this page.
6. Multiply the total cost by the location factor listed on page 7.
7. Add the cost of heating and cooling equipment, fire sprinklers, exterior signs, yard improvements, loading docks, ramps and walk-in boxes if they are an integral part of the building. See pages 201 to 213.

Discount House, Class 3

Discount House, Class 4

Square Foot Area

Quality Class	15,000	20,000	25,000	30,000	35,000	40,000	50,000	75,000	100,000	150,000	200,000
1, Best	36.79	35.54	34.68	34.04	33.54	33.13	32.51	31.54	30.96	30.26	29.84
1 & 2	34.25	33.09	32.29	31.69	31.23	30.85	30.27	29.37	28.82	28.17	27.79
2, Good	31.72	30.64	29.90	29.35	28.91	28.57	28.03	27.19	26.69	26.09	25.73
2 & 3	29.18	28.19	27.51	27.00	26.61	26.29	25.79	25.02	24.56	24.01	23.68
3, Average	26.67	25.76	25.14	24.67	24.31	24.02	23.57	22.86	22.44	21.94	21.63
3 & 4	24.11	23.29	22.73	22.31	21.98	21.72	21.31	20.67	20.29	19.83	19.56
4, Low	21.57	20.84	20.33	19.96	19.66	19.43	19.06	18.49	18.15	17.74	17.50
Wall Height Adjustment *	.38	.32	.29	.26	.24	.23	.21	.18	.16	.14	.13

*Wall Height Adjustment

Add or subtract the amount listed in this column to the square foot of floor cost for each foot of wall height more or less than 20 feet.

Perimeter Wall Adjustment

A common wall exists when two buildings share one wall. Adjust for common walls by deducting the linear foot costs below from the total structure cost. In some structures one or more walls is not owned at all. In this case, deduct the "No Ownership" cost per linear foot of wall not owned.

For Common Wall, deduct $90.00 per linear foot.
For No Wall Ownership, deduct $180.00 per linear foot.

Discount Houses
Wood or Wood and Steel Frame
Quality Classification

	Class 1 Best Quality	Class 2 Good Quality	Class 3 Average Quality	Class 4 Low Quality
Foundation	Reinforced concrete.	Reinforced concrete.	Reinforced concrete.	Reinforced concrete.
Floor Structure	4" reinforced concrete on 6" rock fill.	4" reinforced concrete on 6" rock fill.	4" reinforced concrete on 6" rock fill.	4" reinforced concrete on 6" rock fill.
Wall Structure	2" x 6" - 16" o.c.	2" x 6" - 16" o.c.	2" x 6" - 16" o.c.	2" x 4" - 16" o.c.
Roof Structure	Glu-lams or steel "I" beams on steel intermediate columns, plywood sheathing.	Glu-lams or steel "I" beams on steel intermediate columns, plywood sheathing.	Glu-lams or steel "I" beams on steel intermediate columns, plywood sheathing.	Glu-lams or steel "I" beams on steel intermediate columns, plywood sheathing.
Floor Finish	Terrazzo, sheet vinyl or good carpet in sales area.	Vinyl asbestos tile in sales area with some terrazzo, solid vinyl tile or carpet.	Vinyl asbestos tile in sales area.	Inexpensive vinyl asbestos tile in sales area.
Interior Wall Finish	Gypsum wallboard and vinyl wall cover or interior stucco on inside of walls and on partitions.	Gypsum wallboard, texture and paint or vinyl wall cover on inside of exterior walls and on partitions.	Gypsum wallboard, texture and paint on inside of exterior walls and on partitions.	Gypsum wallboard and paint on inside of exterior walls and on partitions.
Ceiling Finish	Suspended acoustical tile, dropped ceilings over some areas.	Suspended acoustical tile or gypsum board and acoustical texture, dropped ceilings over some areas.	Ceiling tile on roof purlins, dropped ceilings over some areas.	Open.
Front	A large amount of plate glass in good aluminum frames, (18'-22' high for ¾ of width), brick or stone veneer on remainder, 1 pair good automatic doors per 7,000 S.F. of floor area, anodized aluminum sunshade over glass area, 8' canopy across front, 10' to 12' raised walk across front.	A large amount of plate glass in good aluminum frames, (16'-18' high for 2/3 of width), brick or stone veneer on remainder, 1 pair good automatic doors per 10,000 S.F. of floor area, 8' canopy across front, 10' raised walk across front.	A moderate amount of plate glass in average quality aluminum frames, (12'-16' high extending 20' on each side of entrances), exposed aggregate on remainder, 1 pair average automatic doors per 20,000 S.F. of floor area, 6' canopy over glass areas.	Stucco or exposed aggregate with a small amount of plate glass in inexpensive aluminum frames (6'-10' high for ½ of width), 6' canopy at entrances, 6' ground level walk across front.
Exterior Wall Finish	Good wood siding or masonry veneer.	Wood siding, some masonry veneer.	Stucco, some masonry trim.	Stucco.
Roof Cover	5 ply built-up roofing with large rock.	5 ply built-up roofing with tar and rock.	4 ply built-up roofing.	4 ply built-up roofing.
Plumbing	2 rest rooms with 8 fixtures each, floor piping and drains to refrigerated cases.	2 rest rooms with 6 fixtures each, floor piping and drains to refrigerated cases.	2 rest rooms with 4 fixtures each, floor piping and drains to refrigerated cases.	2 rest rooms with 2 fixtures each, floor piping and drains to refrigerated cases.
Electrical	Conduit wiring, recessed 4 tube fluorescent fixtures, 8' o.c., 60 to 80 spotlights.	Conduit wiring, 4 tube fluorescent fixtures with diffusers, 8' o.c., 60 to 80 spotlights.	Conduit wiring, 3 tube fluorescent fixtures, 8' o.c., 20 to 40 spotlights.	Conduit wiring, double tube fluorescent fixtures, 8' o.c.

Square foot costs include the following components: Foundations as required for normal soil conditions. Floor, wall, and roof structures. Interior floor, wall and ceiling finishes. Exterior wall finish and roof cover. Display fronts. Interior partitions. Entry and delivery doors. A canopy and walk across the front of the building as described in the applicable building specifications. Basic lighting and electrical systems. Rough and finish plumbing. Design and engineering fees. Permits and hook-up fees. Contractor's mark-up.

Discount Houses
Wood or Wood and Steel Frame

Estimating Procedure

1. Establish the structure quality class by using the information on page 80.
2. Compute the building floor area. This should include everything within the building exterior walls and all insets outside the main walls but under the main building roof.
3. Add to or subtract from the square foot cost below the appropriate amount from the Wall Height Adjustment Column (at the bottom of this page) if the wall height is more or less than 20 feet.
4. Multiply the adjusted square foot cost by the building area.
5. Deduct, if appropriate, for common walls, using the figures at the bottom of this page.
6. Multiply the total cost by the location factor listed on page 7.
7. Add the cost of heating and cooling equipment, fire sprinklers, exterior signs, yard improvements, loading docks, ramps and walk-in boxes if they are an integral part of the building. See pages 201 to 213.

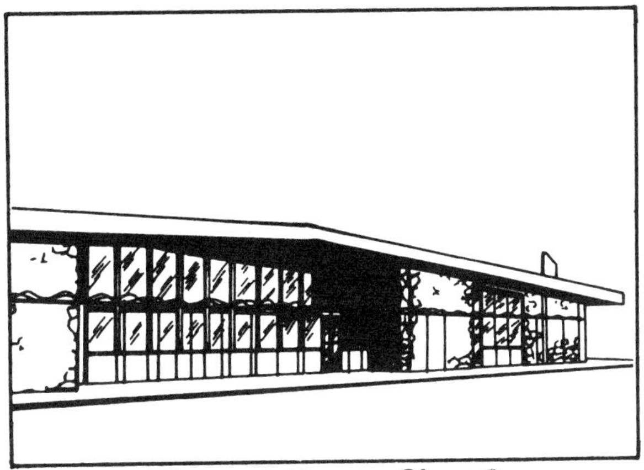

Discount House, Class 2

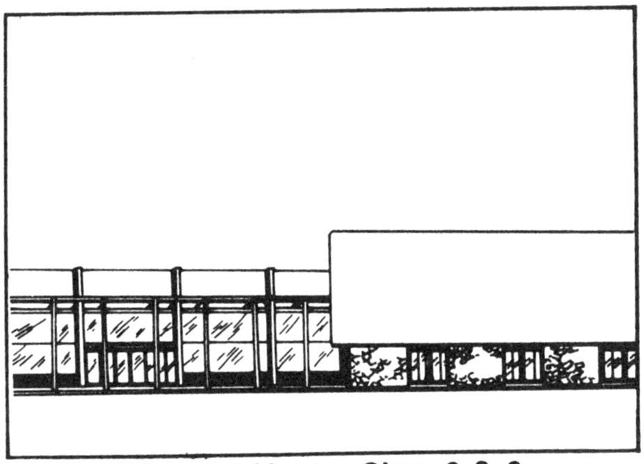

Discount House, Class 2 & 3

Square Foot Area

Quality Class	15,000	20,000	25,000	30,000	35,000	40,000	50,000	75,000	100,000	150,000	200,000
1, Best	32.37	31.57	31.03	30.62	30.31	30.05	29.66	29.05	28.69	28.26	28.00
1 & 2	30.19	29.44	28.94	28.56	28.26	28.03	27.66	27.09	26.75	26.35	26.11
2, Good	27.95	27.26	26.79	26.44	26.17	25.95	25.61	25.08	24.77	24.39	24.17
2 & 3	25.73	25.10	24.66	24.34	24.09	23.89	23.58	23.09	22.80	22.46	22.25
3, Average	23.49	22.91	22.51	22.22	21.99	21.80	21.52	21.08	20.81	20.50	20.31
3 & 4	21.24	20.72	20.36	20.09	19.88	19.72	19.46	19.06	18.82	18.54	18.37
4, Low	19.01	18.54	18.22	17.98	17.80	17.65	17.42	17.06	16.85	16.59	16.44
Wall Height Adjustment *	.19	.18	.17	.16	.15	.14	.13	.11	.10	.09	.08

***Wall Height Adjustment**

Add or subtract the amount listed in this column to the square foot of floor cost for each foot of wall height more or less than 20 feet.

Perimeter Wall Adjustment

A common wall exists when two buildings share one wall. Adjust for common walls by deducting the linear foot costs below from the total structure cost. In some structures one or more walls is not owned at all. In this case, deduct the "No Ownership" cost per linear foot of wall not owned.

For Common Wall, Deduct $61.00 per linear foot.
For No Wall Ownership, deduct $122.00 per linear foot.

Banks and Savings Offices
Masonry or Concrete

	Class 1 Best Quality	Class 2 Good Quality	Class 3 High Average Quality	Class 4 Low Average Quality	Class 5 Low Quality
Foundation	Reinforced concrete.	Reinforced concrete.	Reinforced concrete.	Reinforced concrete.	Reinforced concrete.
Floor Structure	Reinforced concrete.	Reinforced concrete.	Reinforced concrete.	Reinforced concrete.	Reinforced concrete.
Wall Structure	12" reinforced brick, 8" reinforced concrete or concrete columns with noncombustible filler walls.	12" reinforced brick, 8" reinforced concrete or concrete columns with noncombustible filler walls.	6" concrete tilt-up, 12" reinforced brick, 8" reinforced concrete or concrete columns with noncombustible filler walls.	8" decorative concrete block, 6" concrete tilt-up, 8" reinforced brick or concrete columns with noncombustible filler walls.	8" concrete block, 6" reinforced concrete, 6" concrete tilt-up or 8" reinforced brick.
Roof and Cover	5 ply built-up roof, some portions copper or slate.	5 ply built-up roof, some portions heavy shake or mission tile.	4 ply built-up roof, some portions heavy shake or mission tile.	4 ply built-up roof, some portions shake or composition shingles.	4 ply built-up roof.
Exterior Wall Finish	Expensive veneers.	Ornamental brick or stone veneer.	Brick, or ornamental block veneer.	Ornamental rock imbedded in tilt-up panels or stucco with masonry trim.	Exposed aggregate or stucco.
Interior Wall Finish	Ornamental plaster painted with murals or designs. Ornamental moldings, marble and similar expensive finishes.	Ornamental plaster furred out from walls. Ornamental moldings, some hardwood panels or marble finishes.	Hard plaster furred out from walls. Ornamental cove at ceiling. Vinyl wall covering. Some hardwood veneer.	Plaster on lath with putty coat finish. Gypsum wallboard, texture and paint. Some vinyl wallcover.	Plaster on lath with putty coat finish or gypsum wallboard, texture and paint.
Glass	Tinted plate glass in customized frames, (50% of exterior wall area).	Tinted plate glass in bronze frames, (50% of exterior wall area).	Tinted plate glass in heavy anodized aluminum frames or custom wood frames, (50% of exterior wall area).	Moderate amount of plate glass in heavy frames, (25% of exterior wall area).	Small to moderate amount of plate glass in average aluminum frames, (5 to 10% of exterior wall area.).
Overhang	4' closed overhang with copper gutters.	4' closed overhang with copper gutters.	4' closed overhang with painted gutters.	3' closed overhang with painted gutters.	None.
Floor Finish, Public Area	Terrazzo and marble.	Detailed terrazzo.	Detailed terrazzo.	Plain terrazzo.	Sheet vinyl or vinyl asbestos tile.
Floor Finish, Officers Area	Excellent carpet.	Very good carpet.	Very good carpet.	Good carpet.	Average quality carpet.
Work Area	Good sheet vinyl or carpet.	Average sheet vinyl, tile or carpet.	Vinyl asbestos tile.	Vinyl asbestos tile.	Vinyl asbestos tile.
Ceiling Finish	Ornamental plaster with exposed ornamental beams.	Ornamental plaster.	Suspended acoustical tile with gypsum wallboard backing or acoustical plaster.	Suspended acoustical tile with exposed grid system.	Acoustical tile on wood furring.
Lighting	Recessed panelized fluorescent light fixtures, custom light fixtures or chandeliers. Many spotlights.	Recessed panelized fluorescent lighting. Many spotlights.	Recessed panelized fluorescent lighting. Many spotlights.	Recessed panelized fluorescent lighting. Some spotlights.	Nonrecessed panelized lighting.
Plumbing	2 or more rest rooms with special fixtures, good ceramic tile, marble or terrazzo wainscot walls. Hard plaster with putty coat and enamel paint. Marble toilet screens.	2 or more rest rooms with many good fixtures, good ceramic tile, marble or terrazzo wainscot walls. Hard plaster with putty coat and enamel paint. Marble toilet screens.	2 rest rooms with 4 or more fixtures, ceramic tile or terrazzo floors. Ceramic tile or terrazzo wainscot. Hard plaster walls with putty coat and enamel paint. Marble or metal toilet screens.	2 rest rooms with 4 or more fixtures each. Ceramic tile or terrazzo floor. Ceramic tile or terrazzo wainscot plaster walls with putty coat. Enamel paint. Good metal toilet partitions.	2 rest rooms with 4 fixtures each. Ceramic tile or vinyl asbestos tile floors. Plaster walls with enamel paint. Metal toilet screens.

Square foot costs include the following components: Foundations as required for normal soil conditions. Floor, wall and roof structures. Interior floor, wall, and ceiling finishes. Exterior wall finish and roof cover. All glass and glazing. Interior partitions. Roof overhang as described above. Basic electrical systems and lighting fixtures. Rough and finish plumbing. Typical bank vault. Alarm systems. Night depository. Typical record vault and fire doors. Fire exits. Design and engineering fees, permit and hook-up fees, contractor's mark-up.

Banks and Savings Offices
Masonry or Concrete
Length Less Than Twice Width

Estimating Procedure

1. Establish the structure quality class by using the information on page 82.
2. Compute the building floor area. This should include everything within the building exterior walls and all insets outside the main walls but under the main building roof.
3. Add to or subtract from the square foot cost below the appropriate amount from the Wall Height Adjustment Table on page 86 if the wall height is more or less than 16 feet for the first floor or 12 feet for higher floors.
4. Multiply the adjusted square foot cost by the building area.
5. Deduct, if appropriate, for common walls, using the figures on page 86.
6. Multiply the total cost by the location factor listed on page 7.
7. Add the cost of heating and cooling equipment, fire sprinklers, exterior signs, elevators, and yard improvements from pages 201 to 213. Add the cost of mezzanines, bank fixtures, external windows, safe deposit boxes and vault doors from page 92.

Savings Office, Class 3

First Story

Square Foot Area at 16 Foot Wall Height

Quality Class	2,500	3,000	3,500	4,000	4,500	5,000	6,000	7,500	10,000	15,000	20,000
1, Best	134.10	129.97	126.36	123.19	120.38	117.89	113.63	108.57	102.39	94.46	89.43
1 & 2	125.46	121.60	118.22	115.25	112.63	110.29	106.31	101.58	95.80	88.38	83.67
2, Good	116.84	113.24	110.09	107.33	104.89	102.71	99.00	94.60	89.22	82.31	77.92
2 & 3	108.72	105.38	102.45	99.87	97.60	95.58	92.12	88.02	83.02	76.59	72.50
3, High Avg.	100.62	97.53	94.82	92.44	90.33	88.46	85.26	81.47	76.84	70.88	67.10
3 & 4	93.90	91.01	88.48	86.26	84.29	82.54	79.56	76.02	71.70	66.14	62.62
4, Low Avg.	87.17	84.49	82.14	80.08	78.26	76.63	73.87	70.58	66.56	61.41	58.13
4 & 5	79.66	77.21	75.06	73.17	71.51	70.03	67.50	64.49	60.82	56.11	53.12
5, Low	72.17	69.95	68.01	66.30	64.79	63.45	61.15	58.43	55.11	50.84	48.13

Second and Higher Stories

Square Foot Area at 12 Foot Wall Height

Quality Class	500	750	1,000	1,500	2,000	3,000	4,000	5,000	7,500	10,000	20,000
1, Best	109.32	97.93	91.20	83.28	78.59	73.06	69.78	67.54	64.08	62.03	58.15
1 & 2	102.76	92.05	85.72	78.28	73.87	68.67	65.59	63.49	60.24	58.30	54.66
2, Good	96.67	86.59	80.64	73.64	69.49	64.60	61.70	59.73	56.67	54.85	51.42
2 & 3	90.53	81.09	75.52	68.96	65.08	60.50	57.78	55.93	53.07	51.36	48.15
3, High Avg.	84.50	75.70	70.49	64.37	60.75	56.47	53.93	52.21	49.53	47.95	44.95
3 & 4	78.85	70.63	65.78	60.07	56.68	52.69	50.33	48.72	46.22	44.74	41.94
4, Low Avg.	73.32	65.68	61.16	55.85	52.71	49.00	46.80	45.30	42.98	41.60	39.00
4 & 5	68.46	61.32	57.11	52.15	49.21	45.75	43.69	42.29	40.13	38.84	36.41
5, Low	63.57	56.94	53.03	48.42	45.70	42.48	40.57	39.27	37.26	36.07	33.81

Banks and Savings Offices
Masonry or Concrete
Length Between Two and Four Times Width

Estimating Procedure

1. Establish the structure quality class by using the information on page 82.
2. Compute the building floor area. This should include everything within the building exterior walls and all insets outside the main walls but under the main building roof.
3. Add to or subtract from the square foot cost below the appropriate amount from the Wall Height Adjustment Table on page 86 if the wall height is more or less than 16 feet for the first floor or 12 feet for higher floors.
4. Multiply the adjusted square foot cost by the building area.
5. Deduct, if appropriate, for common walls, using the figures on page 86.
6. Multiply the total cost by the location factor listed on page 7.
7. Add the cost of heating and cooling equipment, fire sprinklers, exterior signs, elevators, and yard improvements from pages 201 to 213. Add the cost of mezzanines, bank fixtures, external windows, safe deposit boxes and vault doors from page 92.

Savings Office, Class 4

First Story

Square Foot Area at 16 Foot Wall Height

Quality Class	2,500	3,000	3,500	4,000	4,500	5,000	6,000	7,500	10,000	15,000	20,000
1, Best	142.87	135.49	129.97	125.66	122.17	119.29	114.78	109.96	104.73	98.83	95.48
1 & 2	134.14	127.11	121.84	117.73	114.41	111.66	107.36	102.77	97.78	92.15	88.96
2, Good	127.34	120.53	115.43	111.45	108.24	105.58	101.43	96.99	92.18	86.77	83.69
2 & 3	117.15	110.65	105.83	102.08	99.07	96.59	92.73	88.64	84.25	79.36	76.62
3, High Avg.	108.70	102.75	98.29	94.82	92.01	89.69	86.06	82.19	77.98	73.26	70.57
3 & 4	101.23	95.67	91.52	88.28	85.68	83.52	80.16	76.58	72.70	68.35	65.89
4, Low Avg.	93.83	88.65	84.78	81.78	79.37	77.38	74.27	70.98	67.43	63.47	61.24
4 & 5	86.77	81.97	78.40	75.63	73.42	71.59	68.75	65.75	62.52	58.94	56.94
5, Low	77.13	72.89	69.74	67.31	65.36	63.75	61.26	58.64	55.82	52.71	50.98

Second and Higher Stories

Square Foot Area at 12 Foot Wall Height

Quality Class	500	750	1,000	1,500	2,000	3,000	4,000	5,000	7,500	10,000	20,000
1, Best	116.95	104.51	97.06	88.19	82.88	76.57	72.80	70.22	66.21	63.81	59.27
1 & 2	110.24	98.52	91.49	83.13	78.13	72.18	68.62	66.19	62.41	60.15	55.87
2, Good	103.56	92.55	85.95	78.09	73.39	67.80	64.46	62.18	58.63	56.50	52.48
2 & 3	96.82	86.52	80.35	73.01	68.62	63.39	60.27	58.13	54.81	52.83	49.07
3, High Avg.	89.74	80.19	74.47	67.67	63.59	58.75	55.86	53.88	50.80	48.96	45.48
3 & 4	84.02	75.08	69.73	63.35	59.54	55.00	52.30	50.44	47.56	45.84	42.58
4, Low Avg.	78.12	69.81	64.83	58.90	55.36	51.14	48.62	46.90	44.22	42.62	39.59
4 & 5	72.81	65.07	60.43	54.90	51.60	47.67	45.32	43.72	41.22	39.73	36.90
5, Low	67.56	60.37	56.07	50.94	47.88	44.23	42.05	40.56	38.25	36.86	34.24

Banks and Savings Offices
Masonry or Concrete
Length More Than Four Times Width

Estimating Procedure

1. Establish the structure quality class by using the information on page 82.
2. Compute the building floor area. This should include everything within the building exterior walls and all insets outside the main walls but under the main building roof.
3. Add to or subtract from the square foot cost below the appropriate amount from the Wall Height Adjustment Table on page 86 if the wall height is more or less than 16 feet for the first floor or 12 feet for higher floors.
4. Multiply the adjusted square foot cost by the building area.
5. Deduct, if appropriate, for common walls, using the figures on page 86.
6. Multiply the total cost by the location factor listed on page 7.
7. Add the cost of heating and cooling equipment, fire sprinklers, exterior signs, elevators, and yard improvements from pages 201 to 213. Add the cost of mezzanines, bank fixtures, external windows, safe deposit boxes and vault doors from page 92.

Bank, Class 5

First Story

Square Foot Area at 16 Foot Wall Height

Quality Class	2,500	3,000	3,500	4,000	4,500	5,000	6,000	7,500	10,000	15,000	20,000
1, Best	152.61	143.74	137.15	132.04	127.94	124.57	119.31	113.76	107.80	101.18	97.47
1 & 2	142.89	134.58	128.41	123.63	119.79	116.63	111.71	106.51	100.93	94.73	91.26
2, Good	133.17	125.43	119.68	115.22	111.64	108.70	104.11	99.27	94.06	88.29	85.05
2 & 3	124.05	116.84	111.49	107.33	104.00	101.26	96.98	92.47	87.62	82.24	79.23
3, High Avg.	114.90	108.22	103.26	99.41	96.33	93.79	89.83	85.65	81.16	76.18	73.38
3 & 4	106.89	100.68	96.06	92.48	89.61	87.25	83.57	79.68	75.50	70.87	68.27
4, Low Avg.	98.86	93.12	88.85	85.54	82.88	80.70	77.29	73.70	69.83	65.54	63.14
4 & 5	90.02	84.79	80.90	77.89	75.47	73.48	70.38	67.11	63.59	59.68	57.49
5, Low	81.11	76.40	72.90	70.18	68.00	66.21	63.41	60.46	57.29	53.77	51.80

Second and Higher Stories

Square Foot Area at 12 Foot Wall Height

Quality Class	500	750	1,000	1,500	2,000	3,000	4,000	5,000	7,500	10,000	20,000
1, Best	129.43	114.55	105.70	95.22	88.98	81.58	77.18	74.18	69.51	66.73	61.48
1 & 2	121.81	107.81	99.47	89.61	83.74	76.78	72.64	69.81	65.42	62.81	57.86
2, Good	114.18	101.06	93.25	84.00	78.50	71.97	68.09	65.44	61.33	58.87	54.24
2 & 3	106.60	94.34	87.05	78.42	73.28	67.19	63.57	61.09	57.25	54.96	50.63
3, High Avg.	99.07	87.68	80.91	72.88	68.11	62.45	59.08	56.78	53.21	51.08	47.06
3 & 4	92.17	81.58	75.27	67.81	63.36	58.10	54.96	52.83	49.50	47.52	43.78
4, Low Avg.	85.27	75.47	69.64	62.73	58.62	53.75	50.85	48.87	45.80	43.97	40.50
4 & 5	79.49	70.35	64.91	58.48	54.64	50.10	47.40	45.56	42.69	40.98	37.76
5, Low	73.72	65.25	60.21	54.24	50.68	46.47	43.96	42.25	39.60	38.01	35.02

Banks and Savings Offices Masonry or Concrete Wall Adjustments

Wall Height Adjustment

Add or subtract the amount listed in this table to the square foot of floor cost for each foot of wall height more or less than 16 feet, if adjusting for a first floor, and 12 feet if adjusting for upper floors.

	Square Foot Area							
Quality Class	500	750	1,000	1,500	2,000	2,500	3,000	3,500
1, Best	5.29	4.34	3.77	3.09	2.68	2.40	2.19	2.02
2, Good	4.65	3.82	3.32	2.72	2.35	2.11	1.92	1.78
3, High Average	4.04	3.32	2.88	2.36	2.04	1.83	1.67	1.55
4, Low Average	3.26	2.68	2.33	1.90	1.65	1.48	1.35	1.25
5, Low	2.54	2.08	1.81	1.48	1.29	1.15	1.05	.97

	Square Foot Area							
Quality Class	4,000	4,500	5,000	6,000	7,500	10,000	15,000	20,000
1, Best	1.93	1.82	1.73	1.58	1.41	1.21	.98	.85
2, Good	1.69	1.59	1.51	1.38	1.23	1.06	.86	.74
3, High Average	1.48	1.39	1.32	1.20	1.08	.93	.75	.65
4, Low Average	1.18	1.11	1.06	.96	.86	.74	.60	.52
5, Low	.91	.86	.81	.74	.66	.57	.46	.40

Perimeter Wall Adjustment

First Story

Class	For a Common Wall, Deduct Per Linear Foot	For No Wall Ownership, Deduct Per Linear Foot	For Lack of Exterior Finish, Deduct Per Linear Foot
1	$315.00	$426.00	$210.00
2	267.00	363.00	165.00
3	215.00	319.00	119.00
4	130.00	250.00	79.00
5	107.00	193.00	21.00

Second and Higher Stories

Class	For a Common Wall, Deduct Per Linear Foot	For No Wall Ownership, Deduct Per Linear Foot	For Lack of Exterior Finish, Deduct Per Linear Foot
1	$178.00	$242.00	$94.00
2	148.00	210.00	69.00
3	117.00	182.00	44.50
4	95.00	149.00	27.50
5	70.00	132.00	11.80

Note: First floor costs include the cost of overhang as described on page 82. Second floor costs do not include any allowance for overhang.

Banks and Savings Offices
Wood Frame
Quality Classification

	Class 1 Best Quality	Class 2 Good Quality	Class 3 High Average Quality	Class 4 Low Average Quality	Class 5 Low Quality
Foundation	Reinforced concrete.	Reinforced concrete.	Reinforced concrete.	Reinforced concrete.	Reinforced concrete.
Floor Structure	Reinforced concrete.	Reinforced concrete.	Reinforced concrete.	Reinforced concrete.	Reinforced concrete.
Wall Structure	2" x 6" - 16" o.c. brick or concrete columns with combustible filler walls.	2" x 6" - 16" o.c. brick or concrete columns with combustible filler walls.	2" x 6" - 16" o.c.	2" x 6" - 16" o.c.	2" x 6" - 16" o.c.
Roof and Cover	Copper or slate.	Mission tile or heavy shakes.	5 ply built-up roof. Some portions heavy shake or mission tile.	4 ply built-up roof. Some portions shingle or composition shingles.	4 ply built-up roof.
Exterior Wall Finish	Expensive veneers.	Expensive ornamental veneer filler walls.	Wood siding combined with brick or stone veneers.	Good wood siding.	Good stucco or wood siding.
Interior Wall Finish	Ornamental plaster painted with murals or designs, ornamental moldings, marble and similar expensive finishes.	Ornamental plaster, ornamental moldings, some hardwood panels or marble finishes.	Hard plaster, ornamental cove at ceiling, vinyl wall covering, some hardwood veneer.	Plaster on lath with putty coat finish. Gypsum wallboard, texture and paint. Some vinyl wall cover.	Plaster on lath with putty coat finish. Gypsum wallboard, texture and paint.
Glass	Tinted plate glass in customized frames (50% of exterior wall area).	Tinted plate glass in bronze frames (50% of exterior wall area).	Tinted plate glass in heavy anodized aluminum frames or custom wood frames (50% of exterior wall area).	Moderate amount of plate glass in heavy frames (25% of exterior wall area).	Small to moderate amount of plate glass in average aluminum frames (5 to 10% of exterior wall area).
Overhang	4' closed overhang with copper gutters.	4' closed overhang with copper gutters.	4' closed overhang with painted gutters.	3' closed overhang with painted gutters.	None.
Floor Finish Public Area	Terrazzo and marble.	Detailed terrazzo.	Detailed terrazzo.	Plain terrazzo.	Sheet vinyl or asbestos tile.
Officers Area	Excellent carpet.	Very good carpet.	Very good carpet.	Good carpet.	Average quality carpet.
Work Area	Good sheet vinyl or carpet.	Average sheet vinyl or carpet.	Vinyl asbestos tile.	Vinyl asbestos tile.	Vinyl asbestos tile.
Ceiling Finish	Ornamental plaster with exposed ornamental beams.	Ornamental plaster.	Suspended acoustical tile with gypsum wallboard backing or plain acoustical plaster.	Suspended acoustical tile with exposed grid system.	Acoustical tile on wood strips.
Lighting	Recessed panelized fluorescent light fixtures. Custom light fixtures or chandeliers. Many spotlights.	Recessed panelized fluorescent lighting. Custom chandeliers. Many spotlights.	Recessed panelized fluorescent lighting. Many spotlights.	Recessed panelized fluorescent lighting. Some spotlights.	Nonrecessed panelized lighting.
Plumbing	2 or more rest rooms with special fixtures. Good ceramic tile, marble or terrazzo wainscot walls. Hard plaster with putty coat and enamel paint. Marble toilet screens.	2 or more rest rooms with many good fixtures. Good ceramic tile, marble or terrazzo wainscot walls. Hard plaster with putty coat and enamel paint. Marble toilet screens.	2 or more rest rooms with 4 or more fixtures. Ceramic tile or terrazzo wainscot, hard plaster walls with putty coat and enamel paint. Marble or metal toilet screens.	2 rest rooms with 4 or more fixtures each. Ceramic tile or terrazzo floors. Ceramic tile or terrazzo wainscot plaster walls with putty coat. Enamel paint. Good metal toilet partitions.	2 rest rooms with 4 fixtures each. Metal toilet screens. Ceramic tile or vinyl asbestos tile floors. Plaster walls with enamel paint.

Square foot costs include the following components: Foundations as required for normal soil conditions. Floor, wall and roof structures. Interior floor, wall, and ceiling finishes. Exterior wall finish and roof cover. All glass and glazing. Interior partitions. Roof overhang as described above. Basic electrical systems and lighting fixtures. Rough and finish plumbing. Typical bank vault. Alarm systems. Night depository. Typical record vault and fire doors. Fire exits. Design and engineering fees, permits and hook-up fees, contractor's mark-up.

Banks and Savings Offices
Wood Frame
Length Less Than Twice Width

Estimating Procedure

1. Establish the structure quality class by using the information on page 87.
2. Compute the building floor area. This should include everything within the building exterior walls and all insets outside the main walls but under the main building roof.
3. Add to or subtract from the square foot cost below the appropriate amount from the Wall Height Adjustment Table on page 91 if the wall height is more or less than 16 feet for the first floor or 12 feet for higher floors.
4. Multiply the adjusted square foot cost by the building area.
5. Deduct, if appropriate, for common walls, using the figures on page 91.
6. Multiply the total cost by the location factor listed on page 7.
7. Add the cost of heating and cooling equipment, fire sprinklers, exterior signs, elevators, and yard improvements from pages 201 to 213. Add the cost of mezzanines, bank fixtures, external windows, safe deposit boxes and vault doors from page 92.

Savings Office, Class 5

First Story

Square Foot Area at 16 Foot Wall Height

Quality Class	2,500	3,000	3,500	4,000	4,500	5,000	6,000	7,500	10,000	15,000	20,000
1, Best	126.82	119.94	114.94	111.13	108.13	105.70	101.99	98.19	94.25	90.10	87.89
1 & 2	118.40	111.97	107.31	103.75	100.95	98.68	95.22	91.67	87.99	84.12	82.06
2, Good	110.01	104.04	99.70	96.40	93.80	91.69	88.47	85.17	81.76	78.15	76.24
2 & 3	97.50	92.21	88.37	85.44	83.13	81.27	78.42	75.49	72.46	69.27	67.58
3, High Avg.	94.26	89.14	85.43	82.60	80.37	78.56	75.81	72.98	70.05	66.97	65.33
3 & 4	87.32	82.58	79.14	76.52	74.45	72.78	70.22	67.61	64.89	62.04	60.52
4, Low Avg.	80.49	76.12	72.95	70.53	68.63	67.09	64.73	62.32	59.82	57.18	55.79
4 & 5	72.96	69.00	66.12	63.93	62.21	60.81	58.68	56.49	54.22	51.83	50.57
5, Low	65.29	61.74	59.17	57.21	55.67	54.42	52.51	50.55	48.52	46.38	45.25

Second and Higher Stories

Square Foot Area at 12 Foot Wall Height

Quality Class	500	750	1,000	1,500	2,000	3,000	4,000	5,000	7,500	10,000	20,000
1, Best	87.48	79.37	74.54	68.81	65.39	61.34	58.92	57.27	54.71	53.18	50.29
1 & 2	80.68	73.49	69.21	64.16	61.15	57.59	55.47	54.03	51.78	50.44	47.92
2, Good	74.73	68.31	64.49	59.96	57.27	54.08	52.18	50.88	48.87	47.67	45.40
2 & 3	68.64	62.98	59.63	55.65	53.28	50.48	48.81	47.68	45.91	44.86	42.87
3, High Avg.	62.56	57.71	54.82	51.40	49.37	46.95	45.51	44.53	43.01	42.10	40.38
3 & 4	57.31	53.04	50.50	47.50	45.72	43.62	42.37	41.51	40.19	39.40	37.92
4, Low Avg.	51.99	48.38	46.22	43.67	42.15	40.35	39.27	38.54	37.40	36.72	35.43
4 & 5	47.61	44.49	42.64	40.44	39.14	37.59	36.67	36.05	35.07	34.49	33.39
5, Low	44.13	40.97	39.20	37.20	36.06	34.76	34.02	33.52	32.76	32.32	31.51

Banks and Savings Offices
Wood Frame
Length Between Two and Four Times Width

Estimating Procedure

1. Establish the structure quality class by using the information on page 87.
2. Compute the building floor area. This should include everything within the building exterior walls and all insets outside the main walls but under the main building roof.
3. Add to or subtract from the square foot cost below the appropriate amount from the Wall Height Adjustment Table on page 91 if the wall height is more or less than 16 feet for the first floor or 12 feet for higher floors.
4. Multiply the adjusted square foot cost by the building area.
5. Deduct, if appropriate, for common walls, using the figures on page 91.
6. Multiply the total cost by the location factor listed on page 7.
7. Add the cost of heating and cooling equipment, fire sprinklers, exterior signs, elevators, and yard improvements from pages 201 to 213. Add the cost of mezzanines, bank fixtures, external windows, safe deposit boxes and vault doors from page 92.

Bank, Class 4

First Story

Square Foot Area at 16 Foot Wall Height

Quality Class	2,500	3,000	3,500	4,000	4,500	5,000	6,000	7,500	10,000	15,000	20,000
1, Best	130.08	123.37	118.40	114.55	111.46	108.93	104.99	100.84	96.40	91.49	88.75
1 & 2	121.49	115.22	110.58	106.98	104.10	101.73	98.06	94.18	90.03	85.44	82.88
2, Good	112.89	107.07	102.75	99.41	96.74	94.54	91.12	87.52	83.66	79.40	77.02
2 & 3	104.82	99.41	95.40	92.30	89.81	87.77	84.60	81.26	77.67	73.72	71.51
3, High Avg.	96.77	91.78	88.08	85.21	82.92	81.03	78.10	75.02	71.71	68.06	66.02
3 & 4	89.25	84.65	81.24	78.59	76.48	74.74	72.04	69.19	66.14	62.77	60.89
4, Low Avg.	82.13	77.89	74.76	72.32	70.38	68.78	66.29	63.67	60.86	57.76	56.03
4 & 5	74.21	70.38	67.55	65.35	63.59	62.14	59.90	57.53	54.99	52.19	50.63
5, Low	66.33	62.90	60.37	58.41	56.83	55.54	53.53	51.42	49.15	46.65	45.25

Second and Higher Stories

Square Foot Area at 12 Foot Wall Height

Quality Class	500	750	1,000	1,500	2,000	3,000	4,000	5,000	7,500	10,000	20,000
1, Best	88.35	80.91	76.47	71.22	68.09	64.37	62.16	60.65	58.31	56.91	54.26
1 & 2	82.78	75.81	71.65	66.73	63.80	60.32	58.24	56.83	54.63	53.32	50.84
2, Good	77.20	70.70	66.83	62.23	59.50	56.25	54.32	53.00	50.95	49.73	47.42
2 & 3	71.74	65.70	62.10	57.83	55.29	52.28	50.48	49.25	47.35	46.21	44.06
3, High Avg.	66.31	60.73	57.40	53.46	51.10	48.32	46.66	45.53	43.76	42.72	40.73
3 & 4	61.30	56.14	53.06	49.42	47.24	44.67	43.13	42.09	40.46	39.49	37.65
4, Low Avg.	56.45	51.69	48.86	45.50	43.50	41.13	39.72	38.75	37.26	36.36	34.67
4 & 5	52.36	47.95	45.33	42.21	40.35	38.15	36.84	35.95	34.56	33.73	32.16
5, Low	48.29	44.22	41.80	38.93	37.22	35.19	33.98	33.15	31.87	31.11	29.66

Banks and Savings Offices
Wood Frame
Length More Than Four Times Width

Estimating Procedure

1. Establish the structure quality class by using the information on page 87.
2. Compute the building floor area. This should include everything within the building exterior walls and all insets outside the main walls but under the main building roof.
3. Add to or subtract from the square foot cost below the appropriate amount from the Wall Height Adjustment Table on page 91 if the wall height is more or less than 16 feet for the first floor or 12 feet for higher floors.
4. Multiply the adjusted square foot cost by the building area.
5. Deduct, if appropriate, for common walls, using the figures on page 91.
6. Multiply the total cost by the location factor listed on page 7.
7. Add the cost of heating and cooling equipment, fire sprinklers, exterior signs, elevators, and yard improvements from pages 201 to 213. Add the cost of mezzanines, bank fixtures, external windows, safe deposit boxes and vault doors from page 92.

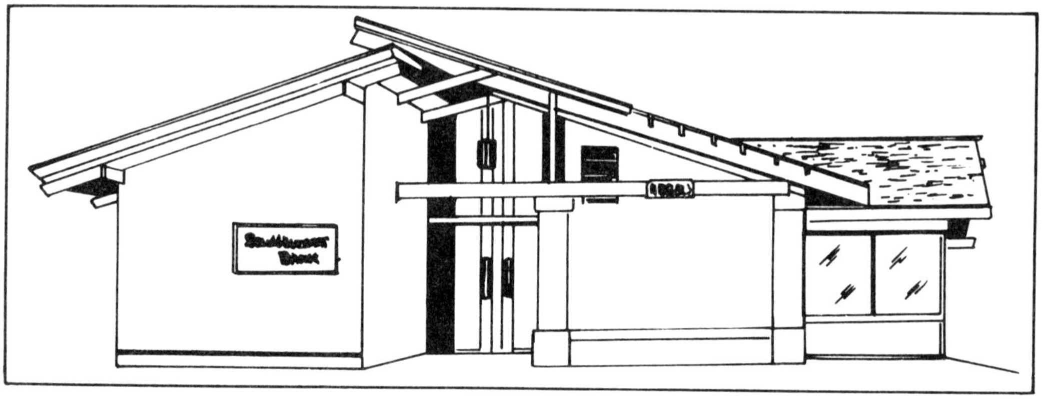

Bank, Class 4 & 5

First Story

Square Foot Area at 16 Foot Wall Height

Quality Class	2,500	3,000	3,500	4,000	4,500	5,000	6,000	7,500	10,000	15,000	20,000
1, Best	136.67	129.16	123.62	119.33	115.91	113.11	108.76	104.20	99.33	93.99	91.03
1 & 2	127.65	120.64	115.46	111.46	108.26	105.64	101.58	97.32	92.78	87.79	85.02
2, Good	118.66	112.14	107.33	103.61	100.64	98.20	94.43	90.47	86.24	81.61	79.04
2 & 3	110.21	104.15	99.68	96.23	93.47	91.20	87.70	84.02	80.09	75.79	73.40
3, High Avg.	101.75	96.16	92.03	88.84	86.30	84.21	80.97	77.57	73.95	69.98	67.77
3 & 4	93.77	88.62	84.82	81.88	79.53	77.60	74.62	71.49	68.15	64.49	62.46
4, Low Avg.	85.83	81.11	77.63	74.94	72.79	71.03	68.30	65.43	62.38	59.02	57.17
4 & 5	77.26	73.01	69.88	67.46	65.52	63.94	61.48	58.90	56.15	53.13	51.46
5, Low	68.71	64.94	62.15	60.00	58.28	56.87	54.68	52.39	49.94	47.25	45.77

Second and Higher Stories

Square Foot Area at 12 Foot Wall Height

Quality Class	500	750	1,000	1,500	2,000	3,000	4,000	5,000	7,500	10,000	20,000
1, Best	101.52	90.78	84.39	76.83	72.33	67.00	63.82	61.66	58.30	56.29	52.51
1 & 2	93.66	84.04	78.33	71.56	67.53	62.76	59.92	57.99	54.98	53.19	49.80
2, Good	85.80	77.32	72.28	66.31	62.76	58.56	56.06	54.36	51.71	50.13	47.15
2 & 3	78.36	70.91	66.49	61.25	58.13	54.44	52.25	50.75	48.43	47.04	44.42
3, High Avg.	70.90	64.52	60.72	56.21	53.53	50.35	48.46	47.17	45.16	43.96	41.70
3 & 4	64.59	58.98	55.65	51.71	49.37	46.59	44.94	43.82	42.07	41.03	39.06
4, Low Avg.	58.26	53.45	50.60	47.23	45.22	42.85	41.45	40.49	38.99	38.11	36.43
4 & 5	52.95	48.84	46.39	43.50	41.79	39.75	38.55	37.72	36.44	35.68	34.24
5, Low	47.72	44.27	42.23	39.82	38.38	36.69	35.68	35.00	33.93	33.30	32.10

Banks and Savings Offices
Wood Frame
Wall Adjustments

Wall Height Adjustment

Add or subtract the amount listed in this table to the square foot of floor cost for each foot of wall height more or less than 16 feet, if adjusting for first floor, and 12 feet if adjusting for upper floors.

Square Foot Area

Quality Class	500	750	1,000	1,500	2,000	2,500	3,000	3,500
1, Best	4.23	3.48	3.02	2.48	2.15	1.92	1.76	1.63
2, Good	3.72	3.06	2.66	2.18	1.89	1.69	1.55	1.43
3, High Average	3.21	2.64	2.29	1.88	1.63	1.46	1.33	1.23
4, Low Average	2.14	1.75	1.53	1.25	1.08	.97	.89	.82
5, Low	1.37	1.13	.98	.80	.70	.62	.57	.53

Square Foot Area

Quality Class	4,000	4,500	5,000	6,000	7,500	10,000	15,000	20,000
1, Best	1.60	1.48	1.39	1.24	1.09	.94	.79	.71
2, Good	1.40	1.29	1.21	1.08	.95	.82	.69	.62
3, High Average	1.20	1.11	1.04	.93	.82	.71	.59	.53
4, Low Average	.80	.74	.69	.62	.55	.47	.39	.36
5, Low	.53	.49	.46	.41	.36	.31	.26	.23

Perimeter Wall Adjustment
First Floor

Class	For a Common Wall, Deduct Per Linear Foot	For No Wall Ownership, Deduct Per Linear Foot	For Lack of Exterior Finish, Deduct Per Linear Foot
1	$200.00	$278.00	$194.00
2	160.00	235.00	153.00
3	129.00	188.00	110.00
4	95.00	144.00	74.00
5	43.00	83.00	25.00

Second and Higher Floors

Class	For a Common Wall, Deduct Per Linear Foot	For No Wall Ownership, Deduct Per Linear Foot	For Lack of Exterior Finish, Deduct Per Linear Foot
1	$124.00	$135.00	$102.00
2	96.00	109.00	76.00
3	71.00	83.00	50.50
4	48.50	62.50	33.00
5	34.40	48.50	14.60

Note: First floor costs include the cost of overhang as described on page 87. Second floor costs do not include any allowance for overhang.

Banks and Savings Offices Additional Costs

Mezzanines

Without Partitions

Quality Class	1, Best	2, Good	3, High Avg.	4, Low Avg.	5, Low
S.F. Cost	$27.20	$26.40	$22.40	$20.70	$18.50

With Partitions

Quality Class	1, Best	2, Good	3, High Avg.	4, Low Avg.	5, Low
S.F. Cost	$31.90	$28.70	$25.60	$24.00	$20.70

Mezzanine costs include: Floor systems, floor finish, stairways, lighting and plumbing. Partitions are included when applicable.

Fixtures, cost per square foot

Shell-type counters, plastic finish, no counter screens or drawers.	$4.70 to $5.40
Counters with drawers, good hardwood, plain counter screens.	5.50 to 7.50
Counters with drawers, good hardwood, formica counter tops, average counter screens.	7.40 to 10.75
Counters with drawers, terrazzo or marble finish, marble counter tops, fancy counter screens.	10.80 to 18.60

Costs include counters, screens, and partitions. The square-foot cost should be applied only to the first floor area. Office areas used for purposes other than conducting bank business related to the immediate site should be excluded.

External Windows, cost per window

Drive-up type	
Flush wall, complete	$5,400 to $7,400
Projected, complete	7,200 to 11,500
Walk-up type	
Flush wall, installed	$3,900

Safe Deposit Boxes, cost per box

Box Sizes	Modular Unit	Custom Built to Order
3" x 5"	$44.50 to $51.50	$62.00 to $69.40
5" x 5"	51.60 to 57.70	70.00 to 66.50
3" x 10"	56.20 to 62.00	76.70 to 89.00
5" x 10"	62.00 to 70.00	95.50 to 110.00
10" x 10"	87.00 to 95.50	130.00 to 155.00

Bank fixtures and safe deposit boxes should be included in the structure cost only if they are affixed to the building.

Vault Doors

Cash and Safe Deposit (rectangular)

Steel thickness	1½"	3½"	7"
In place cost	$13,000 to $17,000	$18,700 to $21,000	$21,000 to $27,300

Record Storage

Steel Thickness	2-11/16"	4"	4-5/8"	6-3/4"
In Place Cost	$1,860	$2,900	$3,300	$4,300

Prices include frames, time clocks, and architraves.

Department Stores
Reinforced Concrete
Quality Classification

	Class 1 Best Quality	Class 2 Good Quality	Class 3 Average Quality	Class 4 Low Quality
Foundation	Reinforced concrete.	Reinforced concrete.	Reinforced concrete.	Reinforced concrete.
Ground Floor Structure	6" reinforced concrete on 6" rock base.	6" reinforced concrete on 6" rock base.	6" reinforced concrete on 6" rock base.	4" reinforced concrete on 6" rock base.
Wall Structure	Reinforced concrete.	Reinforced concrete.	Reinforced concrete.	Reinforced concrete.
Upper Floor Structure	Reinforced concrete.	Reinforced concrete.	Reinforced concrete.	Reinforced concrete.
Roof and Cover	Reinforced concrete with 5 ply built-up roofing and insulation.	Reinforced concrete with 5 ply built-up roofing and insulation.	Reinforced concrete with 4 ply built-up roofing.	Reinforced concrete with 4 ply built-up roofing.
Floor Finish	Terrazzo and very good carpet.	Vinyl asbestos tile with 50% vinyl tile, terrazzo or good carpet.	Vinyl asbestos tile.	Minimum grade tile.
Interior Wall Finish	Gypsum wallboard or lath and plaster finished with good paper or vinyl wall cover on hardwood veneer paneling.	Gypsum wallboard and texture or paper. Some vinyl wall cover or hardwood veneer paneling.	Interior stucco or gypsum wallboard, texture and paint.	Gypsum wallboard, texture and paint.
Ceiling Finish	Suspended good grade acoustical tile with gypsum wallboard backing.	Suspended acoustical tile with concealed grid system.	Suspended acoustical tile with exposed grid system.	Painted.
Lighting	Recessed fluorescent lighting in modular plastic panels. Many spotlights.	Continuous recessed 3 tube fluorescent strips with egg crate diffusers, 8' o.c. Average number of spotlights.	Continuous 3 tube fluorescent strips with egg crate diffusers, 8' o.c. Some spotlights.	Continuous exposed 2 tube fluorescent strips, 8' o.c.
Display Fronts	Very good front as described on page 56. 15 to 25% of the first floor exterior wall is made up of display fronts.	Good quality front as described on page 56. 15 to 25% of the first floor exterior wall is made up of display fronts.	Average quality front as described on page 56. 10 to 20% of the first floor exterior wall is made up of display fronts.	Average quality flat type as described on page 56. 10 to 20% of the first floor exterior wall is made up of display fronts.
Exterior Wall Finish	Ornamental block or brick, some marble veneer.	Decorative block, some stone veneer.	Paint.	Paint.
Plumbing	6 good fixtures per 20,000 square feet of floor area. Metal toilet partitions.	6 standard fixtures per 20,000 square feet of floor area. Metal toilet partitions.	4 standard fixtures per 20,000 square feet of floor area. Metal toilet partitions.	4 standard fixtures per 20,000 square feet of floor area. Wood toilet partitions.

Wall Height Adjustment

Add or subtract the amount listed to the square foot of floor cost for each foot of first and upper story wall height more or less than 20 feet.

Square Foot Area

Quality Class	20,000	25,000	30,000	35,000	40,000	45,000	50,000	60,000	70,000	80,000	100,000
1, Best	.82	.69	.61	.56	.52	.49	.47	.44	.42	.41	.40
2, Good	.55	.46	.41	.37	.35	.33	.31	.29	.28	.27	.26
3, Average	.43	.37	.32	.30	.28	.26	.25	.23	.22	.22	.21
4, Low	.34	.29	.26	.24	.22	.21	.20	.19	.18	.17	.17

Square foot costs include the following components: Foundations as required for normal soil conditions. Floor, wall and roof structures. Interior ceiling, wall and floor finishes (including carpet). Exterior wall finish and roof cover. Display fronts. Interior partitions (including perimeter wall partitions). Entry and delivery doors. Basic lighting and electrical systems. Rough and finish plumbing. Design and engineering fees. Permits and hook-up fees. Contractor's mark-up.

Department Stores
Reinforced Concrete
First Floor

Estimating Procedure

1. Use these square foot costs to estimate the cost of retail stores designed to sell a wider variety of goods. These buildings differ from discount houses in that they have more interior partitions and more elaborate interior and exterior finishes.
2. Establish the structure quality class by applying the information on page 93.
3. Compute the building floor area. This should include everything within the building exterior walls and all inset areas outside the main walls but under the main building roof.
4. Add or subtract from the square foot cost below the appropriate amount from the Wall Height Adjustment Table (near the bottom of page 93) if the wall height is more or less than 20 feet.
5. Multiply the adjusted square foot cost by the building area.
6. Deduct, if appropriate, for common or unfinished walls, using the figures at the bottom of this page.
7. Multiply the total cost by the location factor on page 7.
8. Add the cost of appropriate additional components from page 201: Heating and air conditioning systems, elevators and escalators, fire sprinklers, exterior signs, canopies, and walks, paving and curbing, loading docks and ramps, miscellaneous yard improvements, and mezzanines.
9. Add the cost of second and higher floors and basements from page 95.

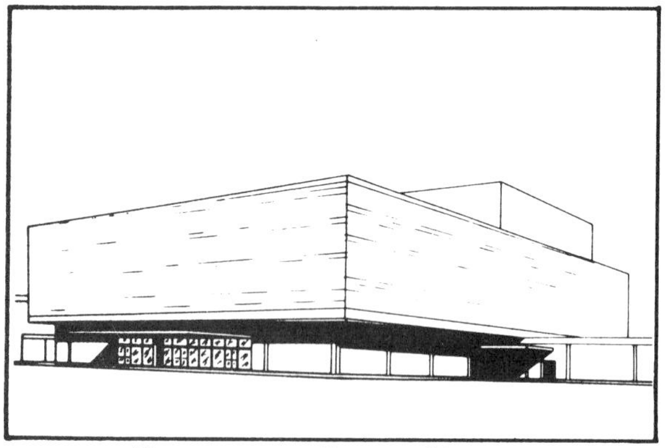

Department Store, Class 1

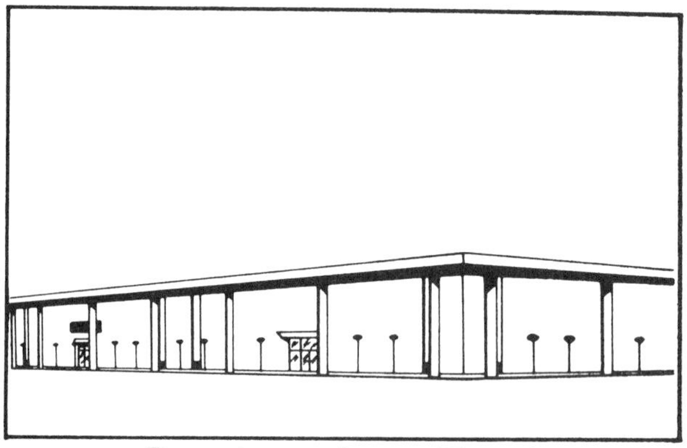

Department Store, Class 2

Square Foot Area

Quality Class	20,000	25,000	30,000	35,000	40,000	45,000	50,000	60,000	70,000	80,000	100,000
1, Best	78.58	75.50	73.00	70.93	69.17	67.66	66.34	64.13	62.35	60.88	58.54
1 & 2	72.32	69.63	67.44	65.63	64.09	62.77	61.62	59.69	58.13	56.84	54.80
2, Good	66.09	63.77	61.89	60.34	59.02	57.89	56.90	55.25	53.92	52.82	51.08
2 & 3	60.49	58.52	56.93	55.62	54.50	53.55	52.72	51.33	50.21	49.28	47.82
3, Average	54.91	53.28	51.97	50.90	50.00	49.23	48.56	47.45	46.56	45.82	44.66
3 & 4	49.63	48.32	47.28	46.43	45.72	45.10	44.57	43.69	42.98	42.40	41.48
4, Low	44.33	43.42	42.67	42.05	41.53	41.07	40.68	40.01	39.48	39.03	38.32

Perimeter Wall Adjustment
First Floor

Class	For a Common Wall, Deduct per Linear Foot	For no Wall Ownership, Deduct per Linear Foot	For Lack of Exterior Finish, Deduct per Linear Foot
1	$287.00	$381.00	$185.00
2	210.00	307.00	108.00
3	148.00	243.00	49.00
4	100.00	204.00	28.00

Department Stores Reinforced Concrete

Estimating Procedure

1. Establish the basement and upper floor quality class. The quality class will usually be the same as the first floor of the building. Square foot costs for unfinished basements will be nearly the same regardless of the structure quality class.
2. Compute the floor area.
3. Add to or subtract from the square foot cost below the appropriate amount from the Wall Height Adjustment Table on page 93 for second and higher floors and from the bottom of this page for basements for each foot of wall height more or less than 20 feet for second and higher floors and 16 feet for basements.
4. Multiply the adjusted square foot cost from one of the 3 tables below by the floor area.
5. Deduct, if appropriate, for common or unfinished upper floor walls, using the costs in the table below titled "Second and Higher Floor Perimeter Wall Adjustments."
6. Multiply the total cost by the location factor on page 7.
7. Add the cost of the appropriate additional components from page 201: Heating and air conditioning systems, elevators and escalators, fire sprinklers, and mezzanines.
8. Add the total from this page to the total from page 94 to find the building cost.

Second and Higher Stories

Square Foot Area

Quality Class	20,000	25,000	30,000	35,000	40,000	45,000	50,000	60,000	70,000	80,000	100,000
1, Best	70.44	67.55	65.19	63.22	61.54	60.09	58.83	56.72	55.00	53.58	51.33
1 & 2	64.99	62.43	60.34	58.60	57.13	55.86	54.75	52.90	51.40	50.15	48.18
2, Good	59.53	57.29	55.48	53.98	52.72	51.62	50.67	49.08	47.80	46.74	45.07
2 & 3	54.66	52.77	51.24	49.97	48.89	47.97	47.17	45.83	44.75	43.85	42.43
3, Average	49.82	48.25	46.98	45.94	45.06	44.31	43.65	42.55	41.67	40.94	39.79
3 & 4	45.02	43.74	42.72	41.88	41.18	40.59	40.07	39.21	38.51	37.94	37.05
4, Low	40.20	39.21	38.44	37.82	37.30	36.86	36.49	35.86	35.36	34.95	34.31

Second and Higher Floor Perimeter Wall Adjustment

Class	For a Common Wall, Deduct per Linear Foot	For no Wall Ownership, Deduct per Linear Foot	For Lack of Exterior Finish, Deduct per Linear Foot
1	$223.00	$307.00	$145.00
2	162.00	245.00	87.00
3	115.00	193.00	41.00
4	77.00	163.00	28.00

Finished Basements

Square Foot Area

Quality Class	20,000	25,000	30,000	35,000	40,000	45,000	50,000	60,000	70,000	80,000	100,000
1, Best	60.02	58.12	56.54	55.20	54.06	53.06	52.19	50.71	49.52	48.52	46.93
1 & 2	55.61	53.93	52.55	51.38	50.39	49.52	48.77	47.50	46.46	45.61	44.24
2, Good	52.10	50.58	49.33	48.28	47.38	46.61	45.93	44.79	43.87	43.11	41.89
2 & 3	48.11	46.82	45.77	44.89	44.14	43.50	42.94	42.00	41.24	40.61	39.61
3, Average	44.85	43.89	43.05	42.32	41.63	41.14	40.65	39.82	39.13	38.55	37.63
3 & 4	41.02	40.12	39.40	38.81	38.31	37.89	37.52	36.91	36.42	36.01	35.37
4, Low	37.09	36.37	35.82	35.38	35.03	34.73	34.47	34.05	33.72	33.45	33.03

Unfinished Basements

Square Foot Area

Area	20,000	25,000	30,000	35,000	40,000	45,000	50,000	60,000	70,000	80,000	100,000
Cost	26.94	26.42	26.04	25.75	25.51	25.31	25.14	24.87	24.66	24.49	24.23

Basement Wall Height Adjustment

Add or subtract the amount listed to the square foot of floor cost for each foot of basement wall height more or less than 16 feet.

Square Foot Area

Area	20,000	25,000	30,000	35,000	40,000	45,000	50,000	60,000	70,000	80,000	100,000
Finished	.37	.31	.28	.26	.25	.23	.23	.22	.21	.21	.20
Unfinished	.32	.28	.25	.23	.22	.21	.20	.19	.18	.18	.18

Department Stores
Masonry or Concrete
Quality Classification

	Class 1 Best Quality	Class 2 Good Quality	Class 3 Average Quality	Class 4 Low Quality
Foundation	Reinforced concrete.	Reinforced concrete.	Reinforced concrete.	Reinforced concrete.
Floor Structure	6" reinforced concrete on 6" rock base.	6" reinforced concrete on 6" rock base.	6" reinforced concrete on 6" rock base.	4" reinforced concrete on 6" rock base.
Wall Structure	8" reinforced decorative concrete block, 6" concrete tilt-up or 8" reinforced brick.	8" reinforced decorative concrete block, 6" concrete tilt-up or 8" reinforced brick.	8" reinforced concrete block, 6" concrete tilt-up or 8" reinforced common brick.	8" reinforced concrete block or 6" concrete tilt-up.
Roof and Cover	Glu-lams or steel beams on steel intermediate columns. Panelized roof system, ½" plywood sheathing, 5 ply built-up roof with insulation.	Glu-lams or steel beams on intermediate columns. Panelized roof system, ½" plywood sheathing, 5 ply built-up roof with insulation.	Glu-lams on steel intermediate columns. Panelized roof system, ½" plywood sheathing, 4 ply built-up roof.	Glu-lams on steel intermediate columns. Panelized roof system, ½" plywood sheathing, 4 ply built-up roof.
Floor Finish	Terrazzo and very good carpet.	Vinyl asbestos tile with 50% sheet vinyl, terrazzo or good carpet.	Vinyl asbestos tile.	Minimum grade tile.
Interior Wall Finish	Gypsum wallboard or lath and plaster, finished with good paper or vinyl wall covers, or hardwood veneer paneling.	Gypsum wallboard and texture or paper, some vinyl wall cover or hardwood veneer paneling.	Interior stucco or gypsum wallboard, texture and paint.	Gypsum wallboard, texture and paint.
Ceiling Finish	Suspended good grade acoustical tile with gypsum wallboard backing.	Suspended acoustical tile with concealed grid system.	Suspended acoustical tile with exposed grid system.	Painted.
Lighting	Recessed fluorescent lighting in modular plastic panels. Many spotlights.	Continuous recessed 3 tube fluorescent strips with egg crate diffusers, 8' o.c. Average number of spotlights.	Continuous 3 tube fluorescent strips with egg crate diffusers, 8' o.c. Some spotlights.	Continuous exposed 2 tube fluorescent strips, 8' o.c.
Display Fronts	Very good fronts as described on page 56. 15 to 25% of the first floor exterior wall is made up of display fronts.	Good quality front as described on page 56. 15 to 25% of the first floor exterior wall is made up of display fronts.	Average quality front as described on page 56. 10 to 20% of the first floor exterior wall is made up of display fronts.	Average quality flat type as described on page 56. 10 to 20% of the first floor exterior wall is made up of display fronts.
Exterior Wall Finish	Decorative block or large rock imbedded in tilt-up panels, some stone veneer.	Exposed aggregate or decorative block. Some stone veneer.	Paint, some exposed aggregate.	Paint.
Plumbing	6 good fixtures per 20,000 square feet of floor area. Metal toilet partitions.	6 standard fixtures per 20,000 square feet of floor area. Metal toilet partitions.	4 standard fixtures per 20,000 square feet of floor area. Metal toilet partitions.	4 standard fixtures per 20,000 square feet of floor area. Wood toilet partitions.

Square foot costs include the following components: Foundations as required for normal soil conditions. Floor, wall and roof structures. Interior ceiling, wall and floor finishes (including carpet). Exterior wall finish and roof cover. Display fronts. Interior partitions (including perimeter wall partitions). Entry and delivery doors. Basic lighting and electrical systems. Rough and finish plumbing. Design and engineering fees. Permits and hook-up fees. Contractor's mark-up.

The in-place cost of these extra components should be added to the basic building cost to arrive at the total structure cost. See the section "Additional Costs for Commercial Structures" on page 201. Heating and air conditioning systems. Elevators and escalators. Fire sprinklers. Exterior signs. Canopies and walks. Paving and curbing. Loading docks or ramps. Miscellaneous yard improvements. Mezzanines.

Department Stores
Masonry or Concrete
First Floor

Estimating Procedure

1. Use these square foot costs to estimate the cost of retail stores designed to sell a wider variety of goods. These buildings differ from discount houses in that they have more interior partitions and more elaborate interior and exterior finishes.
2. Establish the structure quality class by applying the information on page 96.
3. Compute the building floor area. This should include everything within the building exterior walls and all inset areas outside the main walls but under the main building roof.
4. Add to or subtract from the square foot cost below the appropriate amount from the Wall Height Adjustment Column (near the bottom of this page) if the wall height is more or less than 16 feet.
5. Multiply the adjusted square foot cost by the building area.
6. Deduct, if appropriate, for common or unfinished walls, using the figures at the bottom of this page.
7. Multiply the total cost by the location factor on page 7.
8. Add the cost of the appropriate additional components from page 201: Heating and air conditioning systems, elevators and escalators, fire sprinklers, exterior signs, canopies and walks, paving and curbing, loading docks and ramps, miscellaneous yard improvements, and mezzanines.
9. Add the cost of second and higher floors and basements from page 98.

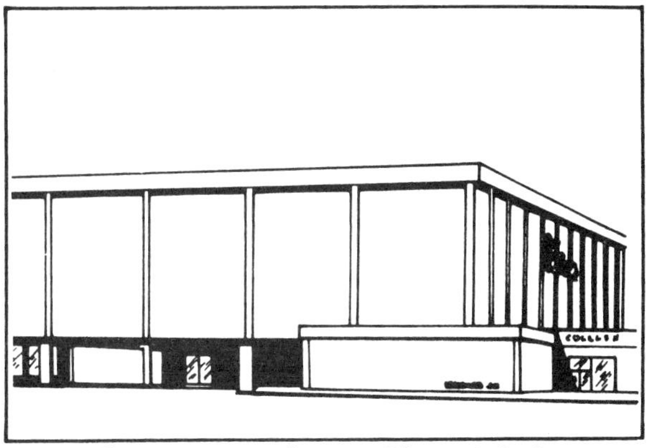

Department Store, Class 1 & 2

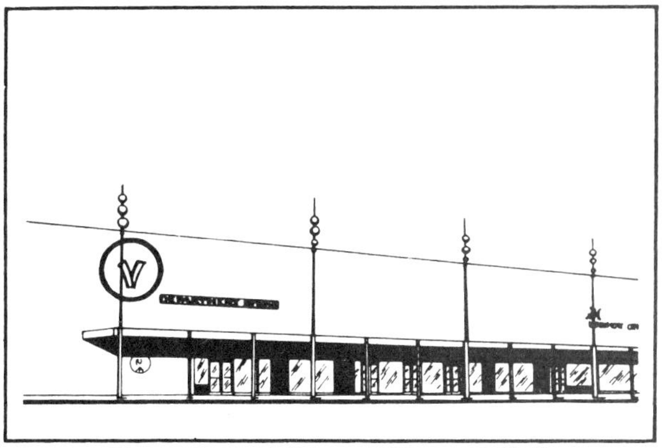

Department Store, Class 2 & 3

Square Foot Area

Quality Class	20,000	25,000	30,000	35,000	40,000	45,000	50,000	60,000	70,000	80,000	100,000
1, Best	61.39	58.68	56.53	54.76	53.28	52.01	50.91	49.08	47.62	46.41	44.51
1 & 2	54.47	52.07	50.15	48.59	47.27	46.15	45.17	43.55	42.25	41.18	39.49
2, Good	47.57	45.47	43.80	42.43	41.28	40.30	39.45	38.03	36.90	35.96	34.49
2 & 3	42.09	40.24	38.76	37.55	36.53	35.66	34.91	33.66	32.65	31.82	30.52
3, Average	36.64	35.02	33.74	32.68	31.80	31.04	30.38	29.29	28.42	27.70	26.56
3 & 4	31.64	30.24	29.13	28.22	27.46	26.81	26.24	25.30	24.54	23.92	22.94
4, Low	26.84	25.65	24.71	23.94	23.29	22.74	22.26	21.46	20.82	20.29	19.46
Wall Height Adjustment *	.33	.29	.26	.24	.22	.22	.21	.20	.20	.19	.19

*Wall Height Adjustment

Add or subtract the amount listed in this column to the square foot of floor cost for each foot of wall height more or less than 16 feet.

Perimeter Wall Adjustment

For a Common Wall, deduct $87.00 per linear foot.
For No Wall Ownership, deduct $169.00 per linear foot.

Department Stores
Masonry or Concrete
Upper Floors and Basements

Estimating Procedure

1. Establish the basement and upper floor quality class. The quality class will usually be the same as the first floor of the building. Square foot costs for unfinished basements will be nearly the same regardless of the structure quality class.
2. Compute the floor area.
3. Add to or subtract from the square foot cost below the appropriate amount from the Wall Height Adjustment Column for each foot of wall height more or less than 12 feet for second and higher floors and 12 feet for basements.
4. Multiply the adjusted square foot cost by the floor area.
5. Deduct, if appropriate, for common or unfinished upper floor walls, using the Perimeter Wall Adjustment costs listed below.
6. Multiply the total cost by the location factor on page 7.
7. Add the cost of the appropriate additional components from page 201: Heating and air conditioning systems, elevators and escalators, fire sprinklers, and mezzanines.
8. Add the total from this page to the total from page 97 to find the building cost.

Second and higher floors

Square Foot Area

Quality Class	20,000	25,000	30,000	35,000	40,000	45,000	50,000	60,000	70,000	80,000	100,000
1, Best	54.27	52.27	50.44	48.81	47.36	46.08	44.92	42.95	41.31	39.92	37.69
1 & 2	47.37	45.62	44.02	42.60	41.34	40.21	39.21	37.48	36.05	34.84	32.89
2, Good	40.65	39.15	37.78	36.56	35.48	34.51	33.65	32.17	30.94	29.90	28.23
2 & 3	35.50	34.19	32.99	31.93	30.98	30.14	29.39	28.10	27.02	26.12	24.65
3, Average	30.37	29.25	28.22	27.31	26.50	25.78	25.14	24.03	23.12	22.34	21.09
3 & 4	24.87	23.96	23.12	22.37	21.71	21.12	20.59	19.69	18.93	18.30	17.27
4, Low	19.98	19.32	18.79	18.35	17.98	17.66	17.38	16.91	16.54	16.22	15.73
Wall Height Adjustment *	.33	.29	.26	.24	.22	.22	.21	.20	.20	.19	.19

***Wall Height Adustment:** Add or subtract the amount listed in this column to the square foot of floor cost for each foot of wall height more or less than 12 feet.

Perimeter Wall Adjustment: For No Wall Ownership, deduct $78.40 per linear foot. For a Common Wall, deduct $50.70 per linear foot.

Finished Basements

Square Foot Area

Quality Class	20,000	25,000	30,000	35,000	40,000	45,000	50,000	60,000	70,000	80,000	100,000
1, Best	48.21	46.28	44.72	43.43	42.34	41.41	40.60	39.25	38.16	37.26	35.84
2, Good	37.08	35.59	34.39	33.40	32.57	31.85	31.23	30.19	29.35	28.66	27.57
3, Average	28.91	27.75	26.81	26.04	25.39	24.83	24.34	23.53	22.88	22.34	21.49
4, Low	20.37	19.55	18.89	18.35	17.89	17.50	17.15	16.58	16.12	15.74	15.14

Unfinished Basements

Square Foot Area

Area	20,000	25,000	30,000	35,000	40,000	45,000	50,000	60,000	70,000	80,000	100,000
Cost	11.75	11.21	10.81	10.50	10.26	10.06	9.89	9.62	9.41	9.24	8.99

Wall Height Adjustment

Add or subtract the amount listed in this table to the square foot of floor cost for each foot of basement wall height more or less than 12 feet.

Square Foot Area

Area	20,000	25,000	30,000	35,000	40,000	45,000	50,000	60,000	70,000	80,000	100,000
Finished	.26	.25	.24	.23	.22	.22	.21	.20	.20	.20	.19
Unfinished	.25	.23	.22	.22	.21	.20	.20	.19	.19	.19	.18

Department Stores
Wood Frame

	Class 1 Best Quality	Class 2 Good Quality	Class 3 Average Quality	Class 4 Low Quality
Foundation	Reinforced concrete.	Reinforced concrete.	Reinforced concrete.	Reinforced concrete.
Floor Structure	6″ reinforced concrete on 6″ rock base.	6″ reinforced concrete on 6″ rock base.	6″ reinforced concrete on 6″ rock base.	4″ reinforced concrete on 6″ rock base.
Wall Structure	2″ x 6″, 16″ o.c.	2″ x 6″, 16″ o.c.	2″ x 6″, 16″ o.c.	2″ x 4″, 16″ o.c.
Roof and Cover	Glu-lams or steel beams on steel intermediate columns. Panelized roof system, ½″ plywood sheathing, 5 ply built-up roof with insulation.	Glu-lams or steel beams on steel intermediate columns. Panelized roof system, ½″ plywood sheathing, 5 ply built-up roof with insulation.	Glu-lams on steel intermediate columns. Panelized roof system, ½″ plywood sheathing, 4 ply built-up roofing.	Glu-lams on steel intermediate columns. Panelized roof system, ½″ plywood sheathing, 4 ply built-up roof.
Floor Finish	Terrazzo and very good carpet.	Vinyl asbestos tile with 50% sheet vinyl, terrazzo or carpet.	Vinyl asbestos tile.	Minimum tile.
Interior Wall Finish	Gypsum wallboard or lath and plaster finished with good paper or vinyl wall cover or hardwood veneer paneling.	Gypsum wallboard, texture or paper, some vinyl wall cover or hardwood veneer paneling.	Interior stucco or gypsum wallboard. Texture and paint.	Gypsum wallboard. Texture and paint.
Ceiling Finish	Suspended good grade acoustical tile with gypsum wallboard backing.	Suspended acoustical tile with concealed grid system.	Suspended acoustical tile with exposed grid system.	Painted.
Lighting	Recessed fluorescent lighting in modular plastic panels. Many spotlights.	Continuous recessed 3 tube fluorescent strips with egg crate diffusers, 8′ o.c. Average number of spotlights.	Continuous 3 tube fluorescent strips with egg crate diffusers, 8′ o.c. Some spotlights.	Continuous exposed 2 tube fluorescent strips, 8′ o.c.
Display Fronts	Very good fronts as described on page 56. 15 to 25% of the first floor exterior wall is made up of display fronts.	Good quality front as described on page 56. 15 to 25% of the first floor exterior wall is made up of display fronts.	Average quality front as described on page 56. 10 to 20% of the first floor exterior wall is made up of display fronts.	Average quality flat type as described on page 56. 10 to 20% of the first floor exterior wall is made up of display fronts.
Exterior Wall Finish	Good wood siding. Extensive stone veneer.	Wood siding. Some brick or stone veneer.	Stucco. Some brick trim.	Stucco.
Plumbing	6 good fixtures per 20,000 square feet of floor area. Metal toilet partitions.	6 standard fixtures per 20,000 square feet of floor area. Metal toilet partitions.	4 standard fixtures per 20,000 square feet of floor area. Metal toilet partitions.	4 standard fixtures per 20,000 square feet of floor area. Wood toilet partitions.

Square foot costs include the following components: Foundations as required for normal soil conditions. Floor, wall and roof structures. Interior ceiling, wall and floor finishes (including carpet). Exterior wall finish and roof cover. Display fronts. Interior partitions (including perimeter wall partitions). Entry and delivery doors. Basic lighting and electrical systems. Rough and finish plumbing. Design and engineering fees. Permits and hook-up fees. Contractor's mark-up.

The in-place cost of these extra components should be added to the basic building cost to arrive at the total structure cost. See the section "Additional Costs for Commercial Structures" on page 201. Heating and air conditioning systems. Elevators and escalators. Fire sprinklers. Exterior signs. Canopies and walks. Paving and curbing. Loading docks or ramps. Miscellaneous yard improvements. Mezzanines.

Department Stores
Wood Frame
First Floor

Estimating Procedure

1. Use these square foot costs to estimate the cost of retail stores designed to sell a wider variety of goods. These buildings differ from discount houses in that they have more interior partitions and more elaborate interior and exterior finishes.
2. Establish the structure quality class by applying the information on page 99.
3. Compute the building floor area. This should include everything within the building exterior walls and all inset areas outside the main walls but under the main building roof.
4. Add to or subtract from the square foot cost below the appropriate amount from the Wall Height Adjustment Column (near the bottom of this page) if the wall height is more or less than 16 feet.
5. Multiply the adjusted square foot cost by the building area.
6. Deduct, if appropriate, for common walls or no wall ownership, using the figures at the bottom of this page.
7. Multiply the total cost by the location factor on page 7.
8. Add the cost of the appropriate additional components from page 201. Heating and air conditioning systems, elevators and escalators, fire sprinklers, exterior signs, canopies and walks, paving and curbing, loading docks and ramps, miscellaneous yard improvements, and mezzanines.
9. Add the cost of second and higher floors and basements from page 101.

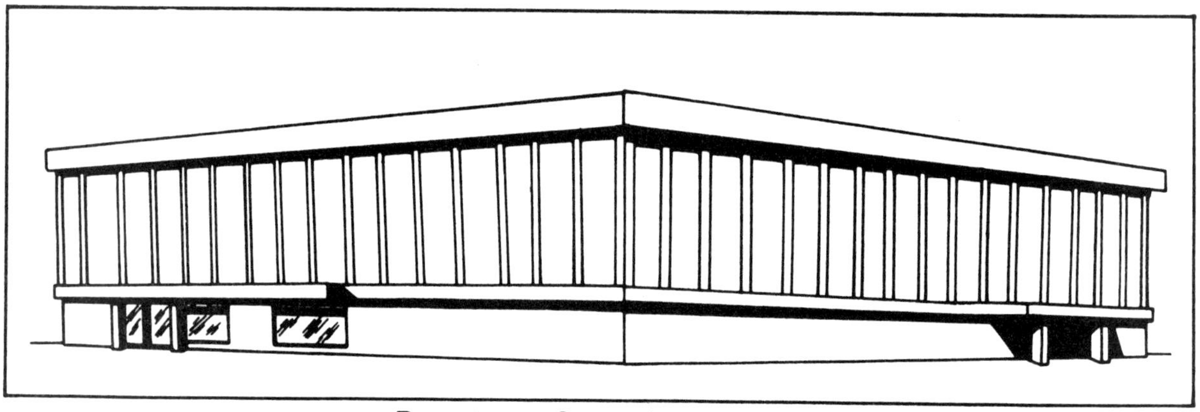

Department Store, Class 1 & 2

Square Foot Area

Quality Class	20,000	25,000	30,000	35,000	40,000	45,000	50,000	60,000	70,000	80,000	100,000
1, Best	58.06	55.71	53.80	52.22	50.89	49.74	48.73	47.06	45.71	44.59	42.82
1 & 2	51.62	49.53	47.83	46.43	45.24	44.22	43.32	41.84	40.64	39.64	38.07
2, Good	44.82	43.00	41.53	40.31	39.28	38.39	37.62	36.33	35.28	34.42	33.05
2 & 3	39.62	38.02	36.72	35.64	34.73	33.94	33.26	32.11	31.19	30.43	29.22
3, Average	34.50	33.10	31.97	31.03	30.24	29.55	28.96	27.96	27.16	26.50	25.45
3 & 4	29.65	28.44	27.47	26.67	25.98	25.40	24.88	24.03	23.34	22.77	21.86
4, Low	24.99	23.98	23.16	22.48	21.90	21.41	20.98	20.26	19.68	19.19	18.43
Wall Height Adjustment *	.15	.15	.14	.13	.13	.12	.12	.11	.10	.09	.08

***Wall Height Adjustment**

Add or subtract the amount listed to the square foot of floor cost for each foot of wall height more or less than 16 feet.

Perimeter Wall Adjustment

For a Common Wall, deduct $17.90 per linear foot.
For No Wall Ownership, deduct $37.00 per linear foot.

Department Stores Wood Frame Upper Floors and Basements

Estimating Procedure

1. Establish the basement and upper floor quality class. The quality class will usually be the same as the first floor of the building. Square foot costs for unfinished basements will be nearly the same regardless of the structure quality class.
2. Compute the floor area.
3. Add to or subtract from the square foot cost below the appropriate amount from the Wall Height Adjustment Column for each foot of wall height more or less than 12 feet.
4. Multiply the adjusted square foot cost by the floor area.
5. Deduct, if appropriate, for common upper floor walls and walls not owned, using the Perimeter Wall Adjustment costs.
6. Multiply the total cost by the location factor on page 7.
7. Add the cost of the appropriate additional components from page 201: Heating and air conditioning systems, elevators and escalators, fire sprinklers, and mezzanines.
8. Add the total from this page to the total from page 100 to find the building cost.

Second and higher floors

Square Foot Area

Quality Class	20,000	25,000	30,000	35,000	40,000	45,000	50,000	60,000	70,000	80,000	100,000
1, Best	48.12	46.42	44.98	43.76	42.70	41.77	40.96	39.58	38.46	37.52	36.02
1 & 2	42.21	40.72	39.45	38.38	37.45	36.64	35.92	34.72	33.73	32.91	31.59
2, Good	36.32	35.03	33.95	33.02	32.22	31.52	30.91	29.87	29.02	28.31	27.18
2 & 3	31.92	30.80	29.84	29.03	28.33	27.71	27.17	26.26	25.51	24.89	23.89
3, Average	27.51	26.54	25.72	25.02	24.41	23.88	23.42	22.63	21.99	21.45	20.59
3 & 4	23.09	22.27	21.58	20.99	20.48	20.04	19.65	18.99	18.45	18.00	17.28
4, Low	18.78	18.11	17.55	17.07	16.66	16.30	15.98	15.44	15.01	14.64	14.05
Wall Height Adjustment*	.13	.12	.11	.10	.10	.09	.08	.08	.07	.06	.05

***Wall Height Adjustment:** Add or subtract the amount listed in this column to the square foot of floor cost for each foot of wall height more or less than 12 feet.

Perimeter Wall Adjustment: For Common Wall, deduct $50.20 per linear foot. For No Wall Ownership, deduct $78.40 per linear foot.

Finished Basements

Square Foot Area

Quality Class	20,000	25,000	30,000	35,000	40,000	45,000	50,000	60,000	70,000	80,000	100,000
1, Best	47.64	46.12	44.67	43.35	42.16	41.10	40.14	38.48	37.10	35.92	34.02
2, Good	36.66	35.49	34.37	33.36	32.45	31.63	30.89	29.61	28.55	27.64	26.18
3, Average	28.79	27.87	27.00	26.20	25.48	24.84	24.26	23.26	22.42	21.71	20.56
4, Low	19.60	18.97	18.37	17.83	17.34	16.91	16.51	15.83	15.26	14.78	13.99

Unfinished Basements

Square Foot Area

Area	20,000	25,000	30,000	35,000	40,000	45,000	50,000	60,000	70,000	80,000	100,000
Cost	11.49	11.04	10.68	10.38	10.12	9.91	9.72	9.40	9.15	8.94	8.61

Wall Height Adjustment

Add or subtract the amount listed in this table to the square foot of floor cost for each foot of basement wall height more or less than 12 feet.

Square Foot Area

Area	20,000	25,000	30,000	35,000	40,000	45,000	50,000	60,000	70,000	80,000	100,000
Finished	.26	.24	.23	.22	.21	.21	.20	.20	.19	.19	.18
Unfinished	.24	.23	.22	.21	.20	.20	.19	.18	.18	.18	.17

General Office Buildings
Masonry or Concrete
Quality Classification

		Class 1 Best Quality	Class 2 Good Quality	Class 3 Average Quality	Class 4 Low Quality
Foundation		Reinforced concrete.	Reinforced concrete.	Reinforced concrete.	Reinforced concrete
First Floor Structure		Reinforced concrete slab on grade or standard wood frame.	Reinforced concrete slab on grade or standard wood frame.	Reinforced concrete slab on grade or 4" x 6" girders with plywood sheathing.	Reinforced concrete slab on grade.
Upper Floor Structure		Standard wood frame, plywood and 1½" lightweight concrete sub floor.	Standard wood frame, plywood and 1½" lightweight concrete sub floor.	Standard wood frame, 5/8" plywood sub floor.	Standard wood frame, 5/8" plywood sub floor.
Walls		8" decorative concrete block or 6" concrete tilt-up.	8" decorative concrete block or 6" concrete tilt-up.	8" reinforced concrete block or 8" reinforced brick or 8" clay tile.	8" reinforced concrete block or 8" clay tile.
Roof Structure		Standard wood frame, flat or low pitch.	Standard wood frame, flat or low pitch.	Standard wood frame, flat or low pitch.	Standard wood frame, flat or low pitch.
Exterior Finish	Walls	Decorative block or large rock imbedded in tilt-up panels with 10 - 20% brick or stone veneer.	Decorative block or exposed aggregate and 10 - 20% brick or stone veneer.	Stucco or colored concrete block.	Painted concrete block or tile.
	Windows	Average number in good aluminum frame. Fixed plate glass in good frame on front side.	Average number in good aluminum frame. Some fixed plate glass in front.	Average number of average aluminum sliding type.	Average number of low cost aluminum sliding type.
	Roof Cover	5 ply built-up roofing on flat roofs. Heavy shake or tile on sloping roofs.	5 ply built-up roofing on flat roofs. Average shake or composition, tar and large rock on sloping roofs.	4 ply built-up roofing on flat roofs. Wood shingle or composition, tar and pea gravel on sloping roofs.	3 ply built-up roofing on flat roofs. Composition shingle on sloping roofs.
	Overhang	3' closed overhang. fully guttered.	2' closed overhang, fully guttered.	None on flat roofs. 18" open on sloping roofs. Fully guttered.	None on flat roofs. 12" to 16" open on sloping roofs. Gutters over entrances.
Floor Finish	Offices	Very good carpet.	Good carpet.	Average grade carpet.	Minimum tile.
	Corridors	Solid vinyl tile or carpet.	Vinyl asbestos tile.	Vinyl asbestos tile.	Minimum tile.
	Bathrooms	Sheet vinyl or ceramic tile.	Sheet vinyl or ceramic tile.	Vinyl asbestos tile.	Minimum tile.
Interior Wall Finish	Offices	Good hardwood veneer.	Hardwood veneer paneling or vinyl wall cover.	Gypsum wallboard, texture and paint.	Gypsum wallboard, texture and paint.
	Corridors	Good hardwood veneer.	Gypsum wallboard and vinyl wall cover.	Gypsum wallboard, texture and paint.	Gypsum wallboard, texture and paint.
	Bathrooms	Gypsum wallboard and enamel with ceramic tile wainscot.	Gypsum wallboard and enamel or vinyl wall covering.	Gypsum wallboard and enamel.	Gypsum wallboard, texture and paint.
Ceiling Finish		Suspended "T" bar and acoustical tile.	Gypsum wallboard and acoustical tile.	Gypsum wallboard and acoustical texture.	Gypsum wallboard and paint.
Utilities	Plumbing	Copper tubing and top quality fixtures.	Copper tubing and good fixtures.	Copper tubing and standard fixtures.	Copper tubing and economy fixtures.
	Lighting	Conduit wiring, good fixtures.	Conduit wiring, good fixtures.	Romex or conduit wiring, average fixtures.	Romex wiring, economy fixtures.

Square foot costs include the following components: Foundations as required for normal soil conditions. Floor, wall, and roof structures. Interior floor, wall and ceiling finishes. Exterior wall finish and roof cover. Interior partitions. Cabinets, doors and windows. Basic electrical systems and lighting fixtures. Rough plumbing and fixtures. Permits and fees. Contractors' mark-up.

General Office Buildings
Masonry or Concrete
Exterior Suite Entrances, Length Less Than Twice Width

Estimating Procedure

1. Use these figures to estimate general office buildings in which access to each suite is through an exterior entrance. Medical and dental offices have smaller rooms and more plumbing fixtures than general offices and should be estimated with figures from the Medical and Dental Buildings Section. See page 118.
2. Establish the building quality class by applying the information on page 102.
3. Compute the first floor area. This should include everything within the exterior walls and all insets outside the main walls but under the main roof.
4. If the first floor wall height is more or less than 10 feet, add or subtract from the first floor square foot cost below the appropriate amount from the Wall Height Adjustment Table on page 109.
5. Multiply the adjusted square foot cost by the first floor area.
6. Deduct, if appropriate, for common walls or no wall finish. Use the figures on page 109.
7. If there are second or higher floors, compute the square foot area on each floor. Locate the appropriate square foot cost from the table at the bottom of this page. Adjust this figure for a wall height more or less than 9 feet, using the figures on page 109. Multiply the adjusted cost by the square foot area on each floor. Use the figures on page 109 to deduct for common walls or no wall finish. Add the result to the cost from step 6 above.
8. Multiply the total cost by the location factor on page 7.
9. Add the cost of heating and air conditioning systems, elevators, fire sprinklers, exterior signs, paving and curbing, miscellaneous yard improvements, covered porches and garages. See pages 201 to 213.

General Offices, Class 2 & 3 (wood frame second floor)

First Story

Square Foot Area

Quality Class	1,000	1,500	2,000	2,500	3,000	4,000	5,000	7,500	10,000	15,000	20,000
Exceptional	86.14	79.05	74.98	72.26	70.29	67.57	65.75	62.96	61.33	59.42	58.30
1, Best	78.31	71.87	68.16	65.70	63.90	61.43	59.77	57.24	55.75	54.02	53.00
1 & 2	70.41	64.62	61.29	59.07	57.46	55.24	53.75	51.47	50.13	48.58	47.66
2, Good	63.12	57.93	54.94	52.95	51.51	49.51	48.18	46.13	44.94	43.54	42.72
2 & 3	57.28	52.57	49.86	48.06	46.75	44.94	43.72	41.87	40.78	39.52	38.77
3, Average	51.40	47.17	44.74	43.12	41.95	40.32	39.23	37.57	36.60	35.46	34.79
3 & 4	46.01	42.23	40.05	38.60	37.55	36.09	35.12	33.63	32.76	31.74	31.14
4, Low	40.51	37.18	35.27	33.99	33.06	31.78	30.92	29.61	28.85	27.95	27.42

Second and Higher Stories

Square Foot Area

Quality Class	1,000	1,500	2,000	2,500	3,000	4,000	5,000	7,500	10,000	15,000	20,000
Exceptional	79.64	73.54	69.98	67.58	65.82	63.38	61.73	59.19	57.69	55.92	54.88
1, Best	72.32	66.78	63.54	61.36	59.77	57.55	56.05	53.75	52.38	50.78	49.83
1 & 2	65.43	60.42	57.49	55.52	54.08	52.07	50.71	48.63	47.39	45.94	45.08
2, Good	59.07	54.55	51.90	50.12	48.82	47.01	45.79	43.90	42.79	41.48	40.70
2 & 3	53.40	49.31	46.92	45.31	44.13	42.49	41.39	39.68	38.68	37.50	36.79
3, Average	47.55	43.91	41.78	40.35	39.30	37.84	36.86	35.34	34.45	33.39	32.77
3 & 4	42.14	38.91	37.02	35.75	34.83	33.53	32.66	31.32	30.52	29.59	29.04
4, Low	37.32	34.46	32.79	31.67	30.84	29.70	28.93	27.73	27.03	26.20	25.71

General Office Buildings
Masonry or Concrete
Exterior Suite Entrances, Length Between 2 and 4 Times Width

Estimating Procedure

1. Use these figures to estimate general office buildings in which access to each suite is through an exterior entrance. Medical and dental offices have smaller rooms and more plumbing fixtures than general offices and should be estimated with figures from the Medical and Dental Buildings Section. See page 118.
2. Establish the building quality class by applying the information on page 102.
3. Compute the first floor area. This should include everything within the exterior walls and all insets outside the main walls but under the main roof.
4. If the first floor wall height is more or less than 10 feet, add or subtract from the first floor square foot cost below the appropriate amount from the Wall Height Adjustment Table on page 109.
5. Multiply the adjusted square foot cost by the first floor area.
6. Deduct, if appropriate, for common walls or no wall finish. Use the figures on page 109.
7. If there are second or higher floors, compute the square foot area on each floor. Locate the appropriate square foot cost from the table at the bottom of this page. Adjust this figure for a wall height more or less than 9 feet, using the figures on page 109. Multiply the adjusted cost by the square foot area on each floor. Use the figures on page 109 to deduct for common walls or no wall finish. Add the result to the cost from step 6 above.
8. Multiply the total cost by the location factor on page 7.
9. Add the cost of heating and air conditioning systems, elevators, fire sprinklers, exterior signs, paving and curbing, miscellaneous yard improvements, covered porches and garages. See pages 201 to 213.

First Story

Square Foot Area

Quality Class	1,000	1,500	2,000	2,500	3,000	4,000	5,000	7,500	10,000	15,000	20,000
Exceptional	93.56	84.48	79.34	75.97	73.54	70.22	68.02	64.69	62.77	60.54	59.24
1, Best	85.06	76.80	72.14	69.07	66.86	63.84	61.84	58.82	57.07	55.04	53.86
1 & 2	76.51	69.08	64.89	62.12	60.14	57.42	55.62	52.90	51.33	49.51	48.45
2, Good	68.62	61.95	58.19	55.71	53.93	51.50	49.88	47.45	46.03	44.40	43.45
2 & 3	62.40	56.34	52.92	50.66	49.04	46.83	45.36	43.15	41.86	40.38	39.51
3, Average	55.99	50.55	47.48	45.46	44.01	42.02	40.70	38.71	37.56	36.23	35.45
3 & 4	50.23	45.35	42.59	40.78	39.48	37.69	36.51	34.73	33.69	32.50	31.80
4, Low	44.35	40.05	37.61	36.01	34.86	33.29	32.24	30.67	29.76	28.70	28.08

Second and Higher Stories

Square Foot Area

Quality Class	1,000	1,500	2,000	2,500	3,000	4,000	5,000	7,500	10,000	15,000	20,000
Exceptional	86.43	78.23	73.60	70.56	68.38	65.40	63.42	60.43	58.71	56.71	55.55
1, Best	78.57	71.12	66.92	64.15	62.16	59.45	57.65	54.94	53.37	51.56	50.50
1 & 2	71.16	64.41	60.60	58.10	56.30	53.84	52.21	49.76	48.34	46.69	45.73
2, Good	64.33	58.22	54.78	52.52	50.89	48.67	47.20	44.98	43.70	42.21	41.35
2 & 3	58.27	52.74	49.62	47.57	46.10	44.09	42.76	40.74	39.58	38.23	37.45
3, Average	51.96	47.03	44.25	42.42	41.11	39.32	38.13	36.33	35.29	34.09	33.40
3 & 4	46.63	42.20	39.71	38.07	36.89	35.28	34.21	32.60	31.67	30.60	29.97
4, Low	41.11	37.21	35.01	33.57	32.53	31.11	30.17	28.75	27.93	26.98	26.42

General Office Buildings
Masonry or Concrete
Exterior Suite Entrances, Length More Than 4 Times Width

Estimating Procedure

1. Use these figures to estimate general office buildings in which access to each suite is through an exterior entrance. Medical and dental offices have smaller rooms and more plumbing fixtures than general offices and should be estimated with figures from the Medical and Dental Buildings Section. See page 118.
2. Establish the building quality class by applying the information on page 102.
3. Compute the first floor area. This should include everything within the exterior walls and all insets outside the main walls but under the main roof.
4. If the first floor wall height is more or less than 10 feet, add or subtract from the first floor square foot cost below the appropriate amount from the Wall Height Adjustment Table on page 109.
5. Multiply the adjusted square foot cost by the first floor area.
6. Deduct, if appropriate, for common walls or no wall finish. Use the figures on page 109.
7. If there are second or higher floors, compute the square foot area on each floor. Locate the appropriate square foot cost from the table at the bottom of this page. Adjust this figure for a wall height more or less than 9 feet, using the figures on page 109. Multiply the adjusted cost by the square foot area on each floor. Use the figures on page 109 to deduct for common walls or no wall finish. Add the result to the cost from step 6 above.
8. Multiply the total cost by the location factor on page 7.
9. Add the cost of heating and air conditioning systems, elevators, fire sprinklers, exterior signs, paving and curbing, miscellaneous yard improvements, covered porches and garages. See pages 201 to 213.

First Story

	Square Foot Area										
Quality Class	1,000	1,500	2,000	2,500	3,000	4,000	5,000	7,500	10,000	15,000	20,000
Exceptional	102.24	91.19	84.99	80.91	77.99	74.01	71.37	67.40	65.10	62.46	60.92
1, Best	92.69	82.68	77.05	73.36	70.70	67.09	64.70	61.10	59.02	56.62	55.23
1 & 2	83.59	74.56	69.48	66.15	63.76	60.50	58.35	55.10	53.23	51.06	49.80
2, Good	74.76	66.68	62.15	59.17	57.03	54.11	52.19	49.28	47.60	45.67	44.54
2 & 3	68.03	60.68	56.55	53.84	51.90	49.25	47.49	44.85	43.32	41.56	40.54
3, Average	61.36	54.73	51.01	48.56	46.81	44.42	42.83	40.45	39.07	37.49	36.56
3 & 4	54.95	49.01	45.68	43.49	41.92	39.78	38.36	36.22	34.99	33.57	32.74
4, Low	48.66	43.40	40.45	38.51	37.12	35.22	33.97	32.08	30.99	29.73	28.99

Second and Higher Stories

	Square Foot Area										
Quality Class	1,000	1,500	2,000	2,500	3,000	4,000	5,000	7,500	10,000	15,000	20,000
Exceptional	93.87	83.83	78.20	74.51	71.86	68.25	65.86	62.27	60.20	57.82	56.43
1, Best	85.33	76.20	71.08	67.72	65.32	62.04	59.87	56.60	54.72	52.55	51.29
1 & 2	77.37	69.10	64.45	61.41	59.23	56.25	54.29	51.33	49.62	47.65	46.51
2, Good	70.06	62.57	58.36	55.61	53.63	50.94	49.15	46.48	44.93	43.15	42.11
2 & 3	63.62	56.82	53.00	50.49	48.70	46.25	44.64	42.20	40.80	39.18	38.24
3, Average	57.00	50.91	47.49	45.24	43.63	41.44	39.99	37.81	36.56	35.11	34.27
3 & 4	51.21	45.74	42.66	40.65	39.20	37.23	35.93	33.97	32.84	31.54	30.78
4, Low	45.39	40.53	37.81	36.02	34.74	33.00	31.84	30.11	29.11	27.95	27.28

General Office Buildings
Masonry or Concrete
Interior Suite Entrances, Length Less Than Twice Width

Estimating Procedure

1. Use these figures to estimate general office buildings in which access to each suite is through an interior corridor. Medical and dental offices have smaller rooms and more plumbing fixtures than general offices and should be estimated with figures from the Medical and Dental Buildings Section. See page 118.
2. Establish the building quality class by applying the information on page 102.
3. Compute the first floor area. This should include everything within the exterior walls and all insets outside the main walls but under the main roof.
4. If the first floor wall height is more or less than 10 feet, add or subtract from the first floor square foot cost below the appropriate amount from the Wall Height Adjustment Table on page 109.
5. Multiply the adjusted square foot cost by the first floor area.
6. Deduct, if appropriate, for common walls or no wall finish. Use the figures on page 109.
7. If there are second or higher floors, compute the square foot area on each floor. Locate the appropriate square foot cost from the table at the bottom of this page. Adjust this figure for a wall height more or less than 9 feet, using the figures on page 109. Multiply the adjusted cost by the square foot area on each floor. Use the figures on page 109 to deduct for common walls or no wall finish. Add the result to the cost from step 6 above.
8. Multiply the total cost by the location factor on page 7.
9. Add the cost of heating and air conditioning systems, elevators, fire sprinklers, exterior signs, paving and curbing, miscellaneous yard improvements, covered porches and garages. See pages 201 to 213.

General Offices, Class 2 & 3

First Story

Square Foot Area

Quality Class	2,000	2,500	3,000	4,000	5,000	7,500	10,000	15,000	20,000	30,000	40,000
Exceptional	72.73	70.05	68.11	65.44	63.65	60.92	59.33	57.47	56.37	55.09	54.33
1, Best	66.10	63.66	61.90	59.47	57.84	55.36	53.91	52.22	51.23	50.06	49.38
1 & 2	59.97	57.77	56.17	53.96	52.48	50.23	48.92	47.38	46.48	45.43	44.80
2, Good	54.36	52.36	50.91	48.91	47.57	45.53	44.34	42.95	42.13	41.18	40.61
2 & 3	49.39	47.57	46.25	44.44	43.22	41.37	40.29	39.02	38.28	37.41	36.89
3, Average	44.31	42.67	41.49	39.86	38.77	37.11	36.14	35.01	34.34	33.56	33.10
3 & 4	39.60	38.14	37.08	35.63	34.65	33.17	32.30	31.29	30.69	29.99	29.58
4, Low	34.80	33.52	32.59	31.31	30.46	29.15	28.39	27.50	26.97	26.36	26.00

Second and Higher Stories

Square Foot Area

Quality Class	2,000	2,500	3,000	4,000	5,000	7,500	10,000	15,000	20,000	30,000	40,000
Exceptional	67.92	65.51	63.77	61.35	59.73	57.24	55.79	54.08	53.07	51.89	51.20
1, Best	61.74	59.56	57.97	55.77	54.30	52.04	50.71	49.16	48.25	47.18	46.54
1 & 2	56.20	54.22	52.77	50.77	49.43	47.37	46.17	44.75	43.92	42.94	42.37
2, Good	51.06	49.26	47.94	46.12	44.90	43.04	41.94	40.66	39.90	39.02	38.49
2 & 3	46.21	44.58	43.39	41.74	40.64	38.95	37.96	36.80	36.11	35.31	34.83
3, Average	41.20	39.74	38.68	37.22	36.23	34.73	33.84	32.81	32.20	31.48	31.06
3 & 4	36.71	35.41	34.46	33.16	32.28	30.94	30.15	29.23	28.68	28.05	27.67
4, Low	32.16	31.03	30.20	29.05	28.29	27.11	26.42	25.61	25.14	24.58	24.25

General Office Buildings
Masonry or Concrete
Interior Suite Entrances, Length Between 2 and 4 Times Width

Estimating Procedure

1. Use these figures to estimate general office buildings in which access to each suite is through an interior corridor. Medical and dental offices have smaller rooms and more plumbing fixtures than general offices and should be estimated with figures from the Medical and Dental Buildings Section. See page 118.
2. Establish the building quality class by applying the information on page 102.
3. Compute the first floor area. This should include everything within the exterior walls and all insets outside the main walls but under the main roof.
4. If the first floor wall height is more or less than 10 feet, add or subtract from the first floor square foot cost below the appropriate amount from the Wall Height Adjustment Table on page 109.
5. Multiply the adjusted square foot cost by the first floor area.
6. Deduct, if appropriate, for common walls or no wall finish. Use the figures on page 109.
7. If there are second or higher floors, compute the square foot area on each floor. Locate the appropriate square foot cost from the table at the bottom of this page. Adjust this figure for a wall height more or less than 9 feet, using the figures on page 109. Multiply the adjusted cost by the square foot area on each floor. Use the figures on page 109 to deduct for common walls or no wall finish. Add the result to the cost from step 6 above.
8. Multiply the total cost by the location factor on page 7.
9. Add the cost of heating and air conditioning systems, elevators, fire sprinklers, exterior signs, paving and curbing, miscellaneous yard improvements, covered porches and garages. See pages 201 to 213.

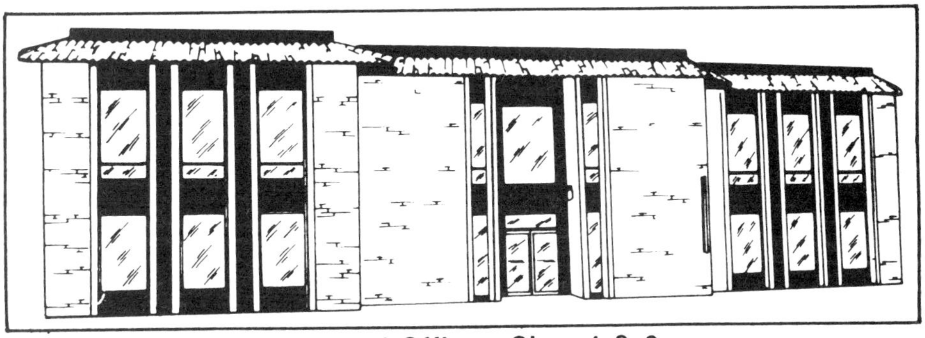

General Offices, Class 1 & 2

First Story

Square Foot Area

Quality Class	2,000	2,500	3,000	4,000	5,000	7,500	10,000	15,000	20,000	30,000	40,000
Exceptional	77.08	73.79	71.42	68.16	66.00	62.71	60.80	58.59	57.29	55.77	54.88
1, Best	70.07	67.08	64.92	61.97	60.00	57.01	55.28	53.26	52.08	50.70	49.89
1 & 2	63.56	60.85	58.90	56.21	54.43	51.72	50.14	48.32	47.25	46.00	45.26
2, Good	57.62	55.16	53.38	50.95	49.33	46.88	45.45	43.79	42.82	41.69	41.03
2 & 3	52.38	50.15	48.53	46.32	44.85	42.62	41.32	39.81	38.93	37.90	37.30
3, Average	47.02	45.02	43.57	41.58	40.26	38.26	37.09	35.74	34.95	34.03	33.48
3 & 4	42.10	40.30	39.01	37.23	36.05	34.25	33.21	32.00	31.29	30.46	29.98
4, Low	37.07	35.49	34.35	32.79	31.74	30.17	29.25	28.18	27.56	26.83	26.40

Second and Higher Stories

Square Foot Area

Quality Class	2,000	2,500	3,000	4,000	5,000	7,500	10,000	15,000	20,000	30,000	40,000
Exceptional	71.71	68.74	66.59	63.66	61.70	58.73	57.00	55.00	53.82	52.45	51.64
1, Best	65.13	62.44	60.49	57.82	56.04	53.35	51.78	49.95	48.89	47.64	46.91
1 & 2	59.35	56.90	55.12	52.69	51.07	48.61	47.18	45.52	44.55	43.41	42.75
2, Good	54.05	51.81	50.20	47.98	46.51	44.27	42.97	41.45	40.57	39.53	38.93
2 & 3	48.92	46.90	45.43	43.43	42.09	40.07	38.89	37.52	36.72	35.78	35.23
3, Average	43.69	41.88	40.58	38.79	37.59	35.78	34.73	33.51	32.79	31.96	31.47
3 & 4	39.04	37.43	36.26	34.66	33.59	31.98	31.04	29.94	29.30	28.56	28.12
4, Low	34.31	32.89	31.87	30.46	29.52	28.10	27.28	26.32	25.75	25.10	24.71

General Office Buildings
Masonry or Concrete
Interior Suite Entrances, Length More Than 4 Times Width

Estimating Procedure

1. Use these figures to estimate general office buildings in which access to each suite is through an interior corridor. Medical and dental offices have smaller rooms and more plumbing fixtures than general offices and should be estimated with figures from the Medical and Dental Buildings Section. See page 118.
2. Establish the building quality class by applying the information on page 102.
3. Compute the first floor area. This should include everything within the exterior walls and all insets outside the main walls but under the main roof.
4. If the first floor wall height is more or less than 10 feet, add or subtract from the first floor square foot cost below the appropriate amount from the Wall Height Adjustment Table on page 109.
5. Multiply the adjusted square foot cost by the first floor area.
6. Deduct, if appropriate, for common walls or no wall finish. Use the figures on page 109.
7. If there are second or higher floors, compute the square foot area on each floor. Locate the appropriate square foot cost from the table at the bottom of this page. Adjust this figure for a wall height more or less than 9 feet, using the figures on page 109. Multiply the adjusted cost by the square foot area on each floor. Use the figures on page 109 to deduct for common walls or no wall finish. Add the result to the cost from step 6 above.
8. Multiply the total cost by the location factor on page 7.
9. Add the cost of heating and air conditioning systems, elevators, fire sprinklers, exterior signs, paving and curbing, miscellaneous yard improvements, covered porches and garages. See pages 201 to 213.

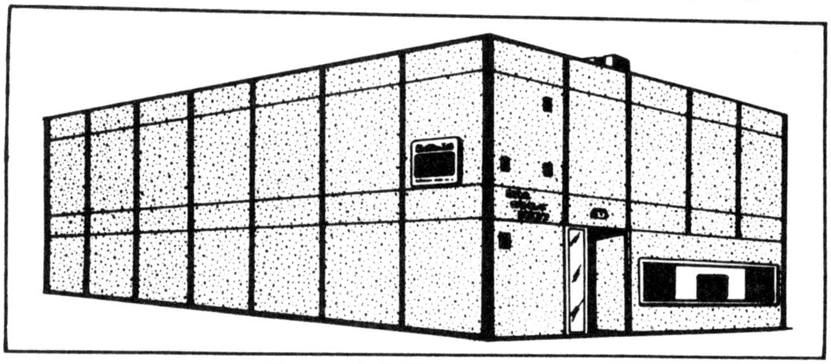

General Office, Class 3

First Story

Square Foot Area

Quality Class	2,000	2,500	3,000	4,000	5,000	7,500	10,000	15,000	20,000	30,000	40,000
Exceptional	82.59	78.49	75.56	71.57	68.94	64.99	62.72	60.10	58.58	56.82	55.79
1, Best	75.08	71.35	68.69	65.07	62.67	59.08	57.02	54.64	53.26	51.65	50.72
1 & 2	68.10	64.72	62.30	59.02	56.85	53.59	51.72	49.56	48.31	46.85	46.00
2, Good	61.71	58.65	56.46	53.48	51.52	48.56	46.87	44.91	43.78	42.46	41.69
2 & 3	56.07	53.29	51.30	48.59	46.81	44.12	42.58	40.80	39.77	38.58	37.88
3, Average	50.47	47.97	46.17	43.74	42.13	39.72	38.33	36.73	35.80	34.72	34.09
3 & 4	45.26	43.02	41.41	39.23	37.78	35.62	34.37	32.94	32.11	31.14	30.58
4, Low	39.97	37.98	36.56	34.64	33.36	31.45	30.35	29.08	28.35	27.50	27.00

Second and Higher Stories

Square Foot Area

Quality Class	2,000	2,500	3,000	4,000	5,000	7,500	10,000	15,000	20,000	30,000	40,000
Exceptional	75.81	72.09	69.49	66.02	63.79	60.53	58.71	56.67	55.52	54.20	53.45
1, Best	68.36	65.01	62.65	59.53	57.52	54.58	52.94	51.10	50.06	48.87	48.20
1 & 2	62.67	59.60	57.44	54.58	52.73	50.04	48.54	46.85	45.89	44.81	44.19
2, Good	57.25	54.44	52.47	49.86	48.17	45.71	44.33	42.79	41.92	40.93	40.36
2 & 3	51.81	49.27	47.48	45.12	43.59	41.36	40.12	38.73	37.94	37.04	36.53
3, Average	46.50	44.22	42.62	40.50	39.13	37.13	36.01	34.76	34.05	33.25	32.79
3 & 4	46.28	44.01	42.42	40.31	38.94	36.95	35.84	34.60	33.89	33.09	32.63
4, Low	36.73	34.93	33.67	31.99	30.91	29.33	28.45	27.46	26.90	26.26	25.90

General Office Buildings
Masonry or Concrete
Wall Height Adjustments

Wall Height Adjustment

The square foot costs for general offices are based on the wall heights of 10 feet for first floors and 9 feet for higher floors. The main or first floor height is the distance from the bottom of the floor slab or joists to the top of the roof slab or ceiling joists. Second and higher floors are measured from the top of the floor slab or floor joists to the top of the roof slab or ceiling joists. Add or subtract the amount listed in this table to the square foot of floor cost for each foot of wall height more or less than 10 feet, if adjusting for a first floor, and 9 feet, if adjusting for upper floors.

	Square Foot Area										
Quality Class	1,000	1,500	2,000	3,000	4,000	5,000	7,500	10,000	15,000	20,000	40,000
1, Best	2.28	1.79	1.51	1.20	1.02	.90	.72	.62	.50	.44	.31
2, Good	1.90	1.49	1.26	1.00	.85	.75	.60	.52	.42	.36	.26
3, Average	1.61	1.26	1.06	.84	.72	.64	.51	.44	.36	.31	.22
4, Low	1.40	1.10	.93	.73	.62	.55	.44	.38	.31	.27	.19

Perimeter Wall Adjustment

A common wall exists when two buildings share one wall. Adjust for common walls by deducting the linear foot cost below from the total structure cost. In some structures one or more walls is not owned at all. In this case, deduct the ''No Ownership'' cost per linear foot of wall not owned. Where a perimeter wall remains unfinished, deduct the ''lack of exterior finish'' cost.

First Story

Class	For a Common Wall, Deduct Per Linear Foot	For no Wall Ownership, Deduct Per Linear Foot	For Lack of Exterior Finish, Deduct Per Linear Foot
1	$119.00	$179.00	$88.00
2	93.00	149.00	56.00
3	69.50	131.00	32.00
4	53.30	114.00	19.30

Second and Higher Stories

Class	For a Common Wall, Deduct Per Linear Foot	For no Wall Ownership, Deduct Per Linear Foot	For Lack of Exterior Finish, Deduct Per Linear Foot
1	$113.00	$155.00	$88.00
2	90.50	139.00	56.30
3	70.60	126.00	32.00
4	65.00	113.00	19.30

General Office Buildings
Wood Frame
Quality Classification

		Class 1 Best Quality	Class 2 Good Quality	Class 3 Average Quality	Class 4 Low Quality
Foundation		Reinforced concrete.	Reinforced concrete.	Reinforced concrete.	Reinforced concrete.
First Floor Structure		Reinforced concrete slab on grade or standard wood frame.	Reinforced concrete slab on grade or standard wood frame.	Reinforced concrete slab on grade or 4" x 6" girders with 2" T&G sub floor.	Reinforced concrete slab on grade.
Upper Floor Structure		Standard wood frame. Plywood and 1½" lightweight concrete sub floor.	Standard wood frame. Plywood and 1½" lightweight concrete sub floor.	Standard wood frame. 5/8" plywood sub floor.	Standard wood frame. 5/8" plywood sub floor.
Walls		Standard wood frame.	Standard wood frame.	Standard wood frame.	Standard wood frame.
Roof Structure		Standard wood frame, flat or low pitch.	Standard wood frame, flat or low pitch.	Standard wood frame, flat or low pitch.	Standard wood frame, flat or low pitch.
Exterior Finish	Walls	Good wood siding with 10 to 20% brick or stone veneer.	Average wood siding or stucco and 10 to 20% brick or stone veneer.	Stucco with some wood trim or cheap wood siding.	Stucco.
	Windows	Average number in good aluminum frame. Fixed plate glass in good frame on front side.	Average number in good aluminum frame. Some fixed plate glass in front.	Average number of average aluminum sliding type.	Average number of low cost aluminum sliding type.
	Roof Cover	5 ply built-up roofing on flat roofs. Heavy shake or tile on sloping roofs.	5 ply built-up roofing on flat roofs. Average shake or composition, tar and large rock on sloping roofs.	4 ply built-up roofing on flat roofs. Wood shingle or composition, tar and pea gravel on sloping roofs.	3 ply built-up roofing on flat roofs. Composition shingle on sloping roofs.
	Overhang	3' closed overhang, fully guttered.	2' closed overhang, fully guttered.	None on flat roofs. 18" on sloping roofs, fully guttered.	None on flat roofs. 12" to 16" open on sloping roofs, gutters over entrances.
Floor Finish	Offices	Very good carpet.	Good carpet.	Average grade carpet.	Minimum tile.
	Corridors	Solid vinyl tile or carpet.	Vinyl asbestos tile.	Vinyl asbestos tile.	Minimum tile.
	Bathrooms	Sheet vinyl or ceramic tile.	Sheet vinyl or ceramic tile.	Vinyl asbestos tile.	Minimum tile.
Interior Wall Finish	Offices	Good hardwood veneer paneling.	Hardwood veneer paneling or vinyl wall cover.	Gypsum wallboard, texture and paint.	Gypsum wallboard, texture and paint.
	Corridors	Good hardwood veneer paneling.	Gypsum wallboard and vinyl wall cover.	Gypsum wallboard, texture and paint.	Gypsum wallboard, texture and paint.
	Bathrooms	Gypsum wallboard and enamel with ceramic tile wainscot.	Gypsum wallboard and enamel or vinyl wall covering.	Gypsum wallboard and enamel.	Gypsum wallboard, texture and paint.
Ceiling Finish		Suspended "T" bar and acoustical tile.	Gypsum wallboard and acoustical tile.	Gypsum wallboard and acoustical texture.	Gypsum wallboard and paint.
Utilities	Plumbing	Copper tubing, top quality fixtures.	Copper tubing, good fixtures.	Copper tubing, standard fixtures.	Copper tubing, economy fixtures.
	Lighting	Conduit wiring, good fixtures.	Conduit wiring, good fixtures.	Romex or conduit wiring, average fixtures.	Romex wiring, economy fixtures.

Square foot costs include the following components: Foundations as required for normal soil conditions. Floor, wall, and roof structures. Interior floor, wall and ceiling finishes. Exterior wall finish and roof cover. Interior partitions. Cabinets, doors and windows. Basic electrical systems and lighting fixtures. Rough plumbing and fixtures. Permits and fees. Contractors' mark-up.

General Office Buildings
Wood Frame

Exterior Suite Entrances, Length Less Than Twice Width

Estimating Procedure

1. Use these figures to estimate general office buildings in which access to each suite is through an exterior entrance. Medical and dental offices have smaller rooms and more plumbing fixtures than general offices and should be estimated with figures from the Medical and Dental Buildings Section. See page 118.
2. Establish the building quality class by applying the information on page 110.
3. Compute the first floor area. This should include everything within the exterior walls and all insets outside the main walls but under the main roof.
4. If the first floor wall height is more or less than 10 feet, add or subtract from the first floor square foot cost below the appropriate amount from the Wall Height Adjustment Table on page 117.
5. Multiply the adjusted square foot cost by the first floor area.
6. Deduct, if appropriate, for common walls or no wall finish. Use the figures on page 117.
7. If there are second or higher floors, compute the square foot area on each floor. Locate the appropriate square foot cost from the table at the bottom of this page. Adjust this figure for a wall height more or less than 9 feet, using the figures on page 117. Multiply the adjusted cost by the square foot area on each floor. Use the figures on page 117 to deduct for common walls or no wall finish. Add the result to the cost from step 6 above.
8. Multiply the total cost by the location factor on page 7.
9. Add the cost of heating and air conditioning systems, elevators, fire sprinklers, exterior signs, paving and curbing, miscellaneous yard improvements, covered porches and garages. See pages 201 to 213.

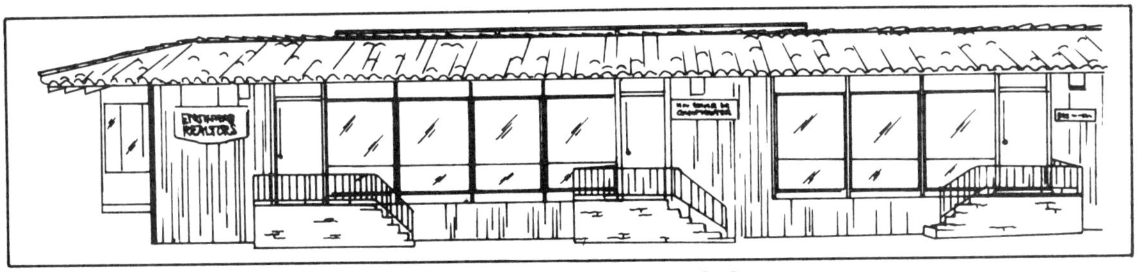

General Offices, Class 1 & 2

First Story

Square Foot Area

Quality Class	1,000	1,500	2,000	2,500	3,000	4,000	5,000	7,500	10,000	15,000	20,000
Exceptional	69.77	65.97	63.79	62.34	61.29	59.84	58.87	57.39	56.53	55.53	54.93
1, Best	63.43	59.97	57.99	56.68	55.72	54.40	53.52	52.18	51.40	50.48	49.94
1 & 2	56.83	53.74	51.96	50.78	49.93	48.75	47.96	46.75	46.05	45.23	44.75
2, Good	50.74	47.97	46.39	45.33	44.57	43.52	42.81	41.74	41.11	40.38	39.95
2 & 3	46.03	43.53	42.09	41.13	40.44	39.48	38.85	37.87	37.30	36.64	36.25
3, Average	41.26	39.01	37.72	36.87	36.25	35.39	34.82	33.94	33.43	32.84	32.49
3 & 4	36.49	34.50	33.36	32.61	32.06	31.30	30.79	30.02	29.57	29.04	28.73
4, Low	31.76	30.03	29.04	28.38	27.90	27.24	26.80	26.13	25.73	25.28	25.01

Second and Higher Stories

Square Foot Area

Quality Class	1,000	1,500	2,000	2,500	3,000	4,000	5,000	7,500	10,000	15,000	20,000
Exceptional	60.94	58.51	57.11	56.18	55.51	54.59	53.97	53.03	52.48	51.83	51.46
1, Best	55.32	53.11	51.84	51.00	50.39	49.55	48.99	48.13	47.63	47.05	46.71
1 & 2	49.97	47.97	46.82	46.06	45.51	44.75	44.25	43.48	43.03	42.50	42.19
2, Good	44.72	42.93	41.91	41.22	40.73	40.05	39.60	38.91	38.50	38.03	37.76
2 & 3	40.11	38.50	37.58	36.97	36.53	35.92	35.51	34.90	34.53	34.11	33.86
3, Average	35.33	33.92	33.11	32.57	32.18	31.64	31.28	30.74	30.42	30.05	29.83
3 & 4	31.08	29.83	29.12	28.65	28.31	27.84	27.52	27.04	26.76	26.43	26.24
4, Low	26.76	25.69	25.08	24.67	24.38	23.97	23.70	23.29	23.04	22.76	22.60

General Office Buildings
Wood Frame
Exterior Suite Entrances, Length Between 2 and 4 Times Width

Estimating Procedure

1. Use these figures to estimate general office buildings in which access to each suite is through an exterior entrance. Medical and dental offices have smaller rooms and more plumbing fixtures than general offices and should be estimated with figures from the Medical and Dental Buildings Section. See page 118.
2. Establish the building quality class by applying the information on page 110.
3. Compute the first floor area. This should include everything within the exterior walls and all insets outside the main walls but under the main roof.
4. If the first floor wall height is more or less than 10 feet, add or subtract from the first floor square foot cost below the appropriate amount from the Wall Height Adjustment Table on page 117.
5. Multiply the adjusted square foot cost by the first floor area.
6. Deduct, if appropriate, for common walls or no wall finish. Use the figures on page 117.
7. If there are second or higher floors, compute the square foot area on each floor. Locate the appropriate square foot cost from the table at the bottom of this page. Adjust this figure for a wall height more or less than 9 feet, using the figures on page 117. Multiply the adjusted cost by the square foot area on each floor. Use the figures on page 117 to deduct for common walls or no wall finish. Add the result to the cost from step 6 above.
8. Multiply the total cost by the location factor on page 7.
9. Add the cost of heating and air conditioning systems, elevators, fire sprinklers, exterior signs, paving and curbing, miscellaneous yard improvements, covered porches and garages. See pages 201 to 213.

General Offices, Class 3

First Story

Square Foot Area

Quality Class	1,000	1,500	2,000	2,500	3,000	4,000	5,000	7,500	10,000	15,000	20,000
Exceptional	73.70	68.87	66.14	64.34	63.05	61.29	60.12	58.35	57.33	56.15	55.45
1, Best	66.99	62.60	60.12	58.49	57.31	55.71	54.65	53.04	52.11	51.03	50.41
1 & 2	60.04	56.10	53.88	52.42	51.37	49.93	48.98	47.54	46.70	45.74	45.18
2, Good	53.57	50.05	48.07	46.77	45.83	44.55	43.70	42.41	41.67	40.81	40.31
2 & 3	48.66	45.47	43.67	42.48	41.63	40.46	39.69	38.52	37.85	37.07	36.61
3, Average	43.66	40.79	39.18	38.12	37.35	36.31	35.61	34.57	33.96	33.26	32.85
3 & 4	38.63	36.10	34.67	33.73	33.05	32.13	31.52	30.59	30.05	29.43	29.07
4, Low	33.67	31.47	30.22	29.40	28.81	28.00	27.47	26.66	26.19	25.65	25.34

Second and Higher Stories

Square Foot Area

Quality Class	1,000	1,500	2,000	2,500	3,000	4,000	5,000	7,500	10,000	15,000	20,000
Exceptional	63.45	60.24	58.46	57.30	56.48	55.36	54.63	53.53	52.91	52.19	51.78
1, Best	57.68	54.76	53.14	52.09	51.34	50.33	49.66	48.66	48.10	47.45	47.07
1 & 2	51.94	49.31	47.85	46.90	46.23	45.31	44.71	43.82	43.31	42.72	42.38
2, Good	46.60	44.24	42.93	42.08	41.47	40.65	40.12	39.31	38.85	38.33	38.02
2 & 3	41.84	39.72	38.55	37.78	37.24	36.50	36.02	35.30	34.89	34.42	34.14
3, Average	36.93	35.06	34.02	33.35	32.87	32.22	31.79	31.16	30.79	30.38	30.14
3 & 4	32.25	30.62	29.71	29.12	28.70	28.14	27.76	27.21	26.89	26.53	26.32
4, Low	28.07	26.65	25.86	25.35	24.98	24.49	24.16	23.68	23.40	23.09	22.90

General Office Buildings
Wood Frame

Exterior Suite Entrances, Length More Than 4 Times Width

Estimating Procedure

1. Use these figures to estimate general office buildings in which access to each suite is through an exterior entrance. Medical and dental offices have smaller rooms and more plumbing fixtures than general offices and should be estimated with figures from the Medical and Dental Buildings Section. See page 118.
2. Establish the building quality class by applying the information on page 110.
3. Compute the first floor area. This should include everything within the exterior walls and all insets outside the main walls but under the main roof.
4. If the first floor wall height is more or less than 10 feet, add or subtract from the first floor square foot cost below the appropriate amount from the Wall Height Adjustment Table on page 117.
5. Multiply the adjusted square foot cost by the first floor area.
6. Deduct, if appropriate, for common walls or no wall finish. Use the figures on page 117.
7. If there are second or higher floors, compute the square foot area on each floor. Locate the appropriate square foot cost from the table at the bottom of this page. Adjust this figure for a wall height more or less than 9 feet, using the figures on page 117. Multiply the adjusted cost by the square foot area on each floor. Use the figures on page 117 to deduct for common walls or no wall finish. Add the result to the cost from step 6 above.
8. Multiply the total cost by the location factor on page 7.
9. Add the cost of heating and air conditioning systems, elevators, fire sprinklers, exterior signs, paving and curbing, miscellaneous yard improvements, covered porches and garages. See pages 201 to 213.

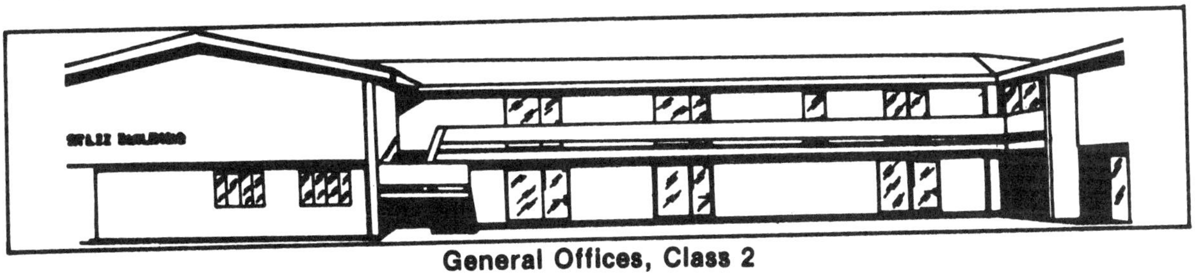

General Offices, Class 2

First Story

Square Foot Area

Quality Class	1,000	1,500	2,000	2,500	3,000	4,000	5,000	7,500	10,000	15,000	20,000
Exceptional	78.42	72.41	69.05	66.85	65.27	63.13	61.71	59.58	58.36	56.94	56.12
1, Best	71.28	65.82	62.77	60.76	59.33	57.38	56.09	54.16	53.04	51.76	51.01
1 & 2	63.87	58.99	56.25	54.45	53.17	51.42	50.27	48.53	47.53	46.38	45.71
2, Good	56.96	52.60	50.16	48.56	47.41	45.85	44.82	43.28	42.39	41.36	40.76
2 & 3	51.79	47.82	45.60	44.15	43.11	41.69	40.76	39.35	38.54	37.61	37.06
3, Average	46.54	42.98	40.98	39.67	38.74	37.47	36.62	35.36	34.63	33.79	33.31
3 & 4	41.23	38.07	36.31	35.15	34.32	33.19	32.45	31.33	30.68	29.94	29.51
4, Low	35.98	33.22	31.68	30.67	29.95	28.96	28.31	27.34	26.77	26.13	25.75

Second and Higher Stories

Square Foot Area

Quality Class	1,000	1,500	2,000	2,500	3,000	4,000	5,000	7,500	10,000	15,000	20,000
Exceptional	66.19	62.39	60.26	58.86	57.85	56.49	55.58	54.22	53.44	52.53	52.01
1, Best	60.18	56.72	54.78	53.51	52.59	51.35	50.53	49.29	48.58	47.76	47.28
1 & 2	54.18	51.07	49.32	48.18	47.36	46.24	45.50	44.38	43.74	43.00	42.57
2, Good	48.66	45.86	44.29	43.26	42.53	41.52	40.86	39.86	39.28	38.62	38.23
2 & 3	43.75	41.23	39.82	38.90	38.24	37.33	36.73	35.84	35.32	34.72	34.37
3, Average	38.70	36.48	35.23	34.41	33.82	33.02	32.49	31.70	31.24	30.71	30.40
3 & 4	34.13	32.17	31.07	30.35	29.83	29.13	28.66	27.96	27.55	27.09	26.82
4, Low	29.52	27.82	26.87	26.25	25.80	25.19	24.79	24.18	23.83	23.43	23.19

General Office Buildings
Wood Frame
Interior Suite Entrances, Length Less Than Twice Width

Estimating Procedure

1. Use these figures to estimate general office buildings in which access to each suite is through an interior corridor. Medical and dental offices have smaller rooms and more plumbing fixtures than general offices and should be estimated with figures from the Medical and Dental Buildings Section. See page 118.
2. Establish the building quality class by applying the information on page 110.
3. Compute the first floor area. This should include everything within the exterior walls and all insets outside the main walls but under the main roof.
4. If the first floor wall height is more or less than 10 feet, add or subtract from the first floor square foot cost below the appropriate amount from the Wall Height Adjustment Table on page 117.
5. Multiply the adjusted square foot cost by the first floor area.
6. Deduct, if appropriate, for common walls or no wall finish. Use the figures on page 117.
7. If there are second or higher floors, compute the square foot area on each floor. Locate the appropriate square foot cost from the table at the bottom of this page. Adjust this figure for a wall height more or less than 9 feet, using the figures on page 117. Multiply the adjusted cost by the square foot area on each floor. Use the figures on page 117 to deduct for common walls or no wall finish. Add the result to the cost from step 6 above.
8. Multiply the total cost by the location factor on page 7.
9. Add the cost of heating and air conditioning systems, elevators, fire sprinklers, exterior signs, paving and curbing, miscellaneous yard improvements, covered porches and garages. See page 201.

General Offices, Class 3 & 4

First Story

Square Foot Area

Quality Class	2,000	2,500	3,000	4,000	5,000	7,500	10,000	15,000	20,000	30,000	40,000
Exceptional	61.55	60.15	59.13	57.72	56.77	55.32	54.47	53.48	52.89	52.20	51.79
1, Best	55.89	54.61	53.69	52.41	51.55	50.23	49.46	48.56	48.02	47.40	47.03
1 & 2	50.62	49.47	48.63	47.47	46.69	45.50	44.80	43.98	43.50	42.93	42.59
2, Good	45.78	44.74	43.98	42.93	42.23	41.15	40.52	39.78	39.34	38.83	38.53
2 & 3	41.55	40.60	39.91	38.96	38.32	37.35	36.77	36.10	35.70	35.24	34.96
3, Average	37.24	36.40	35.78	34.92	34.35	33.47	32.96	32.36	32.00	31.59	31.34
3 & 4	33.03	32.27	31.73	30.97	30.46	29.68	29.23	28.69	28.38	28.01	27.79
4, Low	28.74	28.08	27.60	26.95	26.50	25.83	25.43	24.97	24.69	24.37	24.18

Second and Higher Stories

Square Foot Area

Quality Class	2,000	2,500	3,000	4,000	5,000	7,500	10,000	15,000	20,000	30,000	40,000
Exceptional	54.92	54.02	53.37	52.47	51.86	50.94	50.39	49.76	49.39	48.95	48.69
1, Best	49.93	49.11	48.52	47.70	47.15	46.31	45.81	45.24	44.90	44.50	44.26
1 & 2	45.48	44.74	44.20	43.45	42.95	42.18	41.74	41.21	40.90	40.54	40.32
2, Good	41.02	40.35	39.87	39.19	38.74	38.05	37.64	37.17	36.89	36.56	36.37
2 & 3	36.85	36.25	35.81	35.21	34.80	34.18	33.82	33.39	33.14	32.85	32.67
3, Average	32.59	32.06	31.67	31.14	30.78	30.23	29.90	29.53	29.31	29.05	28.89
3 & 4	28.69	28.22	27.88	27.41	27.09	26.61	26.33	26.00	25.80	25.57	25.44
4, Low	24.77	24.36	24.07	23.66	23.39	22.97	22.73	22.44	22.27	22.07	21.96

General Office Buildings
Wood Frame

Interior Suite Entrances, Length Between 2 and 4 Times Width

Estimating Procedure

1. Use these figures to estimate general office buildings in which access to each suite is through an interior corridor. Medical and dental offices have smaller rooms and more plumbing fixtures than general offices and should be estimated with figures from the Medical and Dental Buildings Section. See page 118.
2. Establish the building quality class by applying the information on page 110.
3. Compute the first floor area. This should include everything within the exterior walls and all insets outside the main walls but under the main roof.
4. If the first floor wall height is more or less than 10 feet, add or subtract from the first floor square foot cost below the appropriate amount from the Wall Height Adjustment Table on page 117.
5. Multiply the adjusted square foot cost by the first floor area.
6. Deduct, if appropriate, for common walls or no wall finish. Use the figures on page 117.
7. If there are second or higher floors, compute the square foot area on each floor. Locate the appropriate square foot cost from the table at the bottom of this page. Adjust this figure for a wall height more or less than 9 feet, using the figures on page 117. Multiply the adjusted cost by the square foot area on each floor. Use the figures on page 117 to deduct for common walls or no wall finish. Add the result to the cost from step 6 above.
8. Multiply the total cost by the location factor on page 7.
9. Add the cost of heating and air conditioning systems, elevators, fire sprinklers, exterior signs, paving and curbing, miscellaneous yard improvements, covered porches and garages. See page 201.

First Story

	Square Foot Area										
Quality Class	2,000	2,500	3,000	4,000	5,000	7,500	10,000	15,000	20,000	30,000	40,000
Exceptional	63.72	62.08	60.88	59.21	58.08	56.33	55.31	54.09	53.37	52.52	52.02
1, Best	58.32	56.83	55.73	54.20	53.16	51.57	50.63	49.52	48.86	48.08	47.62
1 & 2	52.40	51.05	50.06	48.69	47.76	46.33	45.48	44.48	43.89	43.19	42.78
2, Good	47.36	46.14	45.25	44.01	43.17	41.87	41.11	40.21	39.67	39.04	38.67
2 & 3	43.02	41.92	41.11	39.98	39.22	38.04	37.34	36.53	36.04	35.47	35.13
3, Average	38.60	37.61	36.88	35.87	35.19	34.13	33.51	32.77	32.34	31.82	31.52
3 & 4	34.26	33.38	32.73	31.83	31.23	30.29	29.74	29.08	28.70	28.24	27.97
4, Low	29.82	29.05	28.49	27.71	27.18	26.36	25.88	25.31	24.98	24.58	24.34

Second and Higher Stories

	Square Foot Area										
Quality Class	2,000	2,500	3,000	4,000	5,000	7,500	10,000	15,000	20,000	30,000	40,000
Exceptional	56.33	55.21	54.41	53.31	52.58	51.48	50.84	50.11	49.68	49.18	48.89
1, Best	51.22	50.20	49.47	48.47	47.81	46.81	46.23	45.56	45.17	44.72	44.45
1 & 2	46.58	45.66	44.99	44.08	43.48	42.57	42.05	41.44	41.08	40.67	40.43
2, Good	42.10	41.26	40.66	39.84	39.29	38.47	38.00	37.45	37.13	36.75	36.53
2 & 3	37.83	37.08	36.54	35.80	35.31	34.57	34.14	33.65	33.36	33.03	32.83
3, Average	33.50	32.84	32.36	31.71	31.27	30.62	30.24	29.80	29.55	29.25	29.08
3 & 4	29.53	28.95	28.53	27.95	27.57	26.99	26.66	26.27	26.05	25.78	25.63
4, Low	25.44	24.94	24.57	24.08	23.75	23.25	22.96	22.63	22.44	22.21	22.08

General Office Buildings
Wood Frame
Interior Suite Entrances, Length More Than 4 Times Width

Estimating Procedure

1. Use these figures to estimate general office buildings in which access to each suite is through an interior corridor. Medical and dental offices have smaller rooms and more plumbing fixtures than general offices and should be estimated with figures from the Medical and Dental Buildings Section. See page 118.
2. Establish the building quality class by applying the information on page 110.
3. Compute the first floor area. This should include everything within the exterior walls and all insets outside the main walls but under the main roof.
4. If the first floor wall height is more or less than 10 feet, add or subtract from the first floor square foot cost below the appropriate amount from the Wall Height Adjustment Table on page 117.
5. Multiply the adjusted square foot cost by the first floor area.
6. Deduct, if appropriate, for common walls or no wall finish. Use the figures on page 117.
7. If there are second or higher floors, compute the square foot area on each floor. Locate the appropriate square foot cost from the table at the bottom of this page. Adjust this figure for a wall height more or less than 9 feet, using the figures on page 117. Multiply the adjusted cost by the square foot area on each floor. Use the figures on page 117 to deduct for common walls or no wall finish. Add the result to the cost from step 6 above.
8. Multiply the total cost by the location factor on page 7.
9. Add the cost of heating and air conditioning systems, elevators, fire sprinklers, exterior signs, paving and curbing, miscellaneous yard improvements, covered porches and garages. See page 201.

General Offices, Class 2

First Story

Square Foot Area

Quality Class	2,000	2,500	3,000	4,000	5,000	7,500	10,000	15,000	20,000	30,000	40,000
Exceptional	66.78	64.62	63.06	60.95	59.56	57.46	56.26	54.87	54.06	53.12	52.57
1, Best	60.70	58.73	57.32	55.40	54.14	52.23	51.13	49.87	49.13	48.28	47.78
1 & 2	54.91	53.13	51.85	50.12	48.97	47.25	46.26	45.11	44.45	43.68	43.23
2, Good	49.60	47.99	46.84	45.27	44.23	42.68	41.78	40.75	40.15	39.45	39.04
2 & 3	45.11	43.65	42.60	41.18	40.23	38.82	38.00	37.06	36.52	35.88	35.51
3, Average	40.52	39.21	38.27	36.99	36.14	34.87	34.14	33.29	32.80	32.23	31.90
3 & 4	35.93	34.77	33.93	32.80	32.05	30.92	30.27	29.52	29.09	28.58	28.29
4, Low	31.36	30.34	29.62	28.62	27.97	26.99	26.42	25.77	25.39	24.95	24.69

Second and Higher Stories

Square Foot Area

Quality Class	2,000	2,500	3,000	4,000	5,000	7,500	10,000	15,000	20,000	30,000	40,000
Exceptional	57.99	56.62	55.64	54.30	53.42	52.09	51.32	50.44	49.92	49.33	48.98
1, Best	52.84	51.59	50.70	49.48	48.67	47.46	46.76	45.96	45.49	44.95	44.63
1 & 2	48.16	47.02	46.20	45.09	44.35	43.25	42.61	41.88	41.45	40.96	40.67
2, Good	43.46	42.43	41.70	40.69	40.03	39.03	38.46	37.80	37.41	36.97	36.70
2 & 3	39.11	38.18	37.52	36.62	36.02	35.12	34.61	34.01	33.66	33.26	33.03
3, Average	34.68	33.86	33.27	32.47	31.94	31.15	30.69	30.16	29.85	29.50	29.29
3 & 4	30.60	29.88	29.36	28.65	28.19	27.49	27.08	26.61	26.34	26.03	25.85
4, Low	26.48	25.85	25.40	24.79	24.39	23.78	23.43	23.03	22.79	22.52	22.36

General Office Buildings
Wood Frame
Wall Adjustments

The square foot costs for general offices are based on the wall heights of 10 feet for first floors and 9 feet for higher floors. Add or subtract the amount listed in this table to the square foot of floor cost for each foot of wall height more or less than 10 feet, if adjusting for a first floor, and 9 feet, if adjusting for upper floors.

	Square Foot Area										
Quality Class	1,000	1,500	2,000	3,000	4,000	5,000	7,500	10,000	15,000	20,000	40,000
1, Best	.78	.66	.58	.48	.41	.37	.29	.25	.19	.16	.09
2, Good	.65	.55	.49	.40	.35	.31	.25	.21	.16	.13	.08
3, Average	.56	.47	.42	.34	.30	.27	.21	.18	.14	.11	.07
4, Low	.49	.42	.37	.30	.26	.23	.19	.16	.12	.10	.06

Perimeter Wall Adjustment

A common wall exists when two buildings share one wall. Adjust for common walls by deducting the linear foot cost below from the total structure cost. In some structures one or more walls is not owned at all. In this case, deduct the "No Ownership" cost per linear foot of wall not owned. Where a perimeter wall remains unfinished, deduct the "Lack of Exterior Finish" cost.

First Story

Class	For a Common Wall, Deduct Per Linear Foot	For No Wall Ownership, Deduct Per Linear Foot	For Lack of Exterior Finish, Deduct Per Linear Foot
1	$97.00	$115.00	$83.40
2	78.00	92.50	68.00
3	69.00	79.60	54.50
4	50.60	69.00	42.40

Second and Higher Stories

Class	For A Common Wall, Deduct Per Linear Foot	For No Wall Ownership, Deduct Per Linear Foot	For Lack of Exterior Finish, Deduct Per Linear Foot
1	$61.30	$68.60	$50.80
2	49.50	61.30	46.00
3	47.30	55.00	41.80
4	44.00	49.50	34.00

Medical-Dental Buildings
Masonry or Concrete

	Class 1 Best Quality	Class 2 Good Quality	Class 3 Average Quality	Class 4 Low Quality
First Floor Structure	Reinforced concrete slab on grade or standard wood frame.	Reinforced concrete slab on grade or standard wood frame.	Reinforced concrete slab on grade.	Reinforced concrete slab on grade.
Upper Floor Structure	Standard wood frame, plywood and 1½" light weight concrete sub floor.	Standard wood frame, plywood and 1½" light weight concrete sub floor.	Standard wood frame, 5/8" plywood sub floor.	Standard wood frame, 5/8" plywood sub floor.
Walls	8" decorative concrete block or 6" concrete tilt-up.	8" decorative concrete block or 6" concrete tilt-up.	8" reinforced concrete block or 8" reinforced brick.	8" reinforced concrete block or clay tile.
Roof Structure	Standard wood frame, flat or low pitch.	Standard wood frame, flat or low pitch.	Standard wood frame, flat or low pitch.	Standard wood frame, flat or low pitch.
Exterior Finish				
Walls	Decorative block or large rock imbedded in tilt-up panels with 10 to 20% brick or stone veneer.	Decorative block or exposed aggregate and 10 to 20% brick or stone veneer.	Stucco or colored concrete block.	Painted.
Windows	Average number in good aluminum frame. Fixed plate glass in good frame on front.	Average number in good aluminum frame. Some fixed plate glass in front.	Average number of average aluminum sliding type.	Average number of low cost aluminum sliding type.
Roof Cover	5 ply built-up roofing on flat roofs. Heavy shake or tile on sloping roofs.	5 ply built-up roofing on flat roofs. Average shake or composition, tar and large rock on sloping roofs.	4 ply built-up roofing on flat roofs. Wood shingle or composition, tar and pea gravel on sloping roofs.	3 ply built-up roofing on flat roofs. Composition shingle on sloping roofs.
Overhang	3' closed overhang, fully guttered.	2' closed overhang, fully guttered.	None on flat roofs. 18" open on sloping roofs, fully guttered.	None on flat roofs. 12" to 16" open on sloping roofs, gutters over entrances.
Business Offices	Good hardwood veneer paneling. Solid vinyl or carpet.	Hardwood paneling or vinyl wall cover. Vinyl asbestos tile or carpet.	Gypsum wallboard, texture and paint. Vinyl asbestos tile.	Gypsum wallboard texture and paint. Minimum grade tile.
Corridors	Good hardwood veneer paneling. Solid vinyl or carpet.	Gypsum wallboard and vinyl wall cover. Vinyl asbestos tile.	Gypsum wallboard, texture and paint. Vinyl asbestos tile.	Gypsum wallboard, texture and paint. Minimum grade tile.
Waiting Rooms	Good hardwood veneer paneling. Carpet.	Hardwood paneling. Carpet.	Gypsum wallboard, some paneling. Vinyl asbestos tile or carpet.	Gypsum wallboard, texture and paint. Minimum grade tile.
Private Offices	Good hardwood veneer paneling. Carpet.	Textured wall cover and hardwood paneling. Carpet.	Gypsum wallboard and paper, wood paneling. Vinyl asbestos tile or carpet.	Gypsum wallboard, texture and paint. Minimum grade tile.
Treatment Rooms	Gypsum wallboard and vinyl wall covering. Sheet vinyl or carpet.	Gypsum wallboard and enamel. Sheet vinyl.	Gypsum wallboard and enamel. Vinyl asbestos tile.	Gypsum wallboard, texture and paint. Minimum grade tile.
Bathrooms	Gypsum wallboard and enamel with ceramic tile wainscot. Sheet vinyl or ceramic tile.	Gypsum wallboard and enamel or vinyl wall covering. Sheet vinyl or ceramic tile.	Gypsum wallboard and enamel. Vinyl asbestos tile.	Gypsum wallboard, texture and paint. Minimum grade tile.
Ceiling Finish	Suspended "T" bar and acoustical tile.	Gypsum wallboard and acoustical tile.	Gypsum wallboard and acoustical tile.	Gypsum wallboard and paint.
Utilities				
Plumbing	Copper tubing, good fixtures.	Copper tubing, good fixtures.	Copper tubing, average fixtures.	Copper tubing, economy fixtures.
Lighting	Conduit wiring, good fixtures.	Conduit wiring, good fixtures.	Romex or conduit wiring, average fixtures.	Romex wiring, economy fixtures.
Cabinets	Formica faced with formica tops.	Good grade of hardwood with formica tops.	Average amount of painted wood or low grade hardwood with formica top.	Minimum amount of painted wood with formica top.

Square foot costs include the cost of the following components: Foundations as required for normal soil conditions. Floor, wall and roof structures. Interior floor, wall and ceiling finishes. Exterior wall finish and roof cover. Interior partitions. Cabinets, doors and windows. Basic electrical systems and lighting fixtures. Rough plumbing and fixtures. Permits and fees. Contractor's mark-up. In addition to the above components, costs for buildings with more than 10,000 feet include the cost of lead shielding for typical x-ray rooms.

Medical-Dental Buildings
Masonry or Concrete
Exterior Suite Entrances, Length Less Than Twice Width

Estimating Procedure

1. Use these figures to estimate medical, dental, psychiatric, optometry and similar professional buildings in which access to each office suite is through an exterior entrance. Buildings in this section have more plumbing fixtures per square foot of floor and smaller room sizes than general office buildings. Note also that buildings with more than 10,000 square feet are assumed to have lead shielded x-ray rooms.
2. Establish the building quality class by applying the information on page 118.
3. Compute the first floor area. This should include everything within the exterior walls and all insets outside the main walls but under the main roof.
4. If the first floor wall height is more or less than 10 feet, add or subtract from the first floor square foot cost below the appropriate amount from the Wall Height Adjustment Table on page 125.
5. Multiply the adjusted square foot cost by the first floor area.
6. Deduct, if appropriate, for common walls or no wall finish. Use the figures on page 125.
7. If there are second or higher floors, compute the square foot area on each floor. Locate the appropriate square foot cost from the table at the bottom of this page. Adjust this figure for a wall height more or less than 9 feet, using the figures on page 125. Multiply the adjusted cost by the square foot area on each floor. Use the figures on page 125 to deduct for common walls or no wall finish. Add the result to the cost from step 6 above.
8. Multiply the total cost by the location factor on page 7.
9. Add the cost of heating and air conditioning systems, elevators, fire sprinklers, exterior signs, paving and curbing, miscellaneous yard improvements, covered porches and garages. See page 201.

Medical-Dental Building, Class 2

First Floor

Square Foot Area

Quality Class	1,000	1,500	2,000	2,500	3,000	4,000	5,000	7,500	10,000	15,000	20,000
Exceptional	97.43	91.68	88.34	86.10	84.46	82.19	80.67	78.32	76.94	75.32	74.36
1, Best	88.42	83.21	80.17	78.14	76.66	74.60	73.21	71.08	69.83	68.36	67.49
1 & 2	82.87	77.98	75.14	73.23	71.84	69.92	68.62	66.62	65.45	64.07	63.25
2, Good	76.80	72.26	69.63	67.86	66.58	64.79	63.58	61.74	60.65	59.37	58.62
2 & 3	70.64	66.47	64.05	62.43	61.24	59.60	58.49	56.79	55.79	54.61	53.92
3, Average	64.35	60.55	58.35	56.87	55.79	54.29	53.28	51.73	50.82	49.75	49.12
3 & 4	60.04	56.50	54.44	53.06	52.05	50.65	49.71	48.27	47.42	46.42	45.83
4, Low	55.61	52.33	50.42	49.14	48.21	46.91	46.04	44.70	43.91	42.99	42.44

Second and Higher Floors

Square Foot Area

Quality Class	1,000	1,500	2,000	2,500	3,000	4,000	5,000	7,500	10,000	15,000	20,000
Exceptional	87.23	81.40	78.25	76.23	74.82	72.94	71.72	69.94	68.94	67.83	67.19
1, Best	83.53	77.95	74.93	73.00	71.64	69.84	68.67	66.97	66.02	64.95	64.34
1 & 2	78.51	73.26	70.43	68.61	67.34	65.65	64.55	62.95	62.05	61.05	60.47
2, Good	73.52	68.60	65.95	64.25	63.06	61.47	60.44	58.94	58.11	57.16	56.63
2 & 3	67.49	62.98	60.54	58.98	57.89	56.43	55.49	54.11	53.34	52.48	51.98
3, Average	61.50	57.39	55.16	53.74	52.75	51.42	50.56	49.31	48.60	47.82	47.37
3 & 4	57.29	53.46	51.39	50.07	49.14	47.90	47.10	45.93	45.28	44.55	44.13
4, Low	53.11	49.56	47.64	46.41	45.55	44.40	43.66	42.58	41.97	41.29	40.90

Medical-Dental Buildings
Masonry or Concrete

Exterior Suite Entrances, Length Between 2 and 4 Times Width

Estimating Procedure

1. Use these figures to estimate medical, dental, psychiatric, optometry and similar professional buildings in which access to each office suite is through an exterior entrance. Buildings in this section have more plumbing fixtures per square foot of floor and smaller room sizes than general office buildings. Note also that buildings with more than 10,000 square feet are assumed to have lead shielded x-ray rooms.
2. Establish the building quality class by applying the information on page 118.
3. Compute the first floor area. This should include everything within the exterior walls and all insets outside the main walls but under the main roof.
4. If the first floor wall height is more or less than 10 feet, add or subtract from the first floor square foot cost below the appropriate amount from the Wall Height Adjustment Table on page 125.
5. Multiply the adjusted square foot cost by the first floor area.
6. Deduct, if appropriate, for common walls or no wall finish. Use the figures on page 125.
7. If there are second or higher floors, compute the square foot area on each floor. Locate the appropriate square foot cost from the table at the bottom of this page. Adjust this figure for a wall height more or less than 9 feet, using the figures on page 125. Multiply the adjusted cost by the square foot area on each floor. Use the figures on page 125 to deduct for common walls or no wall finish. Add the result to the cost from step 6 above.
8. Multiply the total cost by the location factor on page 7.
9. Add the cost of heating and air conditioning systems, elevators, fire sprinklers, exterior signs, paving and curbing, miscellaneous yard improvements, covered porches and garages. See page 201.

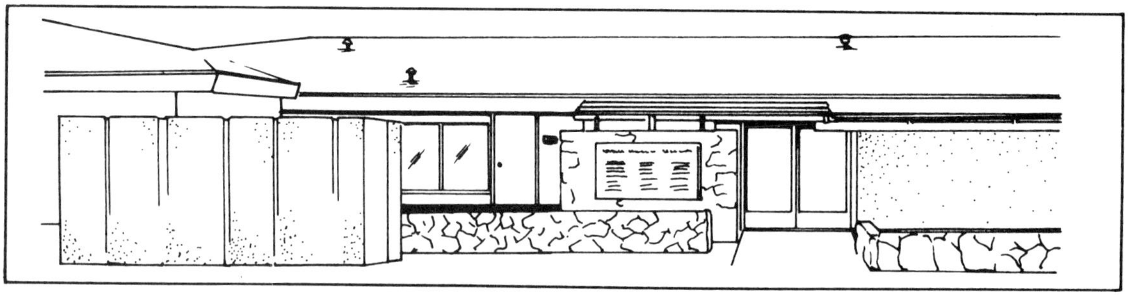

Medical Dental Building, Class 3 & 4

First Story

Square Foot Area

Quality Class	1,000	1,500	2,000	2,500	3,000	4,000	5,000	7,500	10,000	15,000	20,000
Exceptional	105.01	97.24	92.86	89.97	87.90	85.06	83.18	80.34	78.70	76.80	75.69
1, Best	95.48	88.41	84.43	81.80	79.92	77.34	75.63	73.05	71.55	69.83	68.82
1 & 2	88.85	82.28	78.57	76.13	74.37	71.97	70.38	67.98	66.59	64.98	64.04
2, Good	82.27	76.18	72.75	70.49	68.86	66.64	65.17	62.94	61.66	60.17	59.30
2 & 3	75.58	69.99	66.84	64.76	63.27	61.23	59.87	57.83	56.65	55.28	54.48
3, Average	68.87	63.78	60.90	59.01	57.65	55.79	54.56	52.70	51.62	50.37	49.64
3 & 4	64.16	59.41	56.74	54.97	53.70	51.97	50.82	49.09	48.09	46.92	46.25
4, Low	59.42	55.02	52.54	50.91	49.73	48.13	47.07	45.46	44.53	43.45	42.83

Second and Higher Stories

Square Foot Area

Quality Class	1,000	1,500	2,000	2,500	3,000	4,000	5,000	7,500	10,000	15,000	20,000
Exceptional	96.80	89.99	86.15	83.62	81.80	79.31	77.66	75.17	73.73	72.06	71.09
1, Best	87.99	81.81	78.31	76.01	74.36	72.10	70.60	68.34	67.03	65.51	64.63
1 & 2	82.69	76.87	73.59	71.43	69.88	67.75	66.34	64.22	62.99	61.56	60.73
2, Good	77.39	71.95	68.88	66.85	65.40	63.41	62.09	60.10	58.95	57.62	56.84
2 & 3	71.05	66.05	63.23	61.37	60.04	58.21	57.00	55.17	54.12	52.89	52.18
3, Average	64.72	60.17	57.60	55.91	54.69	53.03	51.93	50.26	49.30	48.18	47.53
3 & 4	60.33	56.09	53.69	52.12	50.98	49.43	48.41	46.85	45.95	44.92	44.31
4, Low	55.95	52.02	49.80	48.33	47.28	45.85	44.89	43.45	42.62	41.65	41.09

Medical-Dental Buildings
Masonry or Concrete

Exterior Suite Entrances, Length More Than 4 Times Width

Estimating Procedure

1. Use these figures to estimate medical, dental, psychiatric, optometry and similar professional buildings in which access to each office suite is through an exterior entrance. Buildings in this section have more plumbing fixtures per square foot of floor and smaller room sizes than general office buildings. Note also that buildings with more than 10,000 square feet are assumed to have lead shielded x-ray rooms.
2. Establish the building quality class by applying the information on page 118.
3. Compute the first floor area. This should include everything within the exterior walls and all insets outside the main walls but under the main roof.
4. If the first floor wall height is more or less than 10 feet, add or subtract from the first floor square foot cost below the appropriate amount from the Wall Height Adjustment Table on page 125.
5. Multiply the adjusted square foot cost by the first floor area.
6. Deduct, if appropriate, for common walls or no wall finish. Use the figures on page 125.
7. If there are second or higher floors, compute the square foot area on each floor. Locate the appropriate square foot cost from the table at the bottom of this page. Adjust this figure for a wall height more or less than 9 feet, using the figures on page 125. Multiply the adjusted cost by the square foot area on each floor. Use the figures on page 125 to deduct for common walls or no wall finish. Add the result to the cost from step 6 above.
8. Multiply the total cost by the location factor on page 7.
9. Add the cost of heating and air conditioning systems, elevators, fire sprinklers, exterior signs, paving and curbing, miscellaneous yard improvements, covered porches and garages. See page 201.

Medical-Dental Building, Class 2 & 3

First Story

Square Foot Area

Quality Class	1,000	1,500	2,000	2,500	3,000	4,000	5,000	7,500	10,000	15,000	20,000
Exceptional	113.21	103.39	97.91	94.33	91.78	88.31	86.02	82.60	80.63	78.37	77.06
1, Best	102.91	93.99	89.01	85.76	83.43	80.28	78.20	75.09	73.30	71.25	70.05
1 & 2	95.69	87.39	82.76	79.73	77.57	74.64	72.71	69.81	68.15	66.24	65.13
2, Good	88.76	81.06	76.77	73.97	71.96	69.24	67.45	64.76	63.22	61.45	60.42
2 & 3	81.21	74.17	70.24	67.67	65.84	63.35	61.71	59.25	57.84	56.22	55.28
3, Average	73.91	67.50	63.92	61.59	59.92	57.65	56.16	53.92	52.64	51.17	50.31
3 & 4	68.84	62.87	59.54	57.37	55.81	53.70	52.31	50.23	49.03	47.66	46.86
4, Low	63.74	58.21	55.12	53.11	51.67	49.72	48.43	46.50	45.39	44.12	43.39

Second and Higher Stories

Square Foot Area

Quality Class	1,000	1,500	2,000	2,500	3,000	4,000	5,000	7,500	10,000	15,000	20,000
Exceptional	103.52	95.17	90.46	87.36	85.13	82.09	80.07	77.01	75.25	73.21	72.02
1, Best	93.73	86.17	81.90	79.10	77.08	74.32	72.49	69.73	68.13	66.28	65.20
1 & 2	88.39	81.26	77.24	74.59	72.69	70.09	68.36	65.76	64.25	62.51	61.49
2, Good	82.71	76.04	72.27	69.79	68.01	65.58	63.97	61.53	60.12	58.49	57.54
2 & 3	75.92	69.80	66.34	64.07	62.43	60.20	58.72	56.48	55.19	53.69	52.82
3, Average	69.16	63.58	60.43	58.36	56.87	54.84	53.49	51.45	50.27	48.91	48.11
3 & 4	64.49	59.29	56.35	54.42	53.03	51.13	49.87	47.97	46.87	45.60	44.86
4, Low	59.78	54.96	52.24	50.44	49.16	47.40	46.23	44.47	43.45	42.27	41.58

Medical-Dental Buildings
Masonry or Concrete
Interior Suite Entrances, Length Less Than Twice Width

Estimating Procedure

1. Use these figures to estimate medical, dental, psychiatric, optometry and similar professional buildings in which access to each office suite is through an interior corridor. Buildings in this section have more plumbing fixtures per square foot of floor and smaller room sizes than general office buildings. Note also that buildings with more than 10,000 square feet are assumed to have lead shielded x-ray rooms.
2. Establish the building quality class by applying the information on page 118.
3. Compute the first floor area. This should include everything within the exterior walls and all insets outside the main walls but under the main roof.
4. If the first floor wall height is more or less than 10 feet, add or subtract from the first floor square foot cost below the appropriate amount from the Wall Height Adjustment Table on page 125.
5. Multiply the adjusted square foot cost by the first floor area.
6. Deduct, if appropriate, for common walls or no wall finish. Use the figures on page 125.
7. If there are second or higher floors, compute the square foot area on each floor. Locate the appropriate square foot cost from the table at the bottom of this page. Adjust this figure for a wall height more or less than 9 feet, using the figures on page 125. Multiply the adjusted cost by the square foot area on each floor. Use the figures on page 125 to deduct for common walls or no wall finish. Add the result to the cost from step 6 above.
8. Multiply the total cost by the location factor on page 7.
9. Add the cost of heating and air conditioning systems, elevators, fire sprinklers, exterior signs, paving and curbing, miscellaneous yard improvements, covered porches and garages. See page 201.

Medical-Dental Building, Class 3 & 4

First Story

Square Foot Area

Quality Class	2,000	2,500	3,000	4,000	5,000	7,500	10,000	15,000	20,000	30,000	40,000
Exceptional	82.35	80.12	78.50	76.28	74.78	72.50	71.17	69.61	68.70	67.63	66.99
1, Best	76.97	74.88	73.37	71.29	69.89	67.76	66.52	65.06	64.21	63.21	62.61
1 & 2	71.83	69.89	68.48	66.53	65.23	63.24	62.08	60.72	59.93	58.99	58.44
2, Good	66.70	64.90	63.59	61.78	60.57	58.73	57.65	56.39	55.64	54.78	54.26
2 & 3	61.47	59.80	58.60	56.93	55.82	54.12	53.12	51.96	51.28	50.48	50.01
3, Average	56.20	54.67	53.57	52.05	51.03	49.48	48.57	47.51	46.88	46.15	45.72
3 & 4	52.63	51.21	50.18	48.75	47.80	46.34	45.49	44.49	43.91	43.22	42.82
4, Low	49.06	47.73	46.76	45.44	44.55	43.19	42.40	41.47	40.92	40.29	39.91

Second and Higher Stories

Square Foot Area

Quality Class	2,000	2,500	3,000	4,000	5,000	7,500	10,000	15,000	20,000	30,000	40,000
Exceptional	77.85	75.86	74.42	72.44	71.11	69.08	67.90	66.52	65.70	64.75	64.19
1, Best	72.81	70.96	69.61	67.75	66.51	64.62	63.51	62.21	61.45	60.56	60.04
1 & 2	68.37	66.63	65.36	63.62	62.45	60.67	59.63	58.42	57.70	56.87	56.37
2, Good	63.91	62.28	61.10	59.47	58.38	56.72	55.74	54.61	53.94	53.16	52.70
2 & 3	58.69	57.20	56.11	54.61	53.61	52.09	51.19	50.15	49.54	48.82	48.40
3, Average	53.47	52.11	51.12	49.76	48.84	47.45	46.64	45.69	45.13	44.48	44.09
3 & 4	50.07	48.79	47.86	46.59	45.73	44.43	43.67	42.78	42.25	41.64	41.28
4, Low	46.65	45.46	44.60	43.41	42.61	41.40	40.69	39.86	39.37	38.80	38.46

Medical-Dental Buildings
Masonry or Concrete
Interior Suite Entrances, Length Between 2 and 4 Times Width

Estimating Procedure

1. Use these figures to estimate medical, dental, psychiatric, optometry and similar professional buildings in which access to each office suite is through an interior corridor. Buildings in this section have more plumbing fixtures per square foot of floor and smaller room sizes than general office buildings. Note also that buildings with more than 10,000 square feet are assumed to have lead shielded x-ray rooms.
2. Establish the building quality class by applying the information on page 118.
3. Compute the first floor area. This should include everything within the exterior walls and all insets outside the main walls but under the main roof.
4. If the first floor wall height is more or less than 10 feet, add or subtract from the first floor square foot cost below the appropriate amount from the Wall Height Adjustment Table on page 125.
5. Multiply the adjusted square foot cost by the first floor area.
6. Deduct, if appropriate, for common walls or no wall finish. Use the figures on page 125.
7. If there are second or higher floors, compute the square foot area on each floor. Locate the appropriate square foot cost from the table at the bottom of this page. Adjust this figure for a wall height more or less than 9 feet, using the figures on page 125. Multiply the adjusted cost by the square foot area on each floor. Use the figures on page 125 to deduct for common walls or no wall finish. Add the result to the cost from step 6 above.
8. Multiply the total cost by the location factor on page 7.
9. Add the cost of heating and air conditioning systems, elevators, fire sprinklers, exterior signs, paving and curbing, miscellaneous yard improvements, covered porches and garages. See page 201.

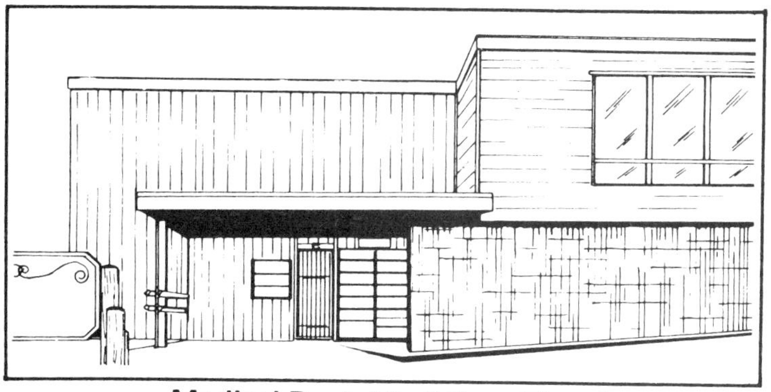

Medical Dental Building, Class 3

First Story

Square Foot Area

Quality Class	2,000	2,500	3,000	4,000	5,000	7,500	10,000	15,000	20,000	30,000	40,000
Exceptional	86.21	83.46	81.48	78.76	76.95	74.21	72.61	70.76	69.67	68.41	67.66
1, Best	80.58	78.01	76.16	73.62	71.92	69.36	67.87	66.13	65.12	63.93	63.24
1 & 2	75.14	72.75	71.02	68.65	67.07	64.68	63.29	61.67	60.73	59.62	58.97
2, Good	69.72	67.50	65.89	63.70	62.23	60.01	58.72	57.22	56.34	55.32	54.72
2 & 3	64.19	62.15	60.67	58.65	57.30	55.26	54.07	52.69	51.88	50.94	50.38
3, Average	58.64	56.77	55.42	53.58	52.34	50.48	49.39	48.13	47.39	46.53	46.02
3 & 4	54.92	53.17	51.91	50.18	49.02	47.27	46.26	45.08	44.39	43.58	43.10
4, Low	51.16	49.53	48.35	46.74	45.67	44.04	43.09	41.99	41.35	40.59	40.15

Second and Higher Stories

Square Foot Area

Quality Class	2,000	2,500	3,000	4,000	5,000	7,500	10,000	15,000	20,000	30,000	40,000
Exceptional	81.26	78.81	77.05	74.62	73.01	70.57	69.15	67.50	66.53	65.41	64.74
1, Best	75.95	73.66	72.00	69.74	68.23	65.95	64.62	63.08	62.18	61.12	60.51
1 & 2	71.29	69.14	67.59	65.46	64.05	61.90	60.66	59.21	58.36	57.38	56.79
2, Good	66.66	64.65	63.20	61.21	59.89	57.88	56.72	55.36	54.57	53.65	53.11
2 & 3	61.18	59.34	58.01	56.18	54.97	53.13	52.06	50.82	50.09	49.24	48.75
3, Average	55.70	54.02	52.81	51.15	50.04	48.37	47.39	46.26	45.60	44.83	44.37
3 & 4	52.19	50.62	49.48	47.93	46.89	45.32	44.41	43.35	42.73	42.01	41.58
4, Low	48.64	47.17	46.12	44.67	43.70	42.24	41.39	40.40	39.82	39.15	38.75

Medical-Dental Buildings
Masonry or Concrete
Interior Suite Entrances, Length More Than 4 Times Width

Estimating Procedure

1. Use these figures to estimate medical, dental, psychiatric, optometry and similar professional buildings in which access to each office suite is through an interior corridor. Buildings in this section have more plumbing fixtures per square foot of floor and smaller room sizes than general office buildings. Note also that buildings with more than 10,000 square feet are assumed to have lead shielded x-ray rooms.
2. Establish the building quality class by applying the information on page 118.
3. Compute the first floor area. This should include everything within the exterior walls and all insets outside the main walls but under the main roof.
4. If the first floor wall height is more or less than 10 feet, add or subtract from the first floor square foot cost below the appropriate amount from the Wall Height Adjustment Table on page 125.
5. Multiply the adjusted square foot cost by the first floor area.
6. Deduct, if appropriate, for common walls or no wall finish. Use the figures on page 125.
7. If there are second or higher floors, compute the square foot area on each floor. Locate the appropriate square foot cost from the table at the bottom of this page. Adjust this figure for a wall height more or less than 9 feet, using the figures on page 125. Multiply the adjusted cost by the square foot area on each floor. Use the figures on page 125 to deduct for common walls or no wall finish. Add the result to the cost from step 6 above.
8. Multiply the total cost by the location factor on page 7.
9. Add the cost of heating and air conditioning systems, elevators, fire sprinklers, exterior signs, paving and curbing, miscellaneous yard improvements, covered porches and garages. See page 201.

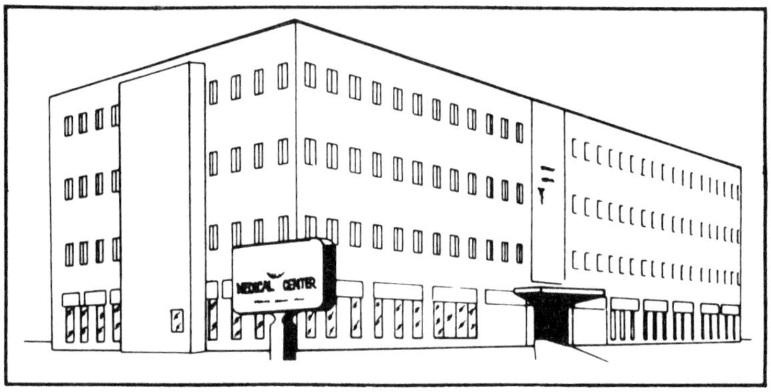

Medical-Dental Building, Class 2 & 3

First Story

Square Foot Area

Quality Class	2,000	2,500	3,000	4,000	5,000	7,500	10,000	15,000	20,000	30,000	40,000
Exceptional	91.15	87.71	85.25	81.90	79.68	76.36	74.45	72.25	70.97	69.49	68.62
1, Best	85.19	81.97	79.67	76.54	74.47	71.37	69.58	67.52	66.33	64.94	64.13
1 & 2	79.38	76.38	74.24	71.32	69.39	66.50	64.84	62.92	61.80	60.51	59.76
2, Good	73.59	70.81	68.82	66.12	64.33	61.65	60.10	58.33	57.29	56.10	55.40
2 & 3	67.69	65.13	63.30	60.81	59.17	56.70	55.28	53.65	52.70	51.60	50.96
3, Average	61.76	59.43	57.76	55.49	53.99	51.74	50.44	48.95	48.08	47.08	46.49
3 & 4	57.82	55.64	54.08	51.95	50.55	48.44	47.23	45.83	45.02	44.08	43.53
4, Low	53.84	51.81	50.36	48.38	47.07	45.11	43.98	42.68	41.92	41.05	40.53

Second and Higher Stories

Square Foot Area

Quality Class	2,000	2,500	3,000	4,000	5,000	7,500	10,000	15,000	20,000	30,000	40,000
Exceptional	85.47	82.41	80.21	77.24	75.27	72.32	70.63	68.68	67.54	66.23	65.47
1, Best	79.92	77.05	75.00	72.22	70.38	67.62	66.04	64.21	63.15	61.93	61.21
1 & 2	74.99	72.30	70.37	67.76	66.04	63.45	61.96	60.25	59.26	58.11	57.43
2, Good	70.09	67.58	65.78	63.34	61.72	59.31	57.92	56.32	55.39	54.31	53.68
2 & 3	64.28	61.97	60.33	58.08	56.61	54.39	53.11	51.65	50.80	49.81	49.23
3, Average	58.60	56.49	54.99	52.95	51.60	49.58	48.42	47.08	46.30	45.41	44.88
3 & 4	54.88	52.92	51.51	49.59	48.33	46.44	45.35	44.10	43.37	42.53	42.04
4, Low	51.15	49.31	48.00	46.22	45.04	43.28	42.26	41.09	40.42	39.63	39.17

Medical-Dental Buildings
Masonry or Concrete
Wall Height Adjustment

The square foot costs for medical-dental buildings are based on the wall heights of 10 feet for first floors and 9 feet for higher floors. The main or first floor height is the distance from the bottom of the floor slab or joists to the top of the roof slab or ceiling joists. Second and higher floors are measured from the top of the floor slab or floor joists to the top of the roof slab or ceiling joists. Add or subtract the amount listed in this table to the square foot of floor cost for each foot of wall height more or less than 10 feet, if adjusting for a first floor, and 9 feet, if adjusting for upper floors.

	Square Foot Area										
Quality Class	1,000	1,500	2,000	3,000	4,000	5,000	7,500	10,000	15,000	20,000	40,000
1, Best	2.41	1.88	1.58	1.25	1.06	.93	.75	.64	.51	.44	.30
2, Good	2.01	1.57	1.32	1.04	.88	.78	.62	.53	.43	.36	.25
3, Average	1.70	1.33	1.12	.88	.75	.66	.53	.45	.36	.31	.21
4, Low	1.47	1.15	.96	.76	.65	.57	.45	.39	.31	.27	.18

Perimeter Wall Adjustment

A common wall exists when two buildings share one wall. Adjust for common walls by deducting the linear foot costs below from the total structure cost. In some structures, one or more walls is not owned at all. In this case, deduct the "No Ownership" cost per linear foot of wall not owned. If a wall has no exterior finish, deduct the "Lack of Exterior Finish" cost.

First Story

Class	For A Common Wall, Deduct Per Linear Foot	For No Wall Ownership, Deduct Per Linear Foot	For Lack Of Exterior Finish, Deduct Per Linear Foot
1	$125.00	$204.00	$100.00
2	95.00	179.00	64.00
3	67.00	158.00	32.00
4	52.00	113.00	21.00

Second and Higher Stories

Class	For A Common Wall, Deduct Per Linear Foot	For No Wall Ownership, Deduct Per Linear Foot	For Lack Of Exterior Finish, Deduct Per Linear Foot
1	$135.00	$173.00	$100.00
2	97.00	149.00	64.00
3	75.00	123.00	32.00
4	65.00	107.00	20.40

Medical-Dental Buildings
Wood Frame

	Class 1 Best Quality	Class 2 Good Quality	Class 3 Average Quality	Class 4 Low Quality
Foundation	Reinforced concrete.	Reinforced concrete.	Reinforced concrete.	Reinforced concrete.
First Floor Structure	Reinforced concrete slab on grade or standard wood frame.	Reinforced concrete slab on grade or standard wood frame.	Reinforced concrete slab on grade or 4" x 6" girders with 2" T&G sub floor.	Reinforced concrete slab on grade.
Upper Floor Structure	Standard wood frame, plywood and 1½" light weight concrete sub floor.	Standard wood frame, plywood and 1½" light weight concrete sub floor.	Standard wood frame, 5/8" plywood sub floor.	Standard wood frame, 5/8" plywood sub floor.
Walls	Standard wood frame.	Standard wood frame.	Standard wood frame.	Standard wood frame.
Roof	Standard wood frame, flat or low pitch.	Standard wood frame, flat or low pitch.	Standard wood frame, flat or low pitch.	Standard wood frame, flat or low pitch.
Exterior Finish Walls	Good wood siding with 10 to 20% brick or stone veneer.	Average wood siding or stucco and 10 to 20% brick or stone veneer.	Stucco with some wood trim or cheap wood siding.	Stucco.
Windows	Average amount in good aluminum frame. Fixed plate glass in good frame on front side.	Average amount in good aluminum frame. Some fixed plate glass in front.	Average amount of average aluminum sliding type.	Average amount of low cost aluminum sliding type.
Roof Cover	5 ply built-up roofing on flat roofs. Heavy shake or tile on sloping roofs.	5 ply built-up roofing on flat roofs. Average shake or composition, tar and large rock on sloping roofs.	4 ply built-up roofing on flat roofs. Wood shingle or composition, tar and pea gravel on sloping roofs.	3 ply built-up roofing on flat roofs. Composition shingle on sloping roofs.
Overhang	3' sealed overhang, fully guttered.	2' sealed overhang, fully guttered.	None on flat roofs. 18" unsealed on sloping roofs, fully guttered.	None on flat roofs. 12" to 16" unsealed on sloping roofs, gutters over entrances.
Floor Finish Business Offices	Solid vinyl tile or carpet.	Vinyl asbestos tile or carpet.	Vinyl asbestos tile.	Minimum tile.
Corridors	Solid vinyl tile or carpet.	Vinyl asbestos tile or carpet.	Vinyl asbestos tile.	Minimum tile.
Waiting Room	Carpet.	Carpet.	Vinyl asbestos tile or carpet.	Minimum tile.
Private Office	Carpet.	Carpet.	Vinyl asbestos tile or carpet.	Minimum tile.
Treatment Room	Sheet vinyl or carpet.	Sheet vinyl.	Vinyl asbestos tile.	Minimum tile.
Bathrooms	Sheet vinyl or ceramic tile.	Sheet vinyl or ceramic tile.	Vinyl asbestos tile.	Minimum tile.
Interior Wall Finish Business Offices	Good hardwood veneer paneling.	Hardwood paneling or vinyl wall cover.	Gypsum wallboard, texture and paint.	Gypsum wallboard, texture and paint.
Corridors	Good hardwood veneer paneling.	Gypsum wallboard and vinyl wall cover.	Gypsum wallboard, texture and paint.	Gypsum wallboard, texture and paint.
Waiting Room	Good hardwood veneer paneling.	Hardwood paneling.	Gypsum wallboard and paper, some wood paneling.	Gypsum wallboard, texture and paint.
Treatment Room	Gypsum wallboard and vinyl wall covering.	Gypsum wallboard and enamel.	Gypsum wallboard and enamel.	Gypsum wallboard, texture and paint.
Bathrooms	Gypsum wallboard and enamel with ceramic tile wainscot.	Gypsum wallboard and enamel or vinyl wall covering.	Gypsum wallboard and enamel.	Gypsum wallboard, texture and paint.
Ceiling Finish	Suspended "T" bar and acoustical tile.	Gypsum wallboard and acoustical tile.	Gypsum wallboard and acoustical tile.	Gypsum wallboard and paint.
Utilities Plumbing	Copper tubing, good fixtures.	Copper tubing, good fixtures.	Copper tubing, average fixtures.	Copper tubing, economy fixtures.
Lighting	Conduit wiring, good fixtures.	Conduit wiring, good fixtures.	Romex or conduit wiring, average fixtures.	Romex wiring, economy fixtures.
Cabinets	Formica faced with formica tops.	Good grade of hardwood with formica tops.	Average amount of painted wood or low grade hardwood with formica top.	Minimum amount of painted wood with formica top.

Square foot costs include the cost of the following components: Foundations as required for normal soil conditions. Floor, wall and roof structures. Interior floor, wall and ceiling finishes. Exterior wall finish and roof cover. Interior partitions. Cabinets, doors and windows. Basic electrical systems and lighting fixtures. Rough plumbing and fixtures. Permits and fees. Contractor's mark-up. In addition to the above components, costs for buildings with more than 10,000 feet include the cost of lead shielding for typical x-ray rooms.

Medical-Dental Buildings
Wood Frame

Exterior Suite Entrances, Length Less Than Twice Width

Estimating Procedure

1. Use these figures to estimate medical, dental, psychiatric, optometry and similar professional buildings in which access to each office suite is through an exterior entrance. Buildings in this section have more plumbing fixtures per square foot of floor and smaller room sizes than general office buildings. Note also that buildings with more than 10,000 square feet are assumed to have lead shielded x-ray rooms.
2. Establish the building quality class by applying the information on page 126.
3. Compute the first floor area. This should include everything within the exterior walls and all insets outside the main walls but under the main roof.
4. If the first floor wall height is more or less than 10 feet, add or subtract from the first floor square foot cost below the appropriate amount from the Wall Height Adjustment Table on page 133.
5. Multiply the adjusted square foot cost by the first floor area.
6. Deduct, if appropriate, for common walls or no wall finish. Use the figures on page 133.
7. If there are second or higher floors, compute the square foot area on each floor. Locate the appropriate square foot cost from the table at the bottom of this page. Adjust this figure for a wall height more or less than 9 feet, using the figures on page 133. Multiply the adjusted cost by the square foot area on each floor. Use the figures on page 133 to deduct for common walls or no wall finish. Add the result to the cost from step 6 above.
8. Multiply the total cost by the location factor on page 7.
9. Add the cost of heating and air conditioning systems, elevators, fire sprinklers, exterior signs, paving and curbing, miscellaneous yard improvements, covered porches and garages. See page 201.

Medical-Dental Building, Class 2 & 3

First Story

Square Foot Area

Quality Class	1,000	1,500	2,000	2,500	3,000	4,000	5,000	7,500	10,000	15,000	20,000
Exceptional	81.70	78.52	76.71	75.51	74.64	73.44	72.64	71.43	70.72	69.89	69.41
1, Best	74.27	71.39	69.74	68.64	67.85	66.76	66.04	64.93	64.29	63.54	63.10
1 & 2	69.28	66.58	65.05	64.03	63.29	62.27	61.60	60.56	59.96	59.26	58.85
2, Good	64.20	61.71	60.28	59.34	58.65	57.71	57.08	56.13	55.57	54.92	54.54
2 & 3	59.27	56.96	55.65	54.77	54.14	53.28	52.70	51.81	51.30	50.70	50.35
3, Average	54.36	52.25	51.04	50.24	49.66	48.87	48.33	47.53	47.05	46.51	46.18
3 & 4	50.52	48.55	47.43	46.69	46.15	45.41	44.92	44.16	43.73	43.22	42.92
4, Low	46.69	44.87	43.84	43.15	42.65	41.97	41.51	40.82	40.41	39.94	39.67

Second and Higher Stories

Square Foot Area

Quality Class	1,000	1,500	2,000	2,500	3,000	4,000	5,000	7,500	10,000	15,000	20,000
Exceptional	72.83	71.00	69.92	69.19	68.65	67.91	67.40	66.62	66.15	65.61	65.28
1, Best	66.30	64.63	63.65	62.98	62.50	61.82	61.35	60.64	60.22	59.72	59.43
1 & 2	61.35	59.81	58.90	58.28	57.83	57.20	56.78	56.12	55.73	55.26	54.99
2, Good	58.56	57.08	56.21	55.63	55.20	54.60	54.19	53.56	53.19	52.75	52.49
2 & 3	53.79	52.43	51.64	51.10	50.70	50.15	49.78	49.20	48.85	48.45	48.21
3, Average	49.05	47.81	47.09	46.60	46.24	45.73	45.39	44.86	44.55	44.18	43.96
3 & 4	45.58	44.43	43.75	43.30	42.96	42.49	42.18	41.69	41.40	41.05	40.85
4, Low	42.18	41.12	40.50	40.07	39.76	39.33	39.04	38.58	38.32	38.00	37.81

Medical-Dental Buildings
Wood Frame
Exterior Suite Entrances, Length Between 2 and 4 Times Width

Estimating Procedure

1. Use these figures to estimate medical, dental, psychiatric, optometry and similar professional buildings in which access to each office suite is through an exterior entrance. Buildings in this section have more plumbing fixtures per square foot of floor and smaller room sizes than general office buildings. Note also that buildings with more than 10,000 square feet are assumed to have lead shielded x-ray rooms.
2. Establish the building quality class by applying the information on page 126.
3. Compute the first floor area. This should include everything within the exterior walls and all insets outside the main walls but under the main roof.
4. If the first floor wall height is more or less than 10 feet, add or subtract from the first floor square foot cost below the appropriate amount from the Wall Height Adjustment Table on page 133.
5. Multiply the adjusted square foot cost by the first floor area.
6. Deduct, if appropriate, for common walls or no wall finish. Use the figures on page 133.
7. If there are second or higher floors, compute the square foot area on each floor. Locate the appropriate square foot cost from the table at the bottom of this page. Adjust this figure for a wall height more or less than 9 feet, using the figures on page 133. Multiply the adjusted cost by the square foot area on each floor. Use the figures on page 133 to deduct for common walls or no wall finish. Add the result to the cost from step 6 above.
8. Multiply the total cost by the location factor on page 7.
9. Add the cost of heating and air conditioning systems, elevators, fire sprinklers, exterior signs, paving and curbing, miscellaneous yard improvements, covered porches and garages. See page 201.

Medical Dental Building, Class 1 & 2

First Story

Square Foot Area

Quality Class	1,000	1,500	2,000	2,500	3,000	4,000	5,000	7,500	10,000	15,000	20,000
Exceptional	85.13	81.12	78.86	77.37	76.30	74.84	73.87	72.41	71.56	70.58	70.01
1, Best	77.38	73.74	71.69	70.34	69.36	68.03	67.15	65.82	65.05	64.16	63.64
1 & 2	72.08	68.69	66.77	65.51	64.61	63.37	62.55	61.31	60.59	59.76	59.28
2, Good	66.76	63.62	61.85	60.68	59.84	58.70	57.94	56.79	56.13	55.36	54.91
2 & 3	61.61	58.71	57.07	55.99	55.22	54.16	53.46	52.40	51.79	51.08	50.67
3, Average	56.48	53.82	52.32	51.33	50.62	49.65	49.01	48.04	47.48	46.83	46.45
3 & 4	52.46	49.99	48.59	47.68	47.02	46.12	45.52	44.62	44.10	43.49	43.14
4, Low	48.47	46.19	44.90	44.05	43.44	42.61	42.06	41.23	40.74	40.19	39.86

Second and Higher Stories

Square Foot Area

Quality Class	1,000	1,500	2,000	2,500	3,000	4,000	5,000	7,500	10,000	15,000	20,000
Exceptional	75.20	72.81	71.43	70.52	69.85	68.94	68.33	67.40	66.85	66.21	65.84
1, Best	68.36	66.18	64.94	64.10	63.50	62.67	62.11	61.27	60.77	60.19	59.85
1 & 2	64.28	62.23	61.05	60.27	59.70	58.92	58.40	57.60	57.14	56.59	56.27
2, Good	60.44	58.51	57.41	56.67	56.14	55.41	54.92	54.17	53.73	53.22	52.92
2 & 3	55.30	53.54	52.53	51.86	51.37	50.70	50.25	49.56	49.16	48.69	48.42
3, Average	50.45	48.84	47.92	47.30	46.86	46.25	45.84	45.21	44.85	44.42	44.17
3 & 4	46.92	45.43	44.57	44.00	43.58	43.01	42.63	42.05	41.71	41.31	41.08
4, Low	43.30	41.92	41.13	40.60	40.22	39.69	39.34	38.81	38.49	38.13	37.91

Medical-Dental Buildings
Wood Frame
Exterior Suite Entrances, Length More Than 4 Times Width

Estimating Procedure

1. Use these figures to estimate medical, dental, psychiatric, optometry and similar professional buildings in which access to each office suite is through an exterior entrance. Buildings in this section have more plumbing fixtures per square foot of floor and smaller room sizes than general office buildings. Note also that buildings with more than 10,000 square feet are assumed to have lead shielded x-ray rooms.
2. Establish the building quality class by applying the information on page 126.
3. Compute the first floor area. This should include everything within the exterior walls and all insets outside the main walls but under the main roof.
4. If the first floor wall height is more or less than 10 feet, add or subtract from the first floor square foot cost below the appropriate amount from the Wall Height Adjustment Table on page 133.
5. Multiply the adjusted square foot cost by the first floor area.
6. Deduct, if appropriate, for common walls or no wall finish. Use the figures on page 133.
7. If there are second or higher floors, compute the square foot area on each floor. Locate the appropriate square foot cost from the table at the bottom of this page. Adjust this figure for a wall height more or less than 9 feet, using the figures on page 133. Multiply the adjusted cost by the square foot area on each floor. Use the figures on page 133 to deduct for common walls or no wall finish. Add the result to the cost from step 6 above.
8. Multiply the total cost by the location factor on page 7.
9. Add the cost of heating and air conditioning systems, elevators, fire sprinklers, exterior signs, paving and curbing, miscellaneous yard improvements, covered porches and garages. See page 201.

Medical-Dental Building, Class 1

First Story

Square Foot Area

Quality Class	1,000	1,500	2,000	2,500	3,000	4,000	5,000	7,500	10,000	15,000	20,000
Exceptional	89.00	84.07	81.31	79.50	78.20	76.43	75.26	73.51	72.49	71.33	70.65
1, Best	81.09	76.60	74.08	72.43	71.25	69.64	68.57	66.97	66.05	64.98	64.37
1 & 2	75.44	71.26	68.92	67.38	66.28	64.78	63.79	62.30	61.45	60.46	59.88
2, Good	69.79	65.92	63.76	62.34	61.32	59.93	59.01	57.64	56.84	55.93	55.40
2 & 3	64.37	60.81	58.81	57.50	56.56	55.28	54.43	53.16	52.43	51.59	51.09
3, Average	58.99	55.72	53.89	52.69	51.82	50.65	49.88	48.71	48.04	47.27	46.82
3 & 4	54.75	51.72	50.02	48.91	48.11	47.02	46.30	45.22	44.60	43.88	43.46
4, Low	50.57	47.76	46.19	45.16	44.43	43.42	42.76	41.76	41.18	40.52	40.14

Second and Higher Stories

Square Foot Area

Quality Class	1,000	1,500	2,000	2,500	3,000	4,000	5,000	7,500	10,000	15,000	20,000
Exceptional	77.73	74.66	72.94	71.81	71.00	69.89	69.16	68.06	67.43	66.70	66.27
1, Best	70.70	67.91	66.34	65.31	64.58	63.57	62.91	61.91	61.33	60.67	60.28
1 & 2	66.45	63.82	62.35	61.38	60.69	59.75	59.12	58.18	57.64	57.02	56.65
2, Good	62.21	59.75	58.37	57.47	56.82	55.94	55.35	54.47	53.96	53.38	53.04
2 & 3	57.16	54.90	53.63	52.80	52.21	51.40	50.86	50.05	49.58	49.05	48.73
3, Average	52.13	50.07	48.91	48.15	47.61	46.87	46.38	45.64	45.22	44.73	44.44
3 & 4	48.44	46.53	45.45	44.75	44.24	43.56	43.10	42.42	42.02	41.57	41.30
4, Low	44.76	42.99	42.00	41.34	40.88	40.24	39.82	39.19	38.82	38.40	38.16

Medical-Dental Buildings
Wood Frame

Interior Suite Entrances, Length Less Than Twice Width

Estimating Procedure

1. Use these figures to estimate medical, dental, psychiatric, optometry and similar professional buildings in which access to each office suite is through an interior corridor. Buildings in this section have more plumbing fixtures per square foot of floor and smaller room sizes than general office buildings. Note also that buildings with more than 10,000 square feet are assumed to have lead shielded x-ray rooms.
2. Establish the building quality class by applying the information on page 126.
3. Compute the first floor area. This should include everything within the exterior walls and all insets outside the main walls but under the main roof.
4. If the first floor wall height is more or less than 10 feet, add or subtract from the first floor square foot cost below the appropriate amount from the Wall Height Adjustment Table on page 133.
5. Multiply the adjusted square foot cost by the first floor area.
6. Deduct, if appropriate, for common walls or no wall finish. Use the figures on page 133.
7. If there are second or higher floors, compute the square foot area on each floor. Locate the appropriate square foot cost from the table at the bottom of this page. Adjust this figure for a wall height more or less than 9 feet, using the figures on page 133. Multiply the adjusted cost by the square foot area on each floor. Use the figures on page 133 to deduct for common walls or no wall finish. Add the result to the cost from step 6 above.
8. Multiply the total cost by the location factor on page 7.
9. Add the cost of heating and air conditioning systems, elevators, fire sprinklers, exterior signs, paving and curbing, miscellaneous yard improvements, covered porches and garages. See page 201.

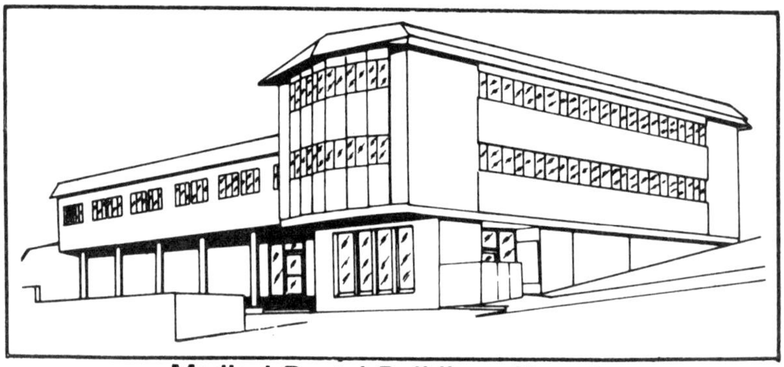

Medical-Dental Building, Class 3

First Story

Square Foot Area

Quality Class	2,000	2,500	3,000	4,000	5,000	7,500	10,000	15,000	20,000	30,000	40,000
Exceptional	70.59	69.43	68.60	67.44	66.67	65.50	64.81	64.01	63.54	62.99	62.66
1, Best	66.24	65.16	64.37	63.29	62.56	61.46	60.82	60.07	59.62	59.11	58.80
1 & 2	61.84	60.83	60.09	59.08	58.41	57.38	56.78	56.08	55.66	55.18	54.90
2, Good	57.48	56.54	55.86	54.92	54.29	53.33	52.77	52.12	51.74	51.29	51.03
2 & 3	53.12	52.25	51.62	50.75	50.17	49.29	48.77	48.17	47.82	47.40	47.16
3, Average	48.78	47.98	47.40	46.60	46.07	45.26	44.78	44.23	43.90	43.52	43.30
3 & 4	45.61	44.86	44.32	43.57	43.08	42.32	41.87	41.36	41.05	40.70	40.49
4, Low	42.45	41.75	41.25	40.55	40.09	39.38	38.97	38.49	38.21	37.87	37.68

Second and Higher Stories

Square Foot Area

Quality Class	2,000	2,500	3,000	4,000	5,000	7,500	10,000	15,000	20,000	30,000	40,000
Exceptional	64.72	64.01	63.50	62.79	62.31	61.58	61.15	60.65	60.35	60.00	59.79
1, Best	60.76	60.10	59.62	58.95	58.50	57.82	57.41	56.94	56.66	56.33	56.14
1 & 2	57.08	56.45	56.00	55.37	54.95	54.31	53.93	53.48	53.22	52.91	52.73
2, Good	53.40	52.82	52.39	51.81	51.41	50.81	50.45	50.04	49.79	49.50	49.33
2 & 3	49.05	48.52	48.13	47.59	47.23	46.67	46.35	45.97	45.74	45.48	45.32
3, Average	44.71	44.22	43.87	43.38	43.05	42.54	42.24	41.90	41.69	41.45	41.31
3 & 4	41.79	41.33	41.00	40.54	40.23	39.76	39.48	39.16	38.97	38.74	38.61
4, Low	38.84	38.42	38.11	37.68	37.40	36.96	36.70	36.40	36.22	36.01	35.89

Medical-Dental Buildings
Wood Frame
Interior Suite Entrances, Length Between 2 and 4 Times Width

Estimating Procedure

1. Use these figures to estimate medical, dental, psychiatric, optometry and similar professional buildings in which access to each office suite is through an interior corridor. Buildings in this section have more plumbing fixtures per square foot of floor and smaller room sizes than general office buildings. Note also that buildings with more than 10,000 square feet are assumed to have lead shielded x-ray rooms.
2. Establish the building quality class by applying the information on page 126.
3. Compute the first floor area. This should include everything within the exterior walls and all insets outside the main walls but under the main roof.
4. If the first floor wall height is more or less than 10 feet, add or subtract from the first floor square foot cost below the appropriate amount from the Wall Height Adjustment Table on page 133.
5. Multiply the adjusted square foot cost by the first floor area.
6. Deduct, if appropriate, for common walls or no wall finish. Use the figures on page 133.
7. If there are second or higher floors, compute the square foot area on each floor. Locate the appropriate square foot cost from the table at the bottom of this page. Adjust this figure for a wall height more or less than 9 feet, using the figures on page 133. Multiply the adjusted cost by the square foot area on each floor. Use the figures on page 133 to deduct for common walls or no wall finish. Add the result to the cost from step 6 above.
8. Multiply the total cost by the location factor on page 7.
9. Add the cost of heating and air conditioning systems, elevators, fire sprinklers, exterior signs, paving and curbing, miscellaneous yard improvements, covered porches and garages. See page 201.

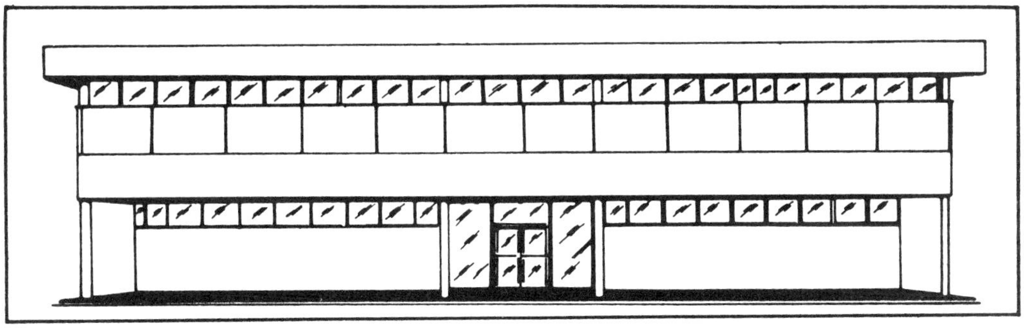

Medical-Dental Building, Class 1 & 2

First Story

Square Foot Area

Quality Class	2,000	2,500	3,000	4,000	5,000	7,500	10,000	15,000	20,000	30,000	40,000
Exceptional	72.36	70.96	69.95	68.56	67.64	66.24	65.43	64.48	63.93	63.29	62.91
1, Best	68.07	66.75	65.80	64.49	63.63	62.31	61.55	60.66	60.14	59.53	59.18
1 & 2	63.51	62.28	61.39	60.17	59.36	58.14	57.42	56.60	56.11	55.55	55.21
2, Good	58.98	57.83	57.01	55.88	55.13	53.99	53.33	52.56	52.11	51.58	51.27
2 & 3	54.48	53.42	52.66	51.62	50.92	49.87	49.26	48.55	48.13	47.65	47.36
3, Average	50.00	49.04	48.34	47.38	46.74	45.78	45.21	44.56	44.18	43.74	43.47
3 & 4	46.70	45.80	45.14	44.25	43.65	42.75	42.23	41.62	41.26	40.85	40.60
4, Low	43.49	42.65	42.04	41.21	40.66	39.82	39.33	38.76	38.43	38.04	37.81

Second and Higher Stories

Square Foot Area

Quality Class	2,000	2,500	3,000	4,000	5,000	7,500	10,000	15,000	20,000	30,000	40,000
Exceptional	66.13	65.16	64.47	63.54	62.93	62.03	61.51	60.93	60.59	60.20	59.97
1, Best	62.09	61.18	60.53	59.66	59.09	58.24	57.76	57.21	56.89	56.53	56.32
1 & 2	58.31	57.45	56.84	56.02	55.49	54.69	54.24	53.72	53.42	53.08	52.88
2, Good	54.53	53.73	53.16	52.39	51.89	51.15	50.72	50.24	49.96	49.64	49.46
2 & 3	50.33	49.59	49.06	48.36	47.89	47.21	46.82	46.37	46.11	45.82	45.65
3, Average	45.67	45.00	44.52	43.88	43.46	42.84	42.48	42.08	41.85	41.58	41.42
3 & 4	42.68	42.06	41.61	41.01	40.62	40.04	39.70	39.33	39.11	38.86	38.71
4, Low	39.68	39.10	38.69	38.13	37.76	37.22	36.91	36.56	36.36	36.13	35.99

Medical-Dental Buildings
Wood Frame
Interior Suite Entrances, Length More Than 4 Times Width

Estimating Procedure

1. Use these figures to estimate medical, dental, psychiatric, optometry and similar professional buildings in which access to each office suite is through an interior corridor. Buildings in this section have more plumbing fixtures per square foot of floor and smaller room sizes than general office buildings. Note also that buildings with more than 10,000 square feet are assumed to have lead shielded x-ray rooms.
2. Establish the building quality class by applying the information on page 126.
3. Compute the first floor area. This should include everything within the exterior walls and all insets outside the main walls but under the main roof.
4. If the first floor wall height is more or less than 10 feet, add or subtract from the first floor square foot cost below the appropriate amount from the Wall Height Adjustment Table on page 133.
5. Multiply the adjusted square foot cost by the first floor area.
6. Deduct, if appropriate, for common walls or no wall finish. Use the figures on page 133.
7. If there are second or higher floors, compute the square foot area on each floor. Locate the appropriate square foot cost from the table at the bottom of this page. Adjust this figure for a wall height more or less than 9 feet, using the figures on page 133. Multiply the adjusted cost by the square foot area on each floor. Use the figures on page 133 to deduct for common walls or no wall finish. Add the result to the cost from step 6 above.
8. Multiply the total cost by the location factor on page 7.
9. Add the cost of heating and air conditioning systems, elevators, fire sprinklers, exterior signs, paving and curbing, miscellaneous yard improvements, covered porches and garages. See page 201.

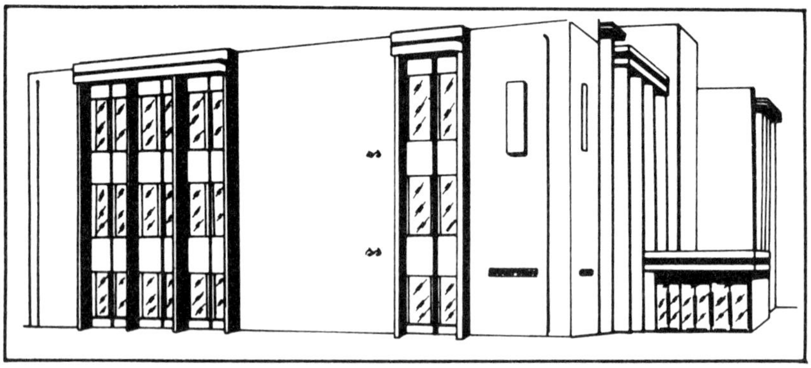

Medical-Dental Building, Class 2

First Story

Square Foot Area

Quality Class	2,000	2,500	3,000	4,000	5,000	7,500	10,000	15,000	20,000	30,000	40,000
Exceptional	75.16	73.50	72.27	70.56	69.40	67.60	66.53	65.27	64.52	63.63	63.10
1, Best	70.57	69.00	67.86	66.25	65.16	63.47	62.47	61.28	60.58	59.74	59.25
1 & 2	65.80	64.34	63.27	61.77	60.75	59.18	58.24	57.14	56.48	55.70	55.24
2, Good	61.03	59.68	58.69	57.30	56.35	54.89	54.03	53.00	52.39	51.67	51.24
2 & 3	56.36	55.11	54.19	52.91	52.04	50.69	49.89	48.94	48.38	47.71	47.32
3, Average	51.72	50.57	49.73	48.56	47.76	46.52	45.78	44.91	44.40	43.79	43.42
3 & 4	49.62	48.52	47.72	46.59	45.82	44.63	43.93	43.09	42.60	42.01	41.66
4, Low	44.92	43.93	43.19	42.17	41.48	40.40	39.76	39.01	38.56	38.03	37.71

Second and Higher Stories

Square Foot Area

Quality Class	2,000	2,500	3,000	4,000	5,000	7,500	10,000	15,000	20,000	30,000	40,000
Exceptional	67.49	66.40	65.61	64.54	63.83	62.76	62.15	61.44	61.03	60.55	60.27
1, Best	63.37	62.34	61.60	60.60	59.93	58.93	58.36	57.69	57.30	56.85	56.59
1 & 2	59.49	58.52	57.83	56.88	56.26	55.32	54.78	54.15	53.79	53.37	53.12
2, Good	55.63	54.73	54.08	53.20	52.61	51.74	51.23	50.64	50.31	49.91	49.68
2 & 3	51.12	50.29	49.69	48.88	48.34	47.54	47.07	46.53	46.22	45.86	45.65
3, Average	46.60	45.84	45.30	44.56	44.07	43.33	42.91	42.42	42.13	41.80	41.61
3 & 4	43.54	42.83	42.32	41.63	41.18	40.49	40.09	39.64	39.37	39.06	38.88
4, Low	40.50	39.84	39.37	38.73	38.30	37.66	37.29	36.87	36.62	36.33	36.16

Medical-Dental Buildings
Wood Frame
Wall Height Adjustment

Wall Height Adjustment

The square foot costs for medical and dental buildings are based on the wall heights of 10 feet for first floors and 9 feet for higher floors. The first floor height is the distance from the bottom of the floor slab or joists to the top of the roof slab or ceiling joists. Second and higher floors are measured from the top of the floor slab or floor joists to the top of the roof slab or ceiling joists. Add or subtract the amount listed in this table to the square foot of floor cost for each foot of wall height more or less than 10 feet, if adjusting for first floor, and 9 feet, if adjusting for upper floors.

	Square Foot Area										
Quality Class	1,000	1,500	2,000	3,000	4,000	5,000	7,500	10,000	15,000	20,000	40,000
1, Best	.78	.62	.53	.42	.36	.32	.26	.22	.18	.16	.11
2, Good	.66	.52	.44	.35	.30	.27	.22	.19	.15	.13	.09
3, Average	.56	.44	.37	.30	.26	.23	.18	.16	.13	.11	.08
4, Low	.49	.38	.33	.26	.22	.20	.16	.14	.11	.10	.07

Perimeter Wall Adjustment

A common wall exists when two buildings share one wall. Adjust for common walls by deducting the linear foot costs below from the total structure cost. In some structures, one or more walls is not owned at all. In this case, deduct the ''No Ownership'' cost per linear foot of wall not owned. If a wall has no exterior finish, deduct the ''Lack or Exterior Finish'' cost.

First Story

Class	For A Common Wall, Deduct Per Linear Foot	For No Wall Ownership, Deduct Per Linear Foot	For Lack Of Exterior Finish, Deduct Per Linear Foot
1	$101.00	$117.00	$90.00
2	76.40	94.00	68.80
3	62.00	75.30	48.40
4	48.40	62.20	33.00

Second and Higher Stories

Class	For A Common Wall, Deduct Per Linear Foot	For No Wall Ownership, Deduct Per Linear Foot	For Lack Of Exterior Finish Deduct Per Linear Foot
1	$63.50	$70.00	$53.50
2	49.50	59.70	47.30
3	47.30	50.00	41.30
4	41.00	47.30	32.50

Convalescent Hospitals Masonry or Concrete Quality Classification

		Class 1 Best Quality	Class 2 Good Quality	Class 3 Average Quality	Class 4 Low Quality
Foundation		Reinforced concrete.	Reinforced concrete.	Reinforced concrete.	Reinforced concrete.
Floor Structure		Reinforced concrete.	Reinforced concrete.	Reinforced concrete.	Reinforced concrete.
Roof		Standard wood frame.	Standard wood frame.	Standard wood frame.	Standard wood frame.
Exterior Finish	**Walls**	Furred out gypsum wallboard interior, colored block exterior.	Furred out gypsum wallboard interior, colored block exterior.	Colored block with stucco interior and painted exterior.	Plain 8" concrete block painted on the inside.
	Front	Brick or stone veneer and plate glass in aluminum frame.	Stone or brick veneer. Small amount of plate glass in aluminum frame.	Rustic siding or some brick or stone veneer.	Stucco and small amount siding.
	Windows	Good aluminum sliding glass doors or anodized aluminum casement windows.	Good aluminum sliding glass doors with aluminum jalousie window for vent.	Average aluminum sliding type. 6' sliding glass door in bedrooms.	Low cost aluminum sliding type.
	Doors	Anodized aluminum entry doors.	Aluminum and glass entry doors.	Good hardwood slab door with side lights or aluminum and glass entry door.	Wood slab door.
	Roof	Composition with large colored rock or tile. 4' to 6' sealed overhang.	Composition with large colored rock or asbestos shingles. 3' to 4' sealed overhang.	Composition and small colored rock. 4' exposed or 3' sealed overhang.	Composition and pea gravel or composition shingles. 2' exposed overhang.
Interior Finish	**Floors**	Vinyl asbestos tile. Good carpet in entry and day room. Quarry tile or epoxy type floor in kitchen.	Vinyl asbestos tile. Good carpet in day room. May have coved sheet vinyl in kitchen.	Vinyl asbestos tile. Day room may have average grade carpet.	Low cost tile. Vinyl asbestos tile in kitchen. Concrete in storage areas.
	Walls	Gypsum wallboard and enamel. Large amount of plastic wall cover. Some wood paneling or glass walls in entry and day rooms. 4' ceramic tile wainscot in kitchen. Gypsum wallboard and paint in storage areas.	Gypsum wallboard and enamel. Gypsum wallboard and some plastic wall cover or wood paneling in entry, bedrooms, dining room, day room, and kitchen. Gypsum wallboard and paint in storage areas.	Gypsum wallboard and enamel. Some plastic wall cover or wood paneling in entry and day room. Gypsum wallboard and paint in storage areas.	Gypsum wallboard and texture. Gypsum wallboard and enamel in kitchen. Unfinished gypsum wallboard in storage areas.
	Ceilings	Acoustical tile. Washable acoustical tile in kitchen. Gypsum wallboard and paint in storage areas.	Gypsum wallboard and enamel. Acoustical tile in corridor. Open beam ceilings with wood decking in entry and day room. Gypsum wallboard and paint in storage areas.	Gypsum wallboard and enamel or acoustical texture. Open beams with wood decking. Gypsum wallboard and paint in storage areas.	Gypsum wallboard and texture. Gypsum wallboard and acoustical texture in entry. Gypsum wallboard and enamel in kitchen. Unfinished gypsum wallboard in storage areas.
Bath Finish Detail	**Main Bath and Showers**	Ceramic tile floors. Ceramic tile walls. Gypsum wallboard and enamel ceilings.	Ceramic tile floors. 6' ceramic tile wainscot. Gypsum wallboard and enamel walls and ceilings.	Ceramic tile floors. 4' ceramic tile wainscot. Gypsum wallboard and enamel walls and ceilings.	Vinyl asbestos tile floors. Gypsum wallboard and enamel walls and ceilings.
	Toilet Rooms	Sheet vinyl or ceramic tile floors. 4" ceramic tile wainscot. Gypsum wallboard and enamel walls and ceilings.	Sheet vinyl floors. 4' plastic or ceramic tile wainscot.	Vinyl asbestos tile or linoleum floors. 4' plastic or marlite wainscot.	Vinyl asbestos tile floors. Gypsum wallboard and enamel walls and ceilings.

Square foot costs include the following components: Foundations as required for normal soil conditions. Floor, wall, and roof structures. Interior floor wall and ceiling finishes. Exterior wall finish and roof cover. Interior partitions. Cabinets, doors and windows. Basic electrical systems and all plumbing. Permits and fees. Contractor's mark-up. Cost of nurses' built-in station desks and cabinets. Nurses' call system. Emergency lighting system. Cubical curtain tracks. Kitchen range hood.

The in-place cost of these extra components should be added to the basic building cost to arrive at the total structure cost. See the section "Additional Costs for Commercial Buildings" on page 201. Heating and air conditioning systems. Elevators. Fire sprinklers. Dumbwaiters. Fireplaces. **Square foot costs do not include the following items:** Drapes. Cubical curtains. Incinerators. Laundry or kitchen equipment, other than the range hood.

Convalescent Hospitals
Masonry or Concrete

Estimating Procedure

1. Use these figures to estimate institutions such as nursing or rest homes which have facilities limited to bed care for people unable to care for themselves.
2. Establish the building quality class by applying the information on page 134.
3. Compute the floor area. This should include everything within the exterior walls and all insets outside the main walls but under the main roof. Areas such as storage rooms which have a substantially inferior finish or any area with a slightly inferior finish and which does not perform a function related to the hospital operation should be estimated at ½ to ¾ of the square foot costs listed.
4. Estimate second stories and basements at 100% of the first floor cost if they are of the same general quality. A storage area of the same or slightly inferior quality located in a basement, second floor or separate building should be estimated at 100% of the building square foot cost.
5. Multiply the appropriate square foot cost below by the floor area.
6. Multiply the total cost by the location factor on page 7.
7. Add the cost of heating and air conditioning systems, fire sprinklers, drapes, therapy facilities, cubical curtains, incinerators, elevators, dumbwaiters, fireplaces, laundry equipment and kitchen equipment other than a range hood. See the section beginning on page 201.

Convalescent Hospital, Class 2

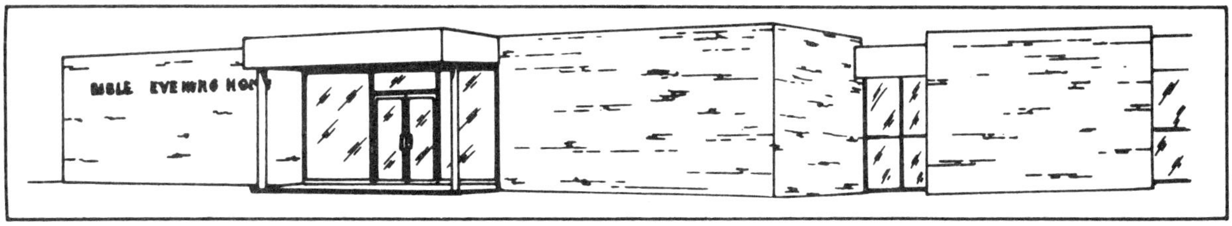

Convalescent Hospital, Class 2

	Square Foot Area											
Quality Class	5,000	6,000	7,000	8,000	9,000	10,000	12,500	15,000	17,500	20,000	25,000	30,000
Exceptional	74.31	71.96	70.29	69.04	68.07	67.30	65.91	64.99	64.34	63.85	63.17	62.73
1, Best	69.46	67.26	65.70	64.53	63.62	62.90	61.61	60.75	60.14	59.68	59.05	58.63
1 & 2	64.78	62.73	61.28	60.19	59.34	58.67	57.46	56.66	56.09	55.67	55.07	54.68
2, Good	59.88	57.99	56.64	55.63	54.85	54.23	53.11	52.37	51.84	51.45	50.91	50.55
2 & 3	54.97	53.24	52.00	51.08	50.36	49.79	48.76	48.08	47.60	47.24	46.74	46.41
3, Average	50.82	49.21	48.07	47.21	46.55	46.02	45.07	44.45	44.00	43.67	43.20	42.90
3 & 4	46.29	44.83	43.79	43.01	42.40	41.92	41.06	40.49	40.08	39.78	39.36	39.08
4, Low	41.92	40.60	39.65	38.95	38.40	37.96	37.18	36.66	36.30	36.02	35.64	35.39

Convalescent Hospitals Wood Frame Quality Classification

	Class 1 Best Quality	Class 2 Good Quality	Class 3 Average Quality	Class 4 Low Quality
Foundation	Reinforced concrete.	Reinforced concrete.	Reinforced concrete.	Reinforced concrete.
Floor Structure	Reinforced concrete.	Reinforced concrete.	Reinforced concrete.	Reinforced concrete.
Wall Structure	Standard wood frame.	Standard wood frame.	Standard wood frame.	Standard wood frame.
Roof	Standard wood frame.	Standard wood frame.	Standard wood frame.	Standard wood frame.
Exterior Finish Walls	Wood siding or stucco with extensive trim.	Stucco or wood siding.	Stucco or good plywood siding.	Stucco.
Front	Brick or stone veneer and plate glass in aluminum frame.	Stone or brick veneer. Small amount of plate glass in aluminum frame.	Rustic siding or some brick or stone veneer.	Stucco and small amount siding.
Windows	Good aluminum sliding glass doors or anodized aluminum casement windows.	Good aluminum sliding glass doors with aluminum jalousie window for vent.	Average aluminum sliding type. 6' sliding glass door in bedrooms.	Low cost aluminum sliding type.
Doors	Anodized aluminum entry doors.	Aluminum and glass entry doors.	Good hardwood slab door with side lights or aluminum and glass entry door.	Wood slab door.
Roof	Composition with large colored rock or tile. 4' to 6' sealed overhang.	Composition with large colored rock or asbestos shingles. 3' to 4' sealed overhang.	Composition and small colored rock. 4' exposed or 3' sealed overhang.	Composition and pea gravel or composition shingles. 2' exposed overhang.
Interior Finish Floors	Vinyl asbestos tile. Good carpet in entry and day room. Quarry tile or epoxy type floor in kitchen.	Vinyl asbestos tile. Good carpet in day room. May have coved vinyl kitchen.	Vinyl asbestos tile. Day room may have average grade carpet.	Low cost tile. Better vinyl asbestos tile in kitchen. Concrete in storage areas.
Walls	Gypsum wallboard and enamel. Large amount of plastic wall cover. Some wood paneling or glass walls in entry and day rooms. 4' ceramic tile wainscot in kitchen. Gypsum wallboard and paint in storage areas.	Gypsum wallboard and enamel. Gypsum wallboard and some plastic wall cover or wood paneling in entry, bedrooms, dining room, day room, and kitchen. Gypsum wallboard and paint in storage areas.	Gypsum wallboard and enamel. Some plastic wall cover or wood paneling in entry and day room. Gypsum wallboard and paint in storage areas.	Gypsum wallboard and texture. Gypsum wallboard and enamel in kitchen. Unfinished gypsum wallboard in storage areas.
Ceilings	Acoustical tile. Washable acoustical tile in kitchen. Gypsum wallboard and paint in storage areas.	Gypsum wallboard and enamel. Acoustical tile in corridor. Open beam ceilings with wood decking in entry and day room. Gypsum wallboard and paint in storage areas.	Gypsum wallboard and enamel or acoustical texture. Open beams with wood decking. Gypsum wallboard and paint in storage areas.	Gypsum wallboard and texture. Gypsum wallboard and acoustical texture in entry. Gypsum wallboard and enamel in kitchen. Unfinished gypsum wallboard in storage areas.
Bath Finish Detail Main Bath and Showers	Ceramic tile floors. Ceramic tile walls. Gypsum wallboard and enamel ceilings.	Ceramic tile floors. 6' ceramic tile wainscot. Gypsum wallboard and enamel walls and ceilings.	Ceramic tile floors. 4' ceramic tile wainscot. Gypsum wallboard and enamel walls and ceilings.	Vinyl asbestos tile floors. Gypsum wallboard and enamel walls and ceilings.
Toilet Rooms	Vinyl or ceramic tile floors. 4' ceramic tile wainscot. Gypsum wallboard and enamel walls and ceilings.	Vinyl floors. 4' plastic or ceramic tile wainscot.	Vinyl asbestos tile or linoleum floors. 4' plastic wainscot.	Vinyl asbestos tile floors. Gypsum wallboard and enamel walls and ceilings.

Square foot costs include the following components: Foundations as required for normal soil conditions. Floor, wall, and roof structures. Interior floor, wall, and ceiling finishes. Exterior wall finish and roof cover. Interior partitions. Cabinets, doors, and windows. Basic electrical systems and plumbing systems. Permits and fees. Contractor's mark-up. Cost of nurses' built-in station desks and cabinets. Nurses' call system. Emergency lighting system. Cubical curtain tracks. Kitchen range hood.

The in-place cost of these extra components should be added to the basic building cost to arrive at the total structure cost. See the section ''Additional Costs for Commercial Buildings'' on page 201. Heating and air conditioning systems. Elevators. Fire sprinklers. Dumbwaiters. Fireplaces. **Square foot costs do not include the following items:** Drapes. Cubical curtains. Incinerators. Laundry or kitchen equipment other than the range hood.

Convalescent Hospitals
Wood Frame

Estimating Procedure

1. Use these figures to estimate institutions such as nursing or rest homes which have facilities limited to bed care for people who are unable to care for themselves.
2. Establish the building quality class by applying the information on page 136.
3. Compute the floor area. This should include everything within the exterior walls and all insets outside the main walls but under the main roof. Areas such as storage rooms which have a substantially inferior finish or any area with a slightly inferior finish which does not perform a function related to the hospital operation should be estimated at ½ to ¾ of the square foot costs listed.
4. Estimate second stories and basements at 100% of the first floor cost if they are of the same general quality. A storage area of the same or slightly inferior quality located in a basement, second floor or separate building should be estimated at 100% of the building square foot cost.
5. Multiply the appropriate square foot cost below by the floor area.
6. Multiply the total cost by the location factor on page 7.
7. Add the cost of heating and air conditioning systems, fire sprinklers, drapes, therapy facilities, cubical curtains, incinerators, elevators, dumbwaiters, fireplaces, laundry equipment and kitchen equipment other than a range hood. See the section beginning on page 201.

Convalescent Hospital, Class 4

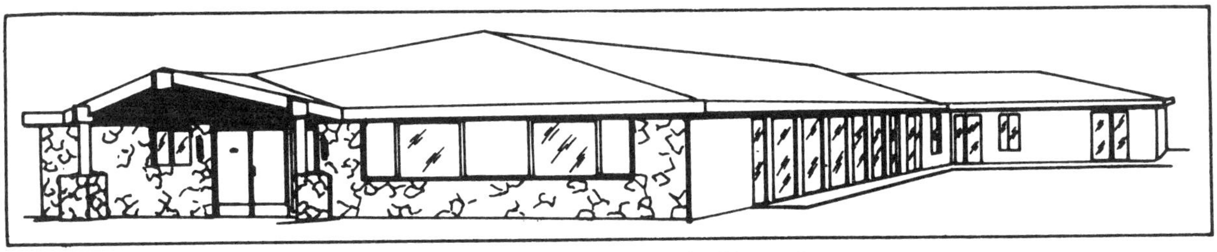

Convalescent Hospital, Class 3

	Square Foot Area											
Quality Class	5,000	6,000	7,000	8,000	9,000	10,000	12,500	15,000	17,500	20,000	25,000	30,000
Exceptional	63.78	61.77	60.32	59.22	58.36	57.66	56.40	55.53	54.91	54.44	53.76	53.30
1, Best	59.74	57.86	56.50	55.47	54.66	54.01	52.82	52.01	51.43	50.99	50.35	49.92
1 & 2	55.55	53.80	52.53	51.58	50.82	50.22	49.11	48.36	47.82	47.41	46.82	46.42
2, Good	51.47	49.85	48.68	47.79	47.10	46.53	45.51	44.82	44.31	43.93	43.38	43.01
2 & 3	47.79	46.29	45.20	44.38	43.73	43.21	42.26	41.61	41.15	40.79	40.28	39.94
3, Average	43.70	42.32	41.33	40.57	39.98	39.51	38.64	38.05	37.62	37.29	36.83	36.52
3 & 4	39.09	37.86	36.97	36.29	35.77	35.34	34.56	34.03	33.65	33.36	32.95	32.66
4, Low	36.22	35.08	34.25	33.63	33.14	32.74	32.02	31.54	31.18	30.91	30.53	30.27

Funeral Homes
Quality Classification

	Class 1 Best Quality	Class 2 Good Quality	Class 3 High Average Quality	Class 4 Low Average Quality	Class 4 Low Quality
Foundations	Reinforced concrete.	Reinforced concrete.	Reinforced concrete.	Reinforced concrete.	Reinforced concrete.
Floor Structure	Standard wood frame or reinforced concrete.	Standard wood frame or reinforced concrete.	Standard wood frame or reinforced concrete.	Standard wood frame or reinforced concrete.	Standard wood frame or reinforced concrete.
Wall Structure *	Standard wood frame.	Standard wood frame.	Standard wood frame.	Standard wood frame.	Standard wood frame.
Roof	Standard wood frame.	Standard wood frame.	Standard wood frame.	Standard wood frame.	Standard wood frame.
Exterior Finish Walls	Brick or stone veneer.	Good wood siding, brick or stone veneer.	Wood siding or stucco with extensive wood trim.	Stucco or wood siding.	Stucco.
Front	Brick or stone veneer. Some tinted plate glass in bronze frames.	Brick or stone veneer. Some tinted plate glass in anodized aluminum frames.	Brick or stone veneer. Some tinted plate glass in anodized aluminum frames.	Stone or brick veneer. Some tinted crystal or plate glass in wood frame.	Wood siding or stucco with some brick veneer. Some colored plastic windows with wood frame (bottle glass type).
Windows	Good anodized aluminum windows. Extensive stained glass in chapel.	Good aluminum sliding type. Stained glass in chapel.	Good aluminum sliding type. Leaded glass in chapel.	Good aluminum sliding type. Mock leaded glass in chapel.	Average aluminum sliding type. Colored plastic in chapel.
Doors	Elaborate hardwood entry doors.	Custom-built hardwood entry doors.	Custom hardwood or anodized aluminum and glass.	Good hardwood or anodized aluminum and glass doors.	Good hardwood slab or aluminum and glass entry doors.
Roof Cover	Slate, 3' sealed overhang.	Bar tile, 3' sealed overhang.	Heavy shake, 3' sealed overhang.	Composition and large colored rock, asbestos shingles or medium shake, 3' exposed overhang.	Composition and small colored rock or good composition shingles, 2' exposed overhang.
Interior Finish Floors	Excellent carpet. Good linoleum in kitchen.	Excellent carpet. Good carpet in living quarters. Good linoleum in kitchen.	Good carpet. Good linoleum in kitchen.	Good carpet. Average carpet in living quarters. Linoleum in kitchen.	Average carpet. Vinyl asbestos tile in hallways and living quarters.
Walls	Hardwood paneling, extensive brick or stone work in lobby and chapel. Gypsum wallboard and texture in living quarters.	Good hardwood veneer with brick or stone work in lobby and chapel. Gypsum wallboard and texture in living quarters.	Gypsum wallboard and embossed wallpaper or fabric textured wallcover. Hardwood in chapel. Gypsum wallboard and texture in living quarters.	Gypsum wallboard and wallpaper. Hardwood veneer in chapel.	Gypsum wallboard and texture. Some hardwood veneer in chapel.
Ceilings	Acoustical plaster. Exposed ornamental beams with ornamental hardwood decking or ornamental acoustical plaster in chapel.	Acoustical plaster. Exposed ornamental beams with ornamental wood decking or acoustical plaster in chapel.	Gypsum wallboard and heavy acoustical texture. Exposed architectural glu lams with redwood deck in chapel. Gypsum wallboard and texture in living quarters.	Gypsum wallboard and acoustical texture. Exposed wood beam with wood deck in chapel.	Gypsum wallboard and acoustical texture. Exposed beam with wood deck in chapel.
Special Lighting Fixtures	Two extensive chandeliers in lobby (4' to 6' diameter). Recessed and spotlighting in chapel.	Two expensive chandeliers in lobby (4' to 6' diameter). Recessed and spotlighting in chapel.	Two good chandeliers in lobby (4' diameter). Recessed and spotlighting in chapel.	Two average chandeliers in lobby (3' diameter).	One average chandelier in lobby.
Bath Detail	Ceramic tile floors and walls. Gypsum wallboard and enamel ceilings. Excellent fixtures.	Ceramic tile floors; 4' ceramic tile wainscot. Gypsum wallboard and enamel ceilings. Excellent fixtures.	Ceramic tile floors. Gypsum wallboard and enamel walls and ceilings. Good fixtures.	Good tile floors. Gypsum wallboard and enamel walls and ceilings. Average fixtures.	Linoleum floors. Gypsum wallboard and enamel walls and ceilings. Average fixtures.

*Add 9% to the square foot cost if the exterior walls are concrete block or brick. Add 13% if the exterior walls are concrete.

Square foot costs include the cost of the following components: Foundations as required for normal soil conditions. Floor, wall, and roof structures. Interior floor, wall and ceiling finishes. Exterior wall finish and roof cover. Interior partitions. Cabinets, doors and windows. Basic electrical system and plumbing system. Special lighting fixtures. Bath detail. Permits and fees. Contractor's mark-up.

Funeral Homes

Estimating Procedure

1. Use these figures to estimate mortuaries and buildings designed for care of the dead and funeral services.
2. Establish the building quality class by applying the information on page 138.
3. Compute the first floor area under the main roof and inside the exterior walls. Include lobby, chapel and living sections, but exclude garages, whether internal, external or attached.
4. Add to the first floor area between ⅓ and ¼ of the actual garage area. Use the higher value for better quality garages.
5. Add to the first floor area all finished basement and second floor area if it is similar in quality to the first floor area. Inferior quality second floor area should be reclassified as to quality and calculated separately. Unfinished funeral home basements should be estimated from the figures on page 202.
6. Finished space within the area formed by the roof but with an exterior wall height less than 7 feet (half story area) should be included at ⅓ to ½ of the actual floor area. Exclude half story area when selecting a square foot cost from the table.
7. Multiply the sum from steps 3 to 6 above by the appropriate square foot cost below.
8. Add 8% if the exterior walls are brick or block and 12% if the exterior walls are concrete. Add 2% if the building has more than 16 corners or if the length is more than twice the width.
9. Multiply the building cost by the location factor listed on page 7.
10. Add the cost of heating and cooling systems, elevators, fireplaces, porches, pews, drapes, signs and yard improvement. See the section beginning on page 201.

Funeral Home, Class 5

	Square Foot Area											
Quality Class	2,000	2,500	3,000	4,000	5,000	7,500	10,000	12,500	15,000	20,000	25,000	30,000
1, Best	--	--	--	--	55.85	53.83	52.74	52.06	51.58	50.94	50.54	50.25
1 & 2	--	--	--	--	51.70	49.82	48.82	48.19	47.74	47.16	46.78	46.52
2, Good	--	--	--	--	48.01	46.27	45.34	44.75	44.34	43.79	43.44	43.20
2 & 3	--	--	49.64	46.51	44.81	42.33	42.14	41.81	41.66	41.58	41.62	41.69
3, High Average	--	--	46.19	43.27	41.69	39.39	39.21	38.90	38.76	38.69	38.72	38.79
3 & 4	47.06	44.54	42.86	40.73	39.44	37.70	36.82	36.28	35.92	35.46	35.18	34.99
4, Low Average	44.03	41.68	40.10	38.11	36.90	35.28	34.45	33.95	33.61	33.18	32.91	32.73
4 & 5	41.08	38.88	37.41	35.56	34.43	32.91	32.14	31.67	31.36	30.95	30.71	30.54
5, Low	37.90	35.87	34.51	32.80	31.76	30.36	29.65	29.22	28.93	28.55	28.33	28.17

Self Service Restaurants Quality Classification

	Class 1 Best Quality	Class 2 Good Quality	Class 3 Average Quality	Class 4 Low Quality
Foundation	Reinforced concrete.	Reinforced concrete.	Reinforced concrete.	Reinforced concrete.
Floor Structure	Reinforced concrete or standard wood frame.	Reinforced concrete or standard wood frame.	Reinforced concrete or standard wood frame.	Reinforced concrete or standard wood frame.
Walls, Wood Frame Const.	Standard wood frame.	Standard wood frame.	Standard wood frame.	Standard wood frame.
Walls, Masonry or Concrete Construction	8" reinforced decorative block or brick.	8" reinforced decorative concrete block or brick.	8" reinforced concrete block.	8" reinforced concrete block.
Roof Structure	Complex angular structure.	Standard wood frame structure, low or medium pitch.	Standard wood frame, low pitch shed or gable type.	Standard wood frame, low pitch shed or gable type.
Roof Cover	Concrete tile, mission tile, or heavy shake.	Composition tar and gravel or medium pitch.	Composition tar and gravel.	Composition or composition shingle.
Floor Finish	Terrazzo in dining areas. Quarry tile in kitchen.	Vinyl tile in dining area. Terrazzo or quarry tile in kitchen.	Vinyl asbestos tile in dining area. Grease proof tile in kitchen area.	Vinyl asbestos tile in dining area. Grease proof tile in kitchen area.
Walls, Exterior Finish	Very good wood siding, ornamental brick or natural stone veneer. Minimum 60% of front and side walls has plate glass in heavy metal frames or rabbeted timbers.	Good wood siding or plywood; ornamental brick or concrete block; good stucco with color finish and with 2' to 4' brick, or natural stone veneer bulkheads. 50% to 75% of front and side walls have plate glass set in good wood or metal frames.	Stucco or average wood siding with 2' to 4' brick or stone veneer bulkheads. Plate glass front from waist high to roof line running along service front. Glassed area from 40% to 70% of front and side walls.	Stucco, low cost wood siding or plywood. 4' to 6' fixed glass waist high with sliding aluminum window opening running along service front. Glassed area from 30% to 40% of front and side walls.
Interior Finish	Plaster with putty coat finish; plastic coated paneling.	Plaster with putty coat finish; plastic coated paneling.	Plaster putty coat finish; textured gypsum wallboard with portions of wall finished with natural wood or plastic coated veneers.	Painted plaster or gypsum wallboard.
Ceilings	Complex angular structure with acoustical plaster.	Multi-level structure with good plaster or acoustical tile.	Flat or low slope with textured gypsum wallboard or acoustical tile.	Flat or low slope with textured gypsum wallboard or acoustical tile.
Plumbing	Good fixtures with metal or marble toilet partitions; full wall height ceramic tile wainscot. Numerous floor drains.	Commercial fixtures with metal toilet partitions; ceramic tile wainscot. Numerous floor drains in kitchen and dishwashing area.	4 to 6 standard commercial fixtures with metal toilet partitions. Floor drains in kitchen and dishwashing area.	4 to 6 standard commercial fixtures with wood toilet partitions. Minimum floor drainage in kitchen area.
Lighting	Triple encased fluorescent fixtures or decorative and ornate incandescent fixtures.	Triple encased fluorescent fixtures or decorative incandescent fixtures.	Triple open or double encased louvered fluorescent fixtures with some decorative incandescent fixtures.	Double open strip fluorescent fixtures or low cost incandescent fixtures.

Square foot costs include the cost of the following components: Foundations as required for normal soil conditions. Floor, wall, and roof structures. Interior floor, wall, and ceiling finishes. Exterior wall finish and roof cover. All glass and glazing. Interior partitions. Basic electrical systems and lighting fixtures. Rough and finish plumbing. Permits and fees. Contractors' mark-up.

The in-place cost of these extra components should be added to the basic building cost to arrive at the total structure cost. See the section beginning on page 201. Heating and air conditioning systems. Booths and counters. Kitchen equipment. Fire sprinklers. Exterior signs. Paving and curbing. Yard improvements. Canopies and overhang.

Self Service Restaurants
Masonry

Estimating Procedure

1. Use these figures to estimate the cost of self-service or drive-in restaurants. These restaurants specialize in rapid service, may or may not have an interior eating area, have a large amount of glass on the front and side walls and usually are well lighted.
2. Establish the structure quality class by applying the information on page 140.
3. Compute the building floor area. This should include everything within the exterior walls and all insets outside the main walls but under the main roof.
4. Multiply the floor area by the appropriate cost below.
5. Multiply the total cost by the factor listed on page 7.
6. Add the cost of heating and air conditioning systems, booths and counters, kitchen equipment, fire sprinklers, exterior signs, paving and curbing, yard improvements and canopies. See the section beginning on page 201.

Self Service Restaurant, Class 2

Self Service Restaurant, Class 3

	Square Foot Area										
Quality Class	400	500	600	700	800	900	1,000	1,100	1,200	1,300	1,400
1, Best	166.58	153.78	144.23	136.75	130.69	125.63	121.34	117.64	114.39	111.52	108.96
1 & 2	143.26	132.26	124.05	117.61	112.40	108.05	104.36	101.17	98.38	95.92	93.71
2, Good	123.20	113.74	106.68	101.15	96.66	92.92	89.75	87.01	84.61	82.49	80.59
2 & 3	105.95	97.81	91.74	86.98	83.12	79.91	77.18	74.82	72.76	70.94	69.31
3, Average	91.13	84.13	78.90	74.81	71.49	68.73	66.38	64.35	62.58	61.01	59.61
3 & 4	78.36	72.34	67.85	64.33	61.48	59.10	57.08	55.34	53.81	52.47	51.26
4, Low	67.40	62.22	58.36	55.33	52.87	50.83	49.09	47.59	46.28	45.12	44.09

	Square Foot Area										
Quality Class	1,500	1,600	1,700	1,800	2,000	2,200	2,400	2,600	2,800	3,000	3,400
1, Best	106.93	105.59	104.32	103.13	100.54	98.94	97.14	95.50	93.99	92.61	90.14
1 & 2	91.99	90.84	89.75	88.73	86.50	85.12	83.57	82.16	80.86	79.67	77.55
2, Good	79.07	78.09	77.15	76.27	74.35	73.17	71.84	70.62	69.51	68.48	66.66
2 & 3	68.01	67.15	66.35	65.59	63.94	62.93	61.78	60.74	59.78	58.90	57.33
3, Average	58.49	57.75	57.06	56.41	54.99	54.12	53.13	52.23	51.41	50.65	49.30
3 & 4	50.30	49.67	49.08	48.51	47.29	46.54	45.70	44.92	44.22	43.56	42.40
4, Low	43.26	42.72	42.20	41.72	40.67	40.03	39.30	38.63	38.02	37.46	36.46

Self Service Restaurants
Wood Frame

Estimating Procedure

1. Use these figures to estimate the cost of self-service or drive-in restaurants. These restaurants specialize in rapid service, may or may not have an interior eating area, have a large amount of glass on the front and side walls and usually are well lighted.
2. Establish the structure quality class by applying the information on page 140.
3. Compute the building floor area. This should include everything within the exterior walls and all insets outside the main walls but under the main roof.
4. Multiply the floor area by the appropriate cost below.
5. Multiply the total cost by the factor listed on page 7.
6. Add the cost of heating and air conditioning systems, booths and counters, kitchen equipment, fire sprinklers, exterior signs, paving and curbing, yard improvements and canopies. See the section beginning on page 201.

Self Service Restaurant, Class 1 & 2

Self Service Restaurant, Class 1 & 2

Square Foot Area

Quality Class	400	500	600	700	800	900	1,000	1,100	1,200	1,300	1,400
1, Best	154.35	142.48	133.64	126.71	121.08	116.41	112.43	109.00	106.00	103.34	100.97
1 & 2	132.74	122.54	114.93	108.97	104.14	100.11	96.69	93.74	91.16	88.88	86.84
2, Good	114.15	105.38	98.83	93.71	89.55	86.09	83.15	80.61	78.39	76.43	74.68
2 & 3	98.13	90.59	84.96	80.56	76.98	74.01	71.48	69.30	67.39	65.70	64.20
3, Average	84.41	77.92	73.09	69.30	66.22	63.66	61.49	59.61	57.97	56.52	55.22
3 & 4	72.63	67.05	62.89	59.63	56.98	54.78	52.91	51.29	49.88	48.63	47.52
4, Low	62.46	57.66	54.08	51.27	49.00	47.10	45.50	44.11	42.89	41.82	40.86

Square Foot Area

Quality Class	1,500	1,600	1,700	1,800	2,000	2,200	2,400	2,600	2,800	3,000	3,400
1, Best	99.49	97.98	96.61	95.36	93.17	91.30	89.69	88.27	87.02	85.91	83.99
1 & 2	85.56	84.26	83.08	82.01	80.12	78.52	77.13	75.92	74.84	73.88	72.23
2, Good	73.58	72.46	71.45	70.52	68.90	67.52	66.33	65.28	64.36	63.53	62.12
2 & 3	63.27	62.31	61.44	60.64	59.25	58.06	57.04	56.14	55.34	54.63	53.42
3, Average	54.42	53.59	52.84	52.15	50.96	49.94	49.05	48.28	47.60	46.98	45.94
3 & 4	46.82	46.10	45.46	44.87	43.84	42.96	42.20	41.54	40.95	40.42	39.52
4, Low	40.26	39.65	39.09	38.59	37.70	36.95	36.29	35.72	35.22	34.76	33.99

Coffee Shop Restaurants
Quality Classification

	Class 1 Best Quality	Class 2 Good Quality	Class 3 Average Quality	Class 4 Low Quality
Foundation	Reinforced concrete.	Reinforced concrete.	Reinforced concrete.	Reinforced concrete.
Floor Structure	Reinforced concrete or standard wood frame.	Reinforced concrete or standard wood frame.	Reinforced concrete or standard wood frame.	Reinforced concrete or standard wood frame.
Walls, Wood Frame Const.	Standard wood frame.	Standard wood frame.	Standard wood frame.	Standard wood frame.
Walls, Masonry or Concrete Construction	8" reinforced decorative concrete block or brick.	8" reinforced decorative concrete block or brick.	8" reinforced concrete block.	8" reinforced concrete block.
Roof Structure	Complex angular structure.	Standard wood frame structure, low or medium pitch.	Standard wood frame structure, low pitch shed or gable type.	Standard wood frame structure, low pitch shed or gable type.
Roof Cover	Concrete tile, mission tile, heavy shake.	Composition tar and gravel, medium shake.	Composition tar and gravel.	Composition or composition shingle.
Floor Finish	Terrazzo or good carpeting in dining area. Quarry tile in kitchen. Terrazzo or natural flagstone in entry vistibule.	Vinyl tile or average carpeting in dining area. Terrazzo or quarry tile in kitchen. Terrazzo in entry vestibule.	Vinyl asbestos in dining area. Grease proof tile in kitchen area. Terrazzo in entry vestibule.	Vinyl asbestos tile in dining area. Grease proof tile in kitchen area.
Exterior Wall Finish	Very good wood siding; ornamental brick or natural stone veneer. Minimum 60% of front and side walls in plate glass in heavy metal frames or rabbeted timbers. Entry vestibule with decorative screens of ornamental concrete block or natural stone panels.	Good wood siding; ornamental brick or concrete blocks. Good stucco with color finish and with 2' to 4' brick, or natural stone veneer bulkheads. 50% to 75% of front and side walls in plate glass set in good wood or metal frames. Double entry doors, recessed entries with decorative screen of concrete block.	Stucco or average wood siding with 2' to 4' brick or stone veneer bulkheads. 50% to 60% of front and side walls in plate glass set in average quality wood or aluminum frames. Double entry doors of metal or wood.	Stucco, low cost wood siding or plywood. 40% to 60% of front and side walls in crystal or plate glass set in low-cost wood or metal frames. Double entry doors.
Interior Finish	Select and matched wood paneling, decorative wallpaper or synthetic fabrics.	Plaster with putty coat finish. Wood or plastic coated veneer paneling. Decorative wallpapers or fabrics.	Plaster putty coat finish textured gypsum wallboard with portions of wall finished with natural wood or plastic coated veneer.	Painted plaster or gypsum wallboard.
Ceilings	Complex angular structure with acoustical plaster.	Multi-level structure with good plaster or acoustical tile.	Flat or low slope with textured gypsum wallboard or acoustical tile.	Flat or low slope with textured gypsum wallboard or acoustical tile.
Plumbing	Good fixtures with metal or marble toilet partitions. Full wall height ceramic tile wainscot. Numerous floor drains.	Commercial fixtures with metal toilet partitions. Ceramic tile wainscot. Numerous floor drains in kitchen and dishwashing area.	4 to 6 standard commercial fixtures with metal toilet partitions. Floor drains in kitchen and dishwashing area.	4 to 6 standard commercial fixtures with wood toilet partitions. Minimum floor drainage in kitchen.
Lighting	Triple encased fluorescent fixtures or decorative and ornate incandescent fixtures.	Triple encased fluorescent fixtures or decorative incandescent fixtures.	Triple open or double encased louvered fluorescent fixtures with some decorative incandescent fixtures.	Double open strip fluorescent fixtures or low-cost incandescent fixtures.

Square foot costs include the cost of the following components: Foundations as required for normal soil conditions. Floor, wall, and roof structures. Interior floor, wall, and ceiling finishes. Exterior wall finish and roof cover. All glass and glazing. Interior partitions. Basic electrical systems and lighting fixtures. Rough and finish plumbing. Permits and fees. Contractors' mark-up.

Coffee Shop Restaurants
Masonry

Estimating Procedure

1. Use these figures to estimate the cost of restaurants with counters, or with counter, booth and table seating. A lounge or bar may be part of the building. Higher quality coffee shops usually have cut-up or complex sloping ceiling and roof structures. Coffee shops have extensive exterior glass and good lighting.
2. Establish the structure quality class by applying the information on page 143.
3. Compute the building floor area. This should include everything within the exterior walls and all insets outside the main walls but under the main roof.
4. Multiply the floor area by the appropriate cost below.
5. Multiply the total cost by the factor listed on page 7.
6. Add the cost of heating and air conditioning systems, booths and counters, kitchen equipment, fire sprinklers, exterior signs, paving and curbing, yard improvements and canopies. See the section beginning on page 201.

Coffee Shop Restaurant, Class 1

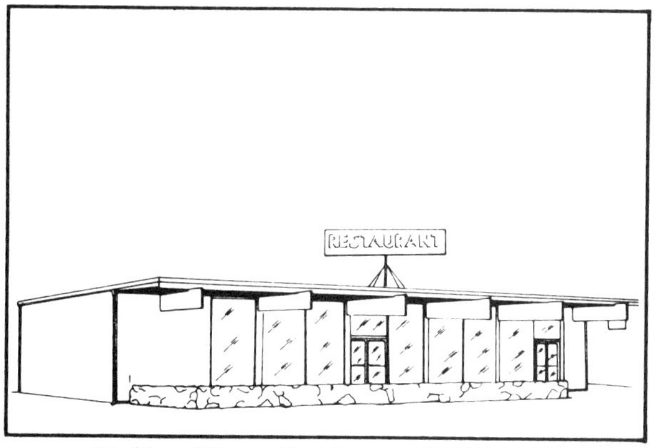

Coffee Shop Restaurant, Class 3

	Square Foot Area										
Quality Class	1,000	1,200	1,400	1,600	1,800	2,000	2,200	2,400	2,600	2,800	3,000
1, Best	115.54	106.98	100.78	96.07	92.36	89.36	86.88	84.80	83.02	81.48	80.13
1 & 2	99.53	92.15	86.81	82.75	79.56	76.98	74.84	73.05	71.51	70.19	69.03
2, Good	85.46	79.12	74.54	71.05	68.31	66.10	64.26	62.72	61.40	60.27	59.27
2 & 3	73.60	68.15	64.20	61.20	58.83	56.92	55.35	54.02	52.88	51.90	51.05
3, Average	63.21	58.53	55.14	52.56	50.53	48.89	47.54	46.39	45.42	44.58	43.84
3 & 4	54.37	50.34	47.43	45.21	43.47	42.05	40.89	39.91	39.07	38.34	37.71
4, Low	46.75	43.29	40.78	38.87	37.37	36.16	35.16	34.31	33.59	32.97	32.42

	Square Foot Area										
Quality Class	3,200	3,400	3,600	3,800	4,000	4,400	4,800	5,200	5,600	6,000	8,000
1, Best	78.02	77.16	76.38	75.66	74.99	73.80	72.77	71.86	71.05	70.33	67.57
1 & 2	67.16	66.42	65.74	65.12	64.55	63.53	62.64	61.86	61.16	60.54	58.16
2, Good	57.75	57.12	56.54	56.01	55.51	54.63	53.87	53.20	52.60	52.06	50.02
2 & 3	49.68	49.13	48.63	48.17	47.75	46.99	46.34	45.76	45.24	44.78	43.03
3, Average	42.72	42.25	41.82	41.43	41.06	40.41	39.85	39.35	38.91	38.51	37.00
3 & 4	36.75	36.34	35.97	35.64	35.32	34.76	34.28	33.85	33.47	33.13	31.83
4, Low	31.60	31.25	30.93	30.64	30.37	29.89	29.47	29.10	28.78	28.48	27.37

Coffee Shop Restaurants
Wood Frame

Estimating Procedure

1. Use these figures to estimate the cost of restaurants with counters, or with counter, booth and table seating. A lounge or bar may be part of the building. Higher quality coffee shops usually have cut-up or complex sloping ceiling and roof structures. Coffee shops have extensive exterior glass and good lighting.
2. Establish the structure quality class by applying the information on page 143.
3. Compute the building floor area. This should include everything within the exterior walls and all insets outside the main walls but under the main roof.
4. Multiply the floor area by the appropriate cost below.
5. Multiply the total cost by the factor listed on page 7.
6. Add the cost of heating and air conditioning systems, booths and counters, kitchen equipment, fire sprinklers, exterior signs, paving and curbing, yard improvements and canopies. See the section beginning on page 201.

Coffee Shop Restaurant, Class 3

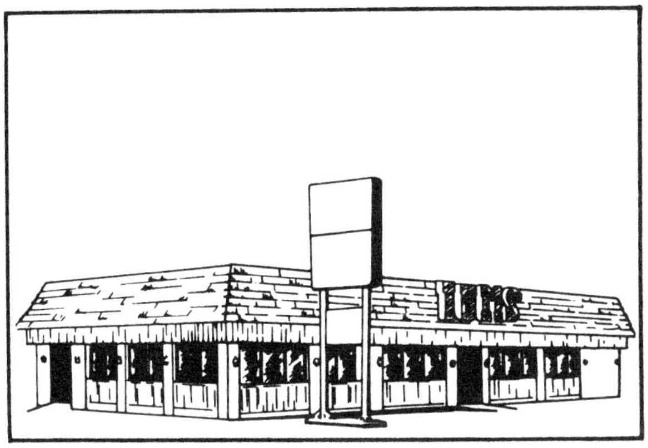

Coffee Shop Restaurant, Class 2

	Square Foot Area										
Quality Class	1,000	1,200	1,400	1,600	1,800	2,000	2,200	2,400	2,600	2,800	3,000
1, Best	110.44	102.18	96.22	91.71	88.16	85.30	82.94	80.96	79.27	77.82	76.54
1 & 2	95.07	87.96	82.83	78.95	75.89	73.43	71.40	69.70	68.24	66.99	65.89
2, Good	81.75	75.64	71.23	67.88	65.26	63.14	61.40	59.93	58.68	57.60	56.66
2 & 3	70.28	65.03	61.24	58.36	56.11	54.29	52.78	51.52	50.45	49.52	48.71
3, Average	60.45	55.93	52.67	50.20	48.26	46.69	45.40	44.32	43.39	42.59	41.90
3 & 4	51.99	48.11	45.30	43.17	41.51	40.16	39.05	38.12	37.32	36.63	36.04
4, Low	44.70	41.36	38.95	37.12	35.69	34.53	33.57	32.77	32.09	31.50	30.98

	Square Foot Area										
Quality Class	3,200	3,400	3,600	3,800	4,000	4,400	4,800	5,200	5,600	6,000	8,000
1, Best	74.61	73.80	73.06	72.37	71.74	70.61	69.62	68.75	67.98	67.28	64.64
1 & 2	64.17	63.48	62.84	62.25	61.70	60.73	59.88	59.13	58.47	57.87	55.59
2, Good	55.18	54.58	54.03	53.53	53.06	52.22	51.49	50.85	50.28	49.76	47.81
2 & 3	47.45	46.94	46.46	46.03	45.63	44.91	44.28	43.73	43.23	42.79	41.11
3, Average	40.81	40.36	39.96	39.58	39.24	38.62	38.08	37.60	37.18	36.80	35.35
3 & 4	35.09	34.71	34.36	34.04	33.74	33.21	32.75	32.34	31.97	31.65	30.40
4, Low	30.18	29.85	29.55	29.27	29.01	28.56	28.16	27.81	27.49	27.21	26.14

Conventional Restuarants Quality Classification

	Class 1 Best Quality	Class 2 Good Quality	Class 3 Average Quality	Class 4 Low Quality
Foundation	Reinforced concrete.	Reinforced concrete	Reinforced concrete.	Reinforced concrete.
Floor Structure	Reinforced concrete or standard wood frame.	Reinforced concrete or standard wood frame.	Reinforced concrete or standard wood frame.	Reinforced concrete or standard wood frame.
Walls, Wood Frame Const.	Standard wood frame.	Standard wood frame.	Standard wood frame.	Standard wood frame.
Walls, Masonry or Concrete Construction	Reinforced decorative concrete block or brick.	Reinforced decorative concrete block or brick.	Reinforced concrete block.	Reinforced concrete block.
Roof Structure	Complex angular structure.	Standard wood frame structure, low or medium pitch.	Standard wood frame, low pitch shed or gable type.	Standard wood frame, low pitch shed or gable type.
Roof Cover	Concrete tile, mission tile, heavy shake.	Composition tar and gravel or medium shake.	Composition tar and gravel.	Composition or composition shingle.
Floor Finish	Terrazzo or good carpeting in dining area. Quarry tile in kitchen. Terrazzo or natural flagstone in entry vestibule.	Vinyl tile or average carpeting in dining area. Terrazzo or quarry tile in kitchen. Terrazzo in entry vestibule.	Vinyl asbestos in dining area. Grease proof tile in kitchen area. Terrazzo in entry vestibule.	Vinyl asbestos tile in dining area. Grease proof tile in kitchen area.
Exterior Wall Finish	Very good wood siding, ornamental brick or natural stone veneer. 10% of front and side walls in plate glass in heavy metal frames or rabbeted timbers. Entry vestibule with decorative screens of ornamental concrete block or natural stone panels.	Good wood siding or plywood; ornamental brick or concrete block. Good stucco with color finish and with 2' to 4' brick, or natural stone veneer bulkheads. 10% to 20% of front and side walls in plate glass set in good wood or metal frames. Double entry doors, recessed entries with decorative screen of concrete block.	Stucco, or average wood siding with 2' to 4' brick or stone veneer bulkheads. 10% to 20% of front and side walls in plate glass set in wood or aluminum extrusions. Double entry doors of metal or wood.	Stucco, low-cost wood siding or plywood. 10% to 20% of front and side walls in crystal plate glass set in low cost wood or metal frames. Double entry doors.
Interior Finish	Select and matched wood paneling, decorative wall papers or synthetic fabrics.	Plaster with putty coat finish; wood or plastic coated veneer paneling; decorative wallpapers or fabrics.	Plaster putty coat finish; textured gypsum wallboard with portions of wall finished with natural wood or plastic coated veneer.	Painted plaster or gypsum wallboard.
Ceilings	Multi-level structure with acoustical plaster.	Multi-level structure with good plaster or acoustical tile.	Flat or low slope with textured gypsum wallboard or acoustical tile.	Flat or low slope with textured gypsum wallboard or acoustical tile.
Plumbing	Good fixtures with metal or marble toilet partitions. Full wall height ceramic tile wainscot. Numerous floor drains.	Commercial fixtures with metal toilet partitions. Ceramic tile wainscot. Numerous floor drains in kitchen and dishwashing area.	4 to 6 standard commercial fixtures with metal toilet partitions. Floor drains in kitchen and dishwashing area.	4 to 6 standard commercial fixtures with wood toilet partitions. Minimum floor drainage in kitchen area.
Lighting	Decorative and ornate incandescent fixtures.	Decorative incandescent fixtures.	Double encased louvered fluorescent fixtures with decorative incandescent fixtures.	Double open strip fluorescent fixtures or low-cost incandescent fixtures.

Square foot costs include the cost of the following components: Foundations as required for normal soil conditions. Floor, wall, and roof structures. Interior floor, wall, and ceiling finishes. Exterior wall finish and roof cover. All glass and glazing. Interior partitions. Basic electrical systems and lighting fixtures. Rough and finish plumbing. Permits and fees. Contractors' mark-up.

Conventional Restaurants
Masonry or Wood Frame

Estimating Procedure

1. Use these figures to estimate the cost of restaurants with limited use of glass as exterior walls and subdued lighting. Roof and ceiling structures are simple and conventional. Seating may be any combination of counter, stools, booths or tables. A lounge or bar is often a part of a building of this type.
2. Establish the structure quality class by applying the information on page 146.
3. Compute the building floor area. This should include everything within the exterior walls and all insets outside the main walls but under the main roof.
4. Multiply the floor area by the appropriate cost below.
5. Multiply the total cost by the factor listed on page 7.
6. Add the cost of heating and air conditioning systems, booths and counters, kitchen equipment, fire sprinklers, exterior signs, paving and curbing, yard improvements and canopies. See the section beginning on page 201.

Conventional Restaurant, Class 2

Masonry

Square Foot Area

Quality Class	1,000	1,500	2,000	2,500	3,000	3,500	4,000	5,000	6,000	8,000	10,000
1, Best	107.49	96.12	89.78	85.63	82.67	80.43	78.65	76.00	74.10	71.49	69.76
1 & 2	92.44	82.66	77.21	73.64	71.10	69.17	67.64	65.36	63.72	61.48	59.99
2, Good	79.51	71.10	66.41	63.35	61.15	59.49	58.18	56.22	54.81	52.88	51.60
2 & 3	68.37	61.14	57.11	54.47	52.59	51.16	50.03	48.35	47.13	45.47	44.37
3, Average	58.80	52.58	49.11	46.85	45.23	44.00	43.03	41.58	40.54	39.11	38.16
3 & 4	50.57	45.22	42.24	40.29	38.89	37.84	37.00	35.76	34.86	33.63	32.82
4, Low	43.50	38.90	36.33	34.65	33.46	32.55	31.83	30.76	29.99	28.93	28.23

Wood Frame

Square Foot Area

Quality Class	1,000	1,500	2,000	2,500	3,000	3,500	4,000	5,000	6,000	8,000	10,000
1, Best	106.33	95.08	88.81	84.71	81.78	79.56	77.81	75.19	73.30	70.72	69.01
1 & 2	91.44	81.77	76.38	72.85	70.33	68.42	66.91	64.66	63.04	60.82	59.35
2, Good	78.64	70.32	65.68	62.65	60.48	58.84	57.54	55.61	54.21	52.30	51.04
2 & 3	67.64	60.49	56.50	53.89	52.02	50.61	49.50	47.83	46.63	44.99	43.90
3, Average	58.17	52.02	48.59	46.35	44.74	43.53	42.57	41.14	40.10	38.69	37.75
3 & 4	50.04	44.75	41.79	39.86	38.49	37.44	36.62	35.38	34.49	33.28	32.47
4, Low	43.03	38.48	35.94	34.28	33.09	32.20	31.49	30.43	29.66	28.62	27.92

"A-Frame" Restaurants Quality Classification

	Class 1 Best Quality	Class 2 Good Quality	Class 3 Average Quality	Class 4 Low Quality
Foundation	Reinforced concrete.	Reinforced concrete.	Reinforced concrete.	Reinforced concrete.
Floor Structure	Reinforced concrete or standard wood frame.	Reinforced concrete or standard wood frame.	Reinforced concrete or standard wood frame.	Reinforced concrete or standard wood frame.
Roof Structure	Wood frame or post and beam, high pitch.	Wood frame or post and beam, high pitch.	Standard wood frame or post and beam, high pitch.	Standard wood frame or post and beam, high pitch.
Roof Cover	Concrete tile, heavy wood shake.	Aluminum shake or medium wood shake.	Wood or composition shingles.	Composition or composition shingles.
Floor Finish	Terrazzo or good carpeting in dining area. Quarry tile in kitchen. Terrazzo or natural flagstone in entry vestibule.	Vinyl tile or average carpeting in dining area. Terrazzo or quarry tile in kitchen, terrazzo in entry vestibule.	Vinyl asbestos in dining area. Grease proof tile in kitchen area. Terrazzo in entry vestibule.	Vinyl asbestos tile in dining area. Grease proof tile in kitchen area.
Interior Finish	Select and matched wood paneling, decorative wallpapers or synthetic fabrics.	Plaster with putty coat finish, wood or plastic coated veneer paneling, decorative wallpapers or fabrics.	Plaster, putty coat finish; textured gypsum wallboard with portions of wall finished with natural wood or plastic coated veneer.	Painted plaster or gypsum wallboard.
End Walls	Very good wood siding; ornamental brick or natural stone veneer. 60% on front in plate glass in heavy metal extrusions or rabbeted timbers. Entry vestibule with decorative screens of ornamental concrete block or natural stone panels.	Good wood siding; ornamental brick or concrete blocks. Good stucco with color finish and with 2' to 4' brick or natural stone veneer bulkheads. 50% of front in plate glass set in good wood or metal frames. Double entry doors, recessed entries with decorative screen of concrete block.	Stucco or average wood siding with 2' to 4' brick or stone veneer bulkheads. 50% to 60% of front in plate glass set in average quality wood or aluminum frames. Double entry doors of metal or wood.	Stucco, low cost wood siding or plywood. 40% to 60% of front in crystal glass set in low cost wood or metal frames.
Ceilings	Decorative wood.	Natural wood, good plaster or acoustical tile.	Painted wood, textured gypsum wallboard or acoustical tile.	Painted wood, textured gypsum wallboard or acoustical tile.
Plumbing	Good fixtures with metal or marble toilet partitions. Full wall height ceramic tile wainscot. Numerous floor drains.	Commercial fixtures with metal toilet partitions. Ceramic tile wainscot. Numerous floor drains in kitchen and dishwashing area.	4 to 6 standard commercial fixtures with metal toilet partitions. Floor drains in kitchen and dishwashing area.	4 to 6 standard commercial fixtures with wood toilet partitions. Minimum floor drainage in kitchen area.
Lighting	Triple encased fluorescent fixtures or decorative and ornate incandescent fixtures.	Triple encased fluorescent or decorative incandescent fixtures.	Triple open, double encased louvered fluorescent fixtures or average incandescent fixtures.	Double open strip fluorescent fixtures or low cost incandescent fixtures.

Square foot costs include the cost of the following components: Foundations as required for normal soil conditions. Floor, wall, and roof structures. Interior floor, wall, and ceiling finishes. Exterior wall finish and roof cover. All glass and glazing. Interior partitions. Basic electrical systems and lighting fixtures. Rough and finish plumbing. Permits and fees. Contractors' mark-up.

The in-place cost of these extra components should be added to the basic building cost to arrive at the total structure cost. See the section beginning on page 201: Heating and air conditioning systems. Booths and counters. Kitchen equipment. Fire sprinklers. Exterior signs. Paving and curbing. Yard improvements. Canopies and overhang.

"A-Frame" Restaurants
Wood Frame

Estimating Procedure

1. Use these figures to estimate the cost of restaurants with a sloping roof that forms two or more exterior walls. Either conventional or self service may be used and a lounge or bar may be part of the building.
2. Establish the structure quality class by applying the information on page 148.
3. Compute the building floor area. This should include everything within the exterior walls and all insets outside the main walls but under the main roof.
4. Multiply the floor area by the appropriate cost below.
5. Multiply the total cost by the factor listed on page 7.
6. Add the cost of heating and air conditioning systems, booths and counters, kitchen equipment, fire sprinklers, exterior signs, paving and curbing, yard improvements and canopies. See the section beginning on page 201.

"A-Frame" Restaurant, Class 1

	Square Foot Area										
Quality Class	400	500	600	800	1,000	1,200	1,400	1,600	1,800	2,000	2,400
1, Best	139.79	124.30	113.83	100.54	92.40	86.87	82.87	79.82	77.42	75.47	72.51
1 & 2	120.20	106.88	97.88	86.45	79.45	74.70	71.25	68.63	66.57	64.90	62.35
2, Good	103.37	91.91	84.17	74.34	68.32	64.24	61.28	59.02	57.25	55.81	53.62
2 & 3	88.90	79.05	72.39	63.94	58.76	55.25	52.70	50.76	49.24	48.00	46.11
3, Average	76.46	67.99	62.26	54.99	50.54	47.52	45.33	43.66	42.35	41.28	39.66
3 & 4	65.68	58.40	53.48	47.23	43.41	40.82	38.93	37.50	36.37	35.46	34.07
4, Low	56.55	50.28	46.05	40.67	37.38	35.14	33.52	32.29	31.32	30.53	29.33

	Square Foot Area										
Quality Class	2,800	3,200	3,600	4,000	4,500	5,000	6,000	7,000	8,000	9,000	10,000
1, Best	70.39	68.78	67.49	66.42	65.32	64.41	62.98	61.91	61.06	60.38	59.81
1 & 2	60.52	59.14	58.03	57.11	56.16	55.38	54.15	53.23	52.50	51.92	51.43
2, Good	52.05	50.86	49.91	49.12	48.30	47.63	46.57	45.78	45.15	44.65	44.23
2 & 3	44.76	43.74	42.92	42.24	41.54	40.96	40.05	39.37	38.83	38.40	38.03
3, Average	38.50	37.62	36.91	36.33	35.73	35.23	34.45	33.86	33.40	33.03	32.71
3 & 4	33.10	32.35	31.74	31.24	30.72	30.29	29.62	29.11	28.72	28.39	28.13
4, Low	28.47	27.82	27.30	26.87	26.42	26.05	25.48	25.04	24.70	24.42	24.19

Theaters
Masonry or Concrete
Quality Classification

	Class 1 Best, with Balcony	Class 2 Very Good, with Balcony	Class 3 Good Quality	Class 4 Average Quality	Class 5 Low Quality
Foundation	Reinforced concrete.	Reinforced concrete.	Reinforced concrete.	Reinforced concrete.	Reinforced concrete.
Floor Structure	6" reinforced concrete on 6" rock fill or 2" x 12" joists, 16" o.c.	4" to 6" reinforced concrete on 6" rock fill or 2" x 10" joists, 16" o.c.	4" reinforced concrete on 6" rock fill or 2" x 10" joists, 16" o.c.	4" reinforced concrete on 6" rock fill or 2" x 8" joists, 16" o.c.	4" reinforced concrete on 6" rock fill or 2" x 6" joists, 16" o.c.
Wall Structure	8" reinforced concrete or 12" common brick.	8" reinforced concrete or 12" common brick.	8" reinforced concrete or 12" common brick.	8" reinforced concrete or 12" common brick.	8" reinforced concrete block or 6" reinforced concrete.
Roof Framing	2" x 10" joists 16" o.c. Trusses on heavy pilasters 20' o.c. 2" x 12" rafters or purlins 16" o.c.	2" x 10" joists 16" o.c. Trusses on heavy pilasters 20' o.c. 2" x 12" rafters or purlins 16" o.c.	2" x 10" joists 16" o.c. Trusses on heavy pilasters 20' o.c. 2" x 12" rafters or purlins 16" o.c.	2" x 10" joists 16" o.c. Trusses on heavy pilasters 20' o.c. 2" x 10" rafters or purlins.	2" x 10" joists 16" o.c. Trusses on pilasters 20' o.c. 2" x 10" rafters or purlins 16" o.c.
Roof Covering	5 ply composition roof on 1" sheathing with insulation.	5 ply composition roof on 1" sheathing with insulation.	5 ply composition roof on 1" sheathing with insulation.	5 ply composition roof on 1" sheathing with insulation.	4 ply composition roof on 1" sheathing.
Front	Highly ornamental stucco or plaster finishes, custom select stone veneers or highly ornamental terra cotta.	Ornamental stucco, custom select brick, natural stone veneers or marble.	Highly ornamental stucco or custom brick or natural stone or terra cotta veneers.	Ornamental stucco or select brick veneers or partial terra cotta veneers.	Plain or colored stucco or common brick or carrara glass veneers.
Floor, Entry	Custom terrazzo with intricate and ornamental designs, custom select natural stone, marble or very good carpet.	Custom terrazzo with highly ornamental designs, natural stone veneers, marble or very good carpet.	Custom designed terrazzo with ornamental designs with portions natural stone veneers, marble or good carpet.	Colored terrazzo with designs, or average carpet.	Colored concrete with some plain colored terrazzo.
Floors, Interior	Concrete with carpet throughout.	Concrete with carpet throughout.	Concrete with carpet throughout.	Concrete with carpet throughout.	Concrete with carpet runners at aisles.
Restrooms	As per code requirement capacity. Custom select ornamental ceramic tile, terrazzo or natural stone on floors and walls.	As per code requirement capacity. Custom ceramic tile or terrazzo on floor and walls.	As per code requirement capacity. Ceramic tile on floors and walls or terrazzo floors and walls.	As per code requirement capacity. Ceramic tiles on walls and floors or terrazzo on floors.	As per code requirement capacity. Ceramic or vinyl tile on floors.
Walls	Painted or custom select canvas backed or custom moulded cloth tapestry wall papers or oriental plaster finished with detailed sirocco type mouldings and trim, walls with selected matched wood veneers.	Painted or custom canvas backed, custom moulded cloth tapestry or ornamental plaster finished with custom mouldings and trimmings. Portions of wall select matched wood veneers.	Painted or finished with custom canvas backed wall papers or moulded tapestry finished wall papers or select wood veneer matched full height at lobby.	Painted or finished with durable canvas or select quality wood veneers on gypsum board or plaster.	Painted and papered with durable canvas materials with portion wood veneers on gypsum board or plaster.
Ceiling	Suspended acoustical with highly ornate mouldings and trim with acoustical baffles.	Suspended acoustical with highly ornate mouldings and trim with acoustical baffles.	Suspended acoustical with ornate plaster cove mouldings and trim with acoustical baffles.	Suspended acoustical with plaster mouldings and sound baffles.	Suspended acoustical tile.
Lighting	Incandescent or recessed lights throughout. Interior with ornate chandelier type fixtures throughout. All dimmer controlled.	Incandescent or recessed lights throughout. Interior with ornate chandelier type fixtures throughout. All dimmer controlled.	Incandescent fixtures with chandelier and recessed at theater area, dimmer controlled.	Incandescent fixtures in lobby with flourescent or chandelier type fixtures in theater area, recessed lighting, dimmer controlled.	Incandescent fixtures with recessed fixtures, dimmer controlled.
Seating	Main floor and balcony seating.	Main floor and balcony seating.	May or may not have balcony.	May or may not have balcony.	May or may not have balcony.

Square foot costs include the cost of the following components: Foundations as required for normal soil conditions. Floor, wall, and roof structures. Interior floor, wall and ceiling finishes. Exterior wall finish and roof cover. Display fronts. Interior partitions. All doors. Ticket booth. Basic lighting and electrical systems. Rough and finish plumbing. A mezzanine floor projection booth. A frame-work for mounting a picture screen. A balcony in auditorium type theaters. Permits and fees. Contractors' mark-up.

Theaters with Balcony
Masonry or Concrete

Estimating Procedure

1. Establish the structure quality class by applying the information on page 150.
2. Compute the building floor area. This should include everything within the main walls and all insets outside the main walls but under the main roof.
3. Add to or subtract from the square foot cost below the appropriate amount from the Wall Height Adjustment Table on page 155 if the wall height is more or less than 28 feet.
4. Multiply the adjusted square foot cost by the building floor area.
5. Deduct, if appropriate, for common walls, using the figures on page 155.
6. Multiply the total cost by the location factor listed on page 7.
7. Add the cost of heating and air conditioning systems, fire extinguishers, exterior signs, paving and curbing. See the section beginning on page 201.

Length less than twice width

	Square Foot Area										
Quality Class	5,000	6,000	7,000	8,000	9,000	10,000	12,000	15 ,000	20,000	24,000	30,000
1, Best	85.58	82.44	80.06	78.18	76.65	75.38	73.36	71.19	68.78	67.46	66.02
1 & 2	81.87	78.87	76.59	74.79	73.33	72.11	70.19	68.10	65.80	64.54	63.16
2, Very Good	78.35	75.47	73.29	71.58	70.18	69.01	67.16	65.17	62.97	61.76	60.45
2 & 3	74.75	72.00	69.93	68.29	66.95	65.84	64.08	62.18	60.08	58.92	57.67
3, Good	71.17	68.56	66.58	65.02	63.75	62.69	61.01	59.20	57.20	56.10	54.91
3 & 4	67.43	64.95	63.08	61.60	60.39	59.39	57.80	56.09	54.19	53.15	52.02
4, Average	63.90	61.56	59.78	58.38	57.24	56.29	54.78	53.16	51.36	50.38	49.30
4 & 5	60.07	57.87	56.20	54.88	53.81	52.91	51.50	49.97	48.28	47.36	46.35
5, Low	56.42	54.35	52.78	51.54	50.53	49.69	48.36	46.93	45.34	44.47	43.53

Length between 2 and 4 times width

	Square Foot Area										
Quality Class	5,000	6,000	7,000	8,000	9,000	10,000	12,000	15,000	20,000	24,000	30,000
1, Best	91.11	87.76	85.23	83.23	81.60	80.24	78.09	75.78	73.22	71.81	70.29
1 & 2	87.18	83.97	81.55	79.63	78.07	76.77	74.72	72.50	70.06	68.71	67.25
2, Very Good	83.45	80.38	78.06	76.22	74.73	73.49	71.52	69.40	67.06	65.77	64.37
2 & 3	79.57	76.65	74.43	72.68	71.26	70.08	68.20	66.18	63.95	62.72	61.38
3, Good	75.78	73.00	70.89	69.22	67.87	66.74	64.95	63.03	60.90	59.73	58.46
3 & 4	71.23	68.61	66.63	65.06	63.79	62.73	61.05	59.24	57.24	56.14	54.95
4, Average	68.09	65.58	63.69	62.19	60.98	59.96	58.36	56.63	54.72	53.67	52.52
4 & 5	63.99	61.64	59.85	58.45	57.31	56.35	54.85	53.22	51.42	50.43	49.36
5, Low	60.08	57.87	56.20	54.88	53.81	52.91	51.50	49.97	48.28	47.36	46.35

Length more than 4 times width

	Square Foot Area										
Quality Class	5,000	6,000	7,000	8,000	9,000	10,000	12,000	15,000	20,000	24,000	30,000
1, Best	97.02	93.48	90.79	88.66	86.93	85.49	83.19	80.72	77.97	76.46	74.81
1 & 2	92.83	89.44	86.87	84.84	83.18	81.79	79.60	77.23	74.60	73.16	71.58
2, Very Good	88.84	85.59	83.13	81.19	79.60	78.27	76.18	73.91	71.39	70.01	68.50
2 & 3	84.71	81.62	79.27	77.41	75.90	74.64	72.64	70.47	68.08	66.76	65.32
3, Good	80.67	77.72	75.49	73.72	72.28	71.08	69.17	67.11	64.83	63.57	62.20
3 & 4	76.45	73.66	71.54	69.87	68.50	67.36	65.56	63.60	61.44	60.25	58.95
4, Average	72.65	70.00	67.99	66.40	65.10	64.02	62.30	60.44	58.39	57.25	56.02
4 & 5	68.11	65.62	63.73	62.24	61.02	60.01	58.40	56.66	54.73	53.67	52.51
5, Low	63.95	61.62	59.85	58.45	57.30	56.35	54.84	53.21	51.40	50.40	49.31

Theaters Without Balcony
Masonry or Concrete
Length Less Than Twice Width

Estimating Procedure

1. Establish the structure quality class by applying the information on page 150.
2. Compute the building floor area. This should include everything within the main walls and all insets outside the main walls but under the main roof.
3. Add to or subtract from the square foot cost below the appropriate amount from the Wall Height Adjustment Table on page 155 if the wall height is more or less than 20 feet.
4. Multiply the adjusted square foot cost by the building floor area.
5. Deduct, if appropriate, for common walls, using the figures on page 155.
6. Multiply the total cost by the location factor listed on page 7.
7. Add the cost of heating and air conditioning systems, fire extinguishers, exterior signs, paving and curbing. See the section beginning on page 201.

Theater Without Balcony, Class 4

	Square Foot Area										
Quality Class	3,000	3,500	4,000	5,000	6,000	7,000	8,000	10,000	12,000	15,000	20,000
3, Good	46.50	45.02	43.87	42.19	41.01	40.13	39.44	38.42	37.70	36.93	36.10
3 & 4	44.37	42.96	41.87	40.26	39.14	38.29	37.64	36.67	35.98	35.25	34.45
4, Average	42.37	41.02	39.98	38.45	37.37	36.57	35.94	35.01	34.36	33.66	32.90
4 & 5	40.68	39.38	38.38	36.91	35.87	35.10	34.50	33.61	32.98	32.31	31.58
5, Low	38.95	37.71	36.75	35.34	34.35	33.61	33.03	32.18	31.58	30.94	30.24

Theaters Without Balcony
Masonry or Concrete
Length Between 2 and 4 Times Width

Estimating Procedure

1. Establish the structure quality class by applying the information on page 150.
2. Compute the building floor area. This should include everything within the main walls and all insets outside the main walls but under the main roof.
3. Add to or subtract from the square foot cost below the appropriate amount from the Wall Height Adjustment Table on page 157 if the wall height is more or less than 20 feet.
4. Multiply the adjusted square foot cost by the building floor area.
5. Deduct, if appropriate, for common walls, using the figures on page 157.
6. Multiply the total cost by the location factor listed on page 7.
7. Add the cost of heating and air conditioning systems, fire extinguishers, exterior signs, paving and curbing. See the section beginning on page 201.

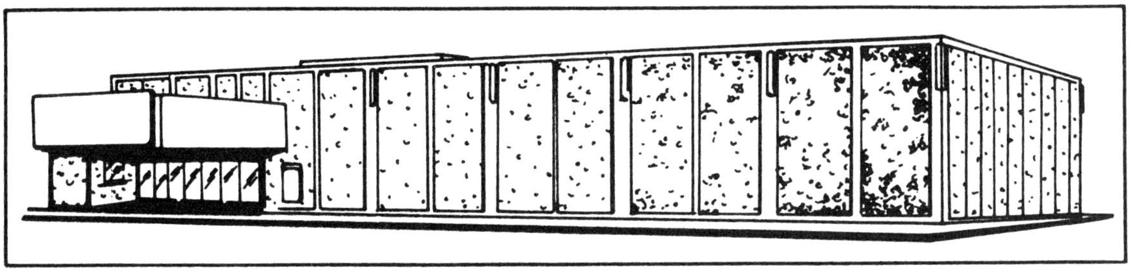

Theater Without Balcony, Class 3

	Square Foot Area										
Quality Class	3,000	3,500	4,000	5,000	6,000	7,000	8,000	10,000	12,000	15,000	20,000
3, Good	49.73	47.62	46.09	44.02	42.73	41.84	41.22	40.40	39.91	39.48	39.13
3 & 4	47.47	45.46	44.00	42.03	40.79	39.95	39.35	38.57	38.10	37.69	37.35
4, Average	45.34	43.42	42.02	40.14	38.95	38.15	37.58	36.84	36.39	35.99	35.67
4 & 5	43.56	41.71	40.37	38.56	37.42	36.65	36.10	35.39	34.96	34.58	34.27
5, Low	41.71	39.94	38.65	36.92	35.83	35.09	34.57	33.88	33.47	33.11	32.81

Theaters Without Balcony
Masonry or Concrete
Length More Than 4 Times Width

Estimating Procedure

1. Establish the structure quality class by applying the information on page 150.
2. Compute the building floor area. This should include everything within the main walls and all insets outside the main walls but under the main roof.
3. Add to or subtract from the square foot cost below the appropriate amount from the Wall Height Adjustment Table on page 157 if the wall height is more or less than 20 feet.
4. Multiply the adjusted square foot cost by the building floor area.
5. Deduct, if appropriate, for common walls, using the figures on page 157.
6. Multiply the total cost by the location factor listed on page 7.
7. Add the cost of heating and air conditioning systems, fire extinguishers, exterior signs, paving and curbing. See the section beginning on page 201.

Theater Without Balcony, Class 3 & 4

Square Foot Area

Quality Class	3,000	3,500	4,000	5,000	6,000	7,000	8,000	10,000	12,000	15,000	20,000
3, Good	52.36	50.65	49.33	47.41	46.07	45.07	44.30	43.17	42.37	41.53	40.62
3 & 4	49.75	48.12	46.87	45.04	43.77	42.82	42.09	41.01	40.26	39.46	38.60
4, Average	47.72	46.16	44.96	43.21	41.99	41.08	40.37	39.34	38.62	37.85	37.02
4 & 5	45.80	44.31	43.15	41.47	40.30	39.43	38.75	37.76	37.07	36.33	35.54
5, Low	43.86	42.43	41.32	39.71	38.59	37.76	37.11	36.16	35.49	34.79	34.03

Theaters
Masonry or Concrete
Wall Adjustments

Wall Height Adjustment

Add or subtract the appropriate amount listed in this table to the square foot of floor cost for each foot of wall height more or less than 28 feet, if adjusting for a theater with balcony, or 20 feet, if adjusting for a theater without a balcony.

	Square Foot Area										
Quality Class	3,000	3,500	4,000	5,000	6,000	7,000	8,000	10,000	12,000	15,000	20,000
1, Best	1.48	1.43	1.40	1.34	1.30	1.28	1.25	1.22	1.20	1.17	1.15
2, Very Good	1.35	1.30	1.27	1.22	1.19	1.16	1.14	1.11	1.09	1.07	1.04
3, Good	1.24	1.20	1.17	1.12	1.09	1.07	1.05	1.02	1.00	.98	.96
4, Average	1.13	1.09	1.07	1.03	1.00	.97	.96	.93	.92	.90	.88
5, Low	1.04	1.01	.98	.94	.92	.90	.88	.86	.84	.83	.81

Perimeter Wall Adjustment

A common wall exists when two buildings share one wall. Adjust for common walls by deducting the linear foot costs below from the total structure cost. In some structures one or more walls is not owned at all. In this case, deduct the ''No Ownership'' cost per linear foot of wall not owned. For Common Wall, deduct $110.00 per linear foot. For No Wall Ownership, deduct $171.00 per linear foot.

Theaters
Wood Frame
Quality Classification

	Class 1 Best Quality	Class 2 Good Quality	Class 3 Average Quality	Class 4 Low Quality
Foundation	Reinforced concrete.	Reinforced concrete.	Reinforced concrete.	Reinforced concrete.
Floor Structure	4" reinforced concrete on 6" rock fill or 2" x 10" joists, 16" o.c.	4" reinforced concrete on 6" rock fill or 2" x 8" joists, 16" o.c.	4" reinforced concrete on 6" rock fill or 2" x 6" joists, 16" o.c.	4" reinforced concrete on 4" rock fill or 2" x 6" joists, 16" o.c.
Wall Structure	2" x 6", 16" o.c.	2" x 4" or 2" x 6", 16" o.c.	2" x 4", 16" o.c. up to 14' high, 2" x 6", 16" o.c. over 14' high.	2" x 4", 16" o.c. up to 14' high, 2" x 6", 16" o.c. over 14' high.
Roof Framing	2" x 10" joists, 16" o.c. Trusses on heavy pilasters, 20' o.c. 2" x 12" rafters or purlins, 16" o.c.	2" x 10" joists, 16" o.c. Trusses on heavy pilasters, 20' o.c. 2" x 10" rafters or purlins.	2" x 10" joists, 16" o.c. Steel trusses on pilasters, 20' o.c. 2" x 10" rafters or purlins, 16" o.c.	2" x 8" joists, 16" o.c. Wood trusses, 2" x 8" purlins, 16" o.c.
Roof Cover	5 ply composition roof on 1" x 6" sheathing with insulation.	5 ply composition roof on 1" x 6" sheathing with insulation.	4 ply composition roof on 1" x 6" sheathing.	4 ply composition roof on 1" x 6" sheathing.
Front	Highly ornamental stucco or custom brick or natural stone or terra cotta veneers.	Ornamental stucco or select brick veneers or partial terra cotta veneers.	Plain or colored stucco or common brick or ornamental wood.	Plain stucco.
Floors, Entry & Lobby	Custom designed terrazzo with ornamental designs with portions natural stone veneers, marble or good carpet.	Colored terrazzo with designs or average carpet.	Colored concrete and portions terrazzo, plain colored.	Plain or colored concrete.
Floors, Interior	Concrete with carpet throughout.	Concrete with carpet throughout.	Concrete with carpet runners at aisles.	Plain or colored concrete.
Restrooms	As per code requirement capacity. Ceramic tile on floors and walls or terrazzo floors and walls.	As per code requirement capacity. Ceramic tile on walls and floors or terrazzo on floors.	As per code requirement capacity. Ceramic or vinyl tile on floors.	As per code requirement capacity. Plain concrete floors and walls, painted.
Walls	Painted and finished with custom backed wallpapers or moulded tapestry finished wallpapers or select wood veneer, matched full height at lobby.	Painted and finished with durable canvas or wood veneers, select quality on gypsum wallboard or plaster.	Painted and papered with durable canvas materials with portion wood veneers on gypsum wallboard or plaster.	Painted with or without stencil type painted moulded designs on gypsum wallboard taped and textured.
Ceilings	Suspended acoustical, ornate cove mouldings and trim with acoustical baffles.	Suspended acoustical with plaster mouldings and sound baffles.	Suspended acoustical tile.	Gypsum wallboard taped, textured and painted.
Lighting	Incandescent fixtures with chandelier fixtures, recessed at theater area, dimmer controlled.	Incandescent fixtures in lobby with fluorescent or chandelier type fixtures in theater area, recessed lighting, dimmer controlled.	Incandescent recessed fixtures, dimmer controlled.	Plain incandescent fixtures with dimmers.

Square foot costs include the cost of the following components: Foundations as required for normal soil conditions. Floor, wall, and roof structures. Interior floor, wall and ceiling finishes. Exterior wall finish and roof cover. Display fronts. Interior partitions. All doors. Ticket booth. Basic lighting and electrical systems. Rough and finish plumbing. A mezzanine floor projection booth. A frame-work for mounting a picture screen. A balcony in auditorium type theaters. Permits and fees. Contractors' mark-up.

Theaters
Wood Frame
Length Less Than Twice Width

Estimating Procedure

1. Establish the structure quality class by applying the information on page 156.
2. Compute the building floor area. This should include everything within the main walls and all insets outside the main walls but under the main roof.
3. Add to or subtract from the square foot cost below the appropriate amount from the Wall Height Adjustment Table on page 160 if the wall height is more or less than 20 feet.
4. Multiply the adjusted square foot cost by the building floor area.
5. Deduct, if appropriate, for common walls, using the figures on page 160.
6. Multiply the total cost by the location factor listed on page 7.
7. Add the cost of heating and air conditioning systems, fire extinguishers, exterior signs, paving and curbing. See the section beginning on page 201.

Theater, Class 4 Front, Class 3 Rear

Square Foot Area

Quality Class	3,000	3,500	4,000	5,000	6,000	7,000	8,000	10,000	12,000	15,000	20,000
1, Best	40.18	38.91	37.92	36.47	35.45	34.69	34.09	33.21	32.59	31.93	31.21
1 & 2	38.69	37.46	36.51	35.11	34.13	33.40	32.82	31.98	31.38	30.74	30.05
2, Good	37.22	36.04	35.12	33.78	32.83	32.13	31.58	30.76	30.19	29.57	28.91
2 & 3	35.86	34.72	33.84	32.54	31.63	30.95	30.42	29.64	29.08	28.49	27.85
3, Average	34.46	33.36	32.51	31.27	30.40	29.74	29.23	28.48	27.95	27.38	26.76
3 & 4	33.20	32.15	31.33	30.13	29.29	28.66	28.17	27.44	26.93	26.38	25.78
4, Low	31.90	30.89	30.10	28.95	28.14	27.54	27.07	26.37	25.88	25.35	24.78

Theaters
Wood Frame
Length Between 2 and 4 Times Width

Estimating Procedure

1. Establish the structure quality class by applying the information on page 156.
2. Compute the building floor area. This should include everything within the main walls and all insets outside the main walls but under the main roof.
3. Add to or subtract from the square foot cost below the appropriate amount from the Wall Height Adjustment Table on page 160 if the wall height is more or less than 20 feet.
4. Multiply the adjusted square foot cost by the building floor area.
5. Deduct, if appropriate, for common walls, using the figures on page 160.
6. Multiply the total cost by the location factor listed on page 7.
7. Add the cost of heating and air conditioning systems, fire extinguishers, exterior signs, paving and curbing. See the section beginning on page 201.

Theater, Class 3

Square Foot Area

Quality Class	3,000	3,500	4,000	5,000	6,000	7,000	8,000	10,000	12,000	15,000	20,000
1, Best	42.82	41.46	40.41	38.86	37.77	36.95	36.32	35.38	34.71	34.00	33.23
1 & 2	41.21	39.90	38.89	37.40	36.35	35.56	34.95	34.05	33.41	32.72	31.98
2, Good	39.63	38.37	37.39	35.96	34.95	34.19	33.60	32.74	32.12	31.46	30.75
2 & 3	38.15	36.94	36.00	34.62	33.65	32.92	32.36	31.52	30.93	30.29	29.61
3, Average	36.68	35.52	34.61	33.29	32.35	31.65	31.11	30.31	29.73	29.13	28.46
3 & 4	35.33	34.20	33.33	32.05	31.16	30.48	29.96	29.18	28.63	28.05	27.41
4, Low	33.91	32.84	32.00	30.77	29.91	29.27	28.76	28.02	27.49	26.93	26.32

Theaters
Wood Frame
Length More Than 4 Times Width

Estimating Procedure

1. Establish the structure quality class by applying the information on page 156.
2. Compute the building floor area. This should include everything within the main walls and all insets outside the main walls but under the main roof.
3. Add to or subtract from the square foot cost below the appropriate amount from the Wall Height Adjustment Table on page 160 if the wall height is more or less than 20 feet.
4. Multiply the adjusted square foot cost by the building floor area.
5. Deduct, if appropriate, for common walls, using the figures on page 160.
6. Multiply the total cost by the location factor listed on page 7.
7. Add the cost of heating and air conditioning systems, fire extinguishers, exterior signs, paving and curbing. See the section beginning on page 201.

Theater, Class 3

	Square Foot Area										
Quality Class	3,000	3,500	4,000	5,000	6,000	7,000	8,000	10,000	12,000	15,000	20,000
1, Best	45.53	44.08	42.96	41.32	40.16	39.30	38.63	37.63	36.93	36.17	35.36
1 & 2	43.84	42.45	41.37	39.79	38.68	37.85	37.20	36.24	35.56	34.83	34.05
2, Good	42.18	40.84	39.80	38.28	37.21	36.41	35.78	34.86	34.21	33.51	32.75
2 & 3	40.60	39.31	38.31	36.85	35.82	35.05	34.44	33.56	32.93	32.26	31.53
3, Average	39.04	37.80	36.84	35.43	34.44	33.70	33.12	32.27	31.66	31.02	30.32
3 & 4	37.60	36.41	35.48	34.13	33.17	32.46	31.90	31.08	30.50	29.88	29.20
4, Low	36.18	35.03	34.14	32.84	31.92	31.23	30.70	29.90	29.34	28.75	28.10

Theaters
Wood Frame
Wall Adjustments

Wall Height Adjustment

Add or subtract the amount listed in this table to the square foot of floor cost for each foot of wall height more or less than 20 feet.

	Square Foot Area										
Quality Class	3,000	3,500	4,000	5,000	6,000	7,000	8,000	10,000	12,000	15,000	20,000
1, Best	1.07	1.04	1.01	.98	.95	.93	.92	.89	.87	.85	.83
2, Good	.99	.96	.94	.91	.88	.86	.85	.82	.81	.79	.77
3, Average	.92	.90	.88	.84	.82	.81	.79	.77	.75	.74	.72
4, Low	.84	.82	.80	.77	.75	.73	.72	.70	.69	.67	.65

Perimeter Wall Adjustment

A common wall exists when two buildings share one wall. Adjust for common walls by deducting the linear foot costs below from the total structure cost. In some structures one or more walls is not owned at all. In this case, deduct the ''No Ownership'' cost per linear foot of wall not owned.

For Common Wall, deduct $60.00 per linear foot.
For No Wall Ownership, deduct $85.00 per linear foot.

Mobile Home Parks

	Class 1 Best Quality	Class 2 Good Quality	Class 3 Average Quality	Class 4 Low Quality
Engineering, Plans, Permits, Surveying	Good planning, necessary permits, good engineering; designed by architect.	Good planning, necessary permits, good engineering; designed by architect.	Average planning, necessary permits, engineered and designed.	Fair planning, necessary permits, minimum surveying.
Grading	Fully graded.	Fully graded.	Fully graded.	Minimum site leveling; grades not engineered; road grading.
Street Paving	2" thick asphalt surface on good base, concrete curbs, 30' width.	2" thick asphalt surface on good base, concrete curbs, 25' width.	20' roads, 2" asphalt on rock base; concrete or wood edging.	Narrow streets, 2" asphalt on ground; no curbs or edging.
Patios and Walks	Patios, 300 to 500 S.F. of good concrete. Walks to utility rooms, pools and recreation areas.	Patios, 200 to 300 S.F. of good concrete. Walks to utility rooms, pools and recreation areas.	Patios, approximately 150 S.F. average concrete or average grade asphalt. Walks to utility buildings.	Some patios, concrete or asphalt paving. No walks.
Trailer Pad and Parking	Concrete or good asphalt pad and driveway.	Asphalt under trailer and extended to one side for driveway.	Gravel under trailer and small asphalt driveway.	Gravel under trailer.
Sewer	8" lines, 10" mains. Meets all code requirements. Storm drain system.	8" lines, 10" mains. Meets all code requirements.	4" to 6" and 8" lines. Meets code requirements in most areas.	3" to 6" lines, inadequate. Below good code requirements.
Water	Engineered system for equalized pressure throughout park. Sprinkler system in common areas.	Adequate line size, designed and properly sized for equalized pressure.	Adequate line size; has required valves at each space.	Small lines; has required valves at each space.
Gas	Supplied to each space, sized to code requirements.	Supplied to each space, sized to code requirements.	None except in utility buildings and recreation buildings.	None except in utility buildings.
Electric	Underground service, designed for larger modern trailers with adequate size to enlarge to take care of future needs. Approximately 100 amp or more. Speaker system, underground television system to each space.	Underground service, designed for larger modern trailers with facilities to enlarge capacity to 100 amp. Approximately 70 amp service or more. Speaker system, underground television system to each space.	Underground services, not designed for more capacity. Approximately 30 amp service or more. Speaker system.	Overhead system wired for 15 amp service at each space.
Outdoor Lighting	Lamp post each five spaces, ornate type.	Lamp post each five spaces, inexpensive type.	Overhead street lights at each corner.	Few overhead street lights.
Telephone	Underground to each space.	Underground to each space.	None.	None.
Signs	Large expensive sign.	Good sign.	Average sign.	Inexpensive sign.
Garbage	Built-in ground.	Built-in ground.	None.	None.
Mail Boxes	Good mail box and post each space.	Inexpensive mail box and post each space.	None.	None.
Fences and Gates	Good wood or cyclone. Ornamental fence or wall in front.	Good wood or cyclone. Block wall in front.	Inexpensive wood or wire.	None or inexpensive wire.
Pools	Good quality.	Good quality, adequate size for park.	Small with few extras.	None or inexpensive
Utility Building	Wood frame and good stucco. Board and batt redwood siding or concrete block exterior. Best composition shingle or tar and rock roofing. Good interior plaster or gypsum wallboard. Well finished concrete floors with vinyl tile. Good lighting. Good heating. Showers ceramic tile or fiberglass walls with ceramic tile floor. Glass shower doors. Good quality plumbing fixtures. Good workmanship throughout.	Wood frame and good stucco or concrete block exterior. Thick-butt composition shingles or tar and gravel roof. Good exterior or plaster or gypsum wallboard. Well finished concrete floors. Good lighting. Good heating. Showers ceramic tile walls with ceramic tile base. Good quality plumbing fixtures. Good workmanship throughout.	Wood frame, average stucco exterior. Composition shingle or roll roofing. Gypsum wallboard taped and textured or plaster interior. Average concrete floors. Average lighting. Fair heating. Metal stall shower or showers with enameled cement plaster walls and tile floor with tile base. Average plumbing fixtures. Average workmanship throughout.	Wood frame, fair stucco or fair siding exterior. Plastic interior. Composition roll roofing. Fair concrete finish. Fair lighting. Inexpensive heating. Showers enameled cement plaster walls and tile floors. Fair plumbing fixtures and fair workmanship throughout.
Recreation Building	Wood frame and stucco. Board and batt redwood siding or concrete block exterior. Best composition shingles or tar and rock roofing. Good interior plaster or gypsum wallboard taped and textured and painted. Well finished concrete floors with vinyl tile. Good heating. Good lighting. Rest room for each sex containing at least one each of the following fixtures: Shower, water closet & lavatory. Showers ceramic tile floors and walls or fiberglass walls and tile floor with glass shower doors. Good quality plumbing fixtures. Kitchen sink, range, refrigerator, cabinets and drainboard of formica or equal material. Large glass area in community room.	Wood frame and good stucco or concrete block exterior. Thick butt composition shingles or tar and gravel roof. Good exterior or plaster or gypsum wallboard. Well finished concrete floors. Good heating. Good lighting. Ceramic tile stall showers with ceramic tile base. Good quality plumbing fixtures. Kitchen with tile drain board and some hardwood cabinets. Small office area. Large glass windows in community room.	Wood frame, average stucco or siding exterior. Composition shingle or roll roofing. Gypsum wallboard taped and textured or plaster interior. Average concrete floors. Average lighting. Average heating. Showers with enameled cement plaster walls and ceramic tile base. Water closets and lavatories. One rest room for each sex with at least 1 each shower, water closet and lavatory. Average grade of plumbing fixtures. Ceiling of gypsum wallboard. Average workmanship throughout.	None.

Mobile Home Parks

Estimating Procedure

1. Establish the park quality class by applying the information on page 161.
2. Compute the square foot area per home space. This should include the mobile home space, streets, recreation and other community use areas but exclude excess land not improved or not in use as a part of the park operation. Divide this total area by the number of home spaces. The result is the average area per home space.
3. Multiply the appropriate cost below by the number of home spaces.
4. Determine the quality class and area of recreation and utility buildings. Compute the total cost of these buildings and add this amount to or subtract it from the total from step 3 to adjust for more or less buildings than included in the quality specification.
5. Multiply the total of space and building costs by the location factor on page 7.
6. Add the cost of septic tank systems, wells, and covered areas built at the individual spaces.

Space costs without community facilities include the cost of the following components: Grading associated with a level site under normal soil conditions. Street paving and curbs. Patios and walks. Pads and parking paving. Sewer, electrical, gas and water systems including normal hook-up costs. Outdoor lighting. Signs. Mail boxes. Fences and gates. Contractors' mark-up.

Space costs with community facilities and buildings include the cost of all the above components plus these components in amounts proportionate to the size of the park: Recreation, administrative and utility buildings adequate for the size of the park. Recreation facilities such as pools, shuffle board courts, playground equipment, fire pits, etc. Telephones. Restrooms.

The cost of the following components are not included in the basic building cost: Septic tank systems. Wells. Structures or covered areas on individual spaces. The cost of grading beyond that associated with a level site.

Parks Without Community Facilities or Buildings

	Square Foot Area Per Space										
Quality Class	1,500	2,000	2,500	3,000	3,500	4,000	4,500	5,000	5,500	6,000	6,500
1, Best	--	--	--	--	--	5,110	5,200	5,500	5,600	5,770	6,200
2, Good	--	--	--	4,580	4,690	4,690	4,900	5,010	5,230	5,230	5,230
3, Average	--	--	3,500	3,600	3,790	3,870	3,970	4,030	4,030	4,030	--
4, Low	2,050	2,380	2,485	2,640	2,760	2,830	2,830	2,830	2,830	--	--

Parks With Community Facilities and Buildings

	Square Foot Area Per Space										
Quality Class	1,500	2,000	2,500	3,000	3,500	4,000	4,500	5,000	5,500	6,000	6,500
1, Best	--	--	--	--	--	6,650	6,790	7,030	7,170	7,400	7,600
2, Good	--	--	--	5,810	5,940	6,060	6,180	6,300	6,540	6,540	6,540
3, Average	--	--	4,520	4,800	4,870	4,900	5,040	5,160	5,160	5,160	--
4, Low	3,160	3,200	3,280	3,440	3,560	3,680	3,680	3,680	3,680	--	--

Square Foot Costs for Buildings Alone

	Recreational Buildings	Utility Buildings
1, Best	$26.00 to $37.70	$24.70 to $30.00
2, Good	22.00 to 30.00	18.30 to 24.70
3, Average	18.20 to 26.00	16.90 to 18.20
4, Low	---	15.50 to 18.00

Service Stations
Wood, Masonry or Painted Steel
Quality Classification

	Wood Frame	Masonry or Concrete	Painted Steel, Good	Painted Steel, Average	Painted Steel, Low
Foundation & Floor	Reinforced concrete.	Reinforced concrete.	Reinforced concrete.	Reinforced concrete.	Reinforced concrete.
Walls	Wood frame 2 x 4, 16" o.c.	8" concrete block.	Steel frame.	Steel frame.	Steel frame.
Roof Structure	Light wood frame, flat or shed type.	Light wood frame, flat or shed type.	Steel frame, flat or shed type.	Steel frame, flat or shed type.	Steel frame, flat or shed type.
Exterior Finish	Painted wood siding or stucco.	Painted concrete block.	Painted steel.	Painted steel.	Painted steel.
Roof Cover	Composition.	Composition.	Steel deck.	Steel deck.	Steel deck.
Glass Area	Small area, painted wood frames.	Small area, painted steel frames.	Large area, painted steel frames.	Average area, painted steel frames.	Small area, painted steel frames.
Lube Room Doors	Folding steel gate.	Folding steel gate.	Painted steel sectional roll up.	Painted steel sectional roll up.	Folding steel gate.
Floor Finish	Concrete.	Concrete.	Concrete, colored concrete in office.	Concrete.	Concrete.
Interior Wall Finish	Exposed studs, painted.	Concrete block, painted.	Exposed structure painted. Painted steel panels in office.	Exposed structure painted. Painted steel panels in office.	Exposed structure painted. Painted steel panels in office.
Ceiling Finish	Exposed structure painted.	Exposed structure painted.	Exposed structure painted. Painted steel panels in office.	Exposed structure painted. Painted steel panels in office.	Exposed structure painted.
Rest Room Finish	Wall board and paint walls and ceilings.	Concrete block and paint walls, wall board and paint ceilings.	Ceramic tile floors, 4' ceramic tile wainscot, painted steel ceilings.	Ceramic tile floors, 4' ceramic tile wainscot, painted steel ceilings.	Concrete floors, painted steel walls, painted steel ceilings.
Rest Room Fixtures	4 low cost fixtures.	4 low cost fixtures.	5 average cost fixtures.	5 average cost fixtures.	4 low cost fixtures.
Exterior Appointments	None.	None.	2' overhang on 3 sides, 3' raised walk on on 3 sides, fluorescent sof-fit lights on 3 sides.	1' overhang on 2 sides, 3' raised walk on 2 sides.	None.

Square foot costs include the cost of the following components: Foundations as required for normal soil conditions. Floor, wall and roof structure. Interior floor, wall and ceiling finishes as described above. Interior partitions. Exterior finish and roof cover. A built-in work bench, tire rack and shelving. Electrical services and fixtures contained within the building. Air and water lines within the building. That portion of rough plumbing serving the building and plumbing fixtures within the building. Roof overhangs and raised walks as described above. Lube room doors. Permits and fees. Contractors' mark-up.

The in-place cost of these extra components should be added to the basic building cost to arrive at the total structure cost. See the section "Additional Costs for Service Stations" on page 169. Canopies. Pumps, dispensers and turbines. Air and water services outside the building. Island lighters. Gasoline storage tanks. Hoists. Compressors. Yard lights. Signs. Paving. Curbs and fences. Miscellaneous equipment and accessories. Island office and storage buildings. Site improvements. Heating and cooling systems.

Land improvement costs: Most service station sites require an expenditure of $7,000 or more for items such as leveling, excavation, curbs, driveways, relocation of power poles, replacement of sidewalks with reinforced walks and street paving.

Service Stations
Wood, Masonry or Painted Steel

Estimating Procedure

1. Establish the structure quality class by applying the information on page 163.
2. Compute the building floor area.
3. Multiply the square foot cost by the building floor area.
4. Multiply the total cost by the location factor listed on page 7.
5. Add the cost of appropriate equipment and fixtures from the section "Additional Costs for Service Stations" beginning on page 169.

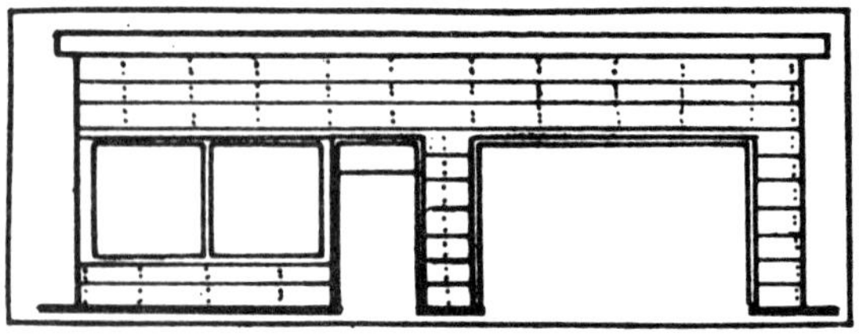

Wood Frame

Masonry

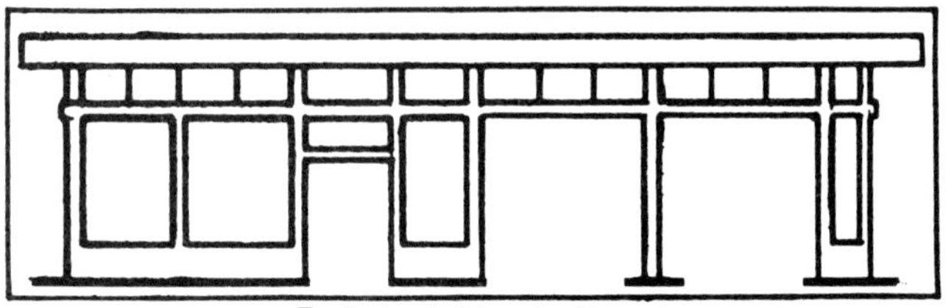

Painted Steel, Good

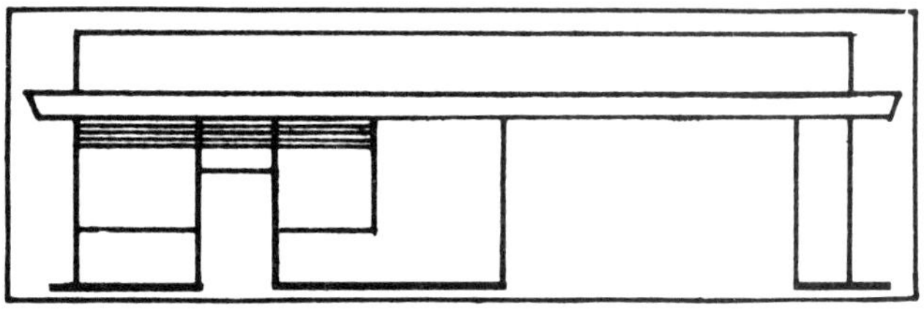

Painted Steel, Average

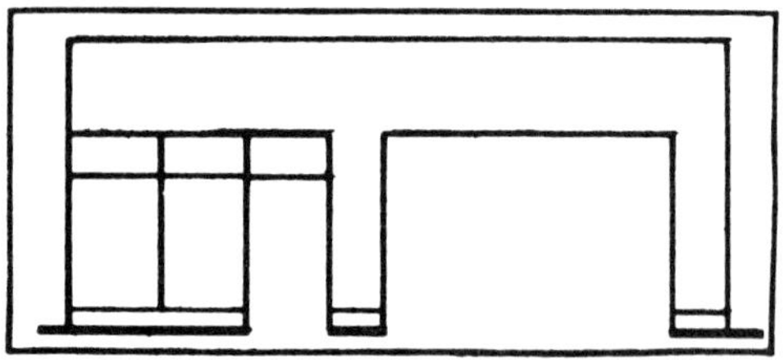

Painted Steel, Low

Square Foot Area

Quality Class	500	600	700	800	900	1,000	1,100	1,200	1,300	1,400	1,800
Wood Frame	36.42	33.44	31.31	29.71	28.47	27.48	26.66	25.99	25.41	24.92	23.51
Masonry or Concrete	42.86	39.35	36.84	34.96	33.50	32.33	31.38	30.58	29.91	29.33	27.66
Painted Steel, Good	64.12	58.86	55.11	52.30	50.11	48.37	46.94	45.74	44.74	43.87	41.38
Painted Steel, Average	56.53	51.90	48.59	46.11	44.19	42.64	41.38	40.33	39.44	38.68	36.48
Painted Steel, Low	49.22	45.19	42.31	40.15	38.47	37.13	36.03	35.12	34.34	33.68	31.77

Service Stations
Porcelain Finished Steel
Quality Classification

	Good Quality	Average Quality	Low Quality
Foundation and Floor	Reinforced concrete.	Reinforced concrete.	Reinforced concrete.
Walls	Steel frame.	Steel frame.	Steel frame.
Roof Structure	Steel frame, flat or shed type.	Steel frame, flat or shed type.	Steel frame, flat or shed type.
Exterior Finish	Porcelain and steel.	Porcelain and steel.	Porcelain and steel.
Roof Cover	Steel deck.	Steel deck.	Steel deck.
Glass Area	Large area, aluminum frames.	Large area, aluminum frames.	Average area, painted steel frames.
Lube Room Doors	Aluminum and glass sectional roll up.	Aluminum and glass sectional roll up.	Painted steel and glass sectional roll up.
Floor Finish	Concrete floors, ceramic tile in office.	Concrete floors, colored concrete in office.	Concrete floors, colored concrete in office.
Interior Wall Finish	Porcelain steel panels.	Exposed structure painted. Painted steel panels in office.	Exposed structure painted. Painted steel panels in office.
Ceiling Finish	Exposed structure painted. Porcelain steel panels in office.	Exposed structure painted. Painted steel panels in office.	Exposed structure painted. Painted steel panels in office.
Rest Room Finish	Ceramic tile floors, 8' ceramic tile or porcelain panels. Painted steel ceiling.	Ceramic tile floors, 5' ceramic tile wainscot. Painted steel ceiling.	Ceramic tile floors, 5' ceramic tile wainscot. Painted steel ceiling.
Rest Room Fixtures	5 good fixtures.	5 good fixtures.	5 good fixtures.
Exterior Appointments	3' to 4' overhang on 3 sides, 6' x 8' sign pylon, 3' raised walk on 3 sides, fluorescent soffit lights on 3 sides.	3' to 4' overhang on 3 sides, 6' x 8' sign pylon, 3' raised walk on 3 sides, fluorescent soffit lights on 3 sides.	3' raised walk on 3 sides, fluorescent soffit lights on 3 sides.

Square foot costs include the cost of the following components: Foundations as required for normal soil conditions. Floor, wall and roof structure. Interior floor, wall and ceiling finishes as described above. Interior partitions. Exterior finish and roof cover. A built-in work bench, tire rack and shelving. Electrical services and fixtures contained within the building. Air and water lines within the building. That portion of rough plumbing serving the building and plumbing fixtures within the building. Roof overhangs and raised walks as described above. Lube room doors. Permits and fees. Contractors' mark-up.

The in-place cost of these extra components should be added to the basic building cost to arrive at the total structure cost. See the section "Additional Costs for Service Stations" on page 169. Canopies. Pumps, dispensers and turbines. Air and water services outside the building. Island lighters. Gasoline storage tanks. Hoists. Compressors. Yard lights. Signs. Paving. Curbs and fences. Miscellaneous equipment and accessories. Island office and storage buildings. Site improvements. Heating and cooling systems.

Land improvement costs: Most service station sites require an expenditure of $7,000 or more for items such as leveling, excavation, curbs, driveways, relocation of power poles, replacement of sidewalks with reinforced walks and street paving.

Service Stations
Porcelain Finished Steel

Estimating Procedure

1. Establish the structure quality class by applying the information on page 165.
2. Compute the building floor area.
3. Multiply the square foot cost by the building floor area.
4. Multiply the total cost by the location factor listed on page 7.
5. Add the cost of appropriate equipment and fixtures from the section "Additional Costs for Service Stations" beginning on page 169.

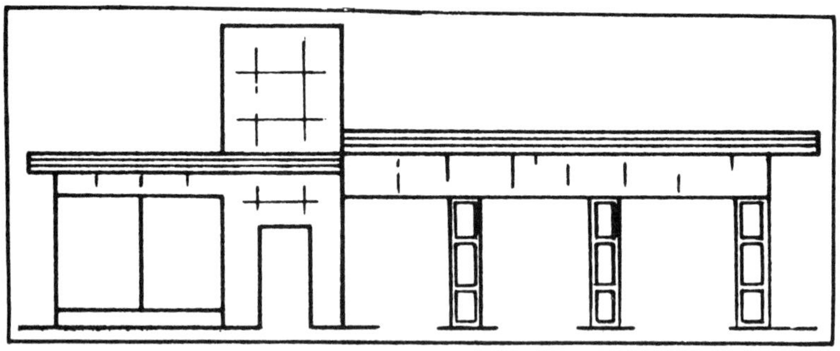

Good Quality

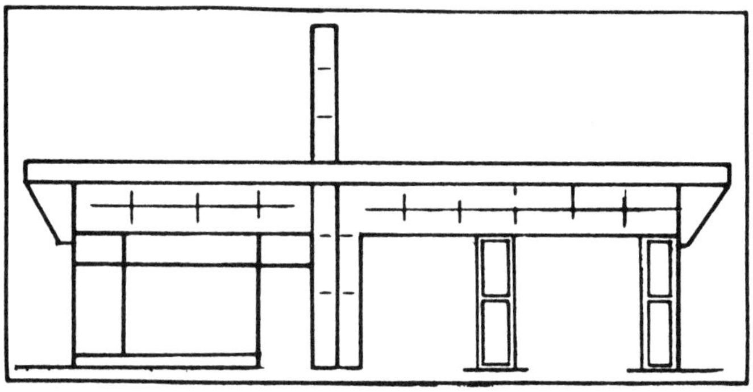

Average Quality

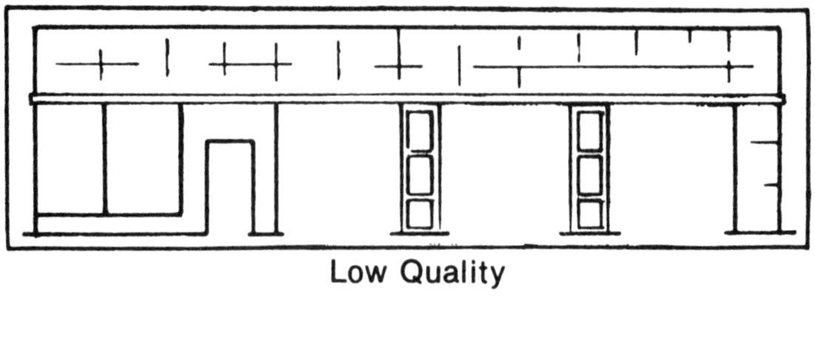

Low Quality

Square Foot Area

Quality Class	1,000	1,100	1,200	1,300	1,400	1,500	1,600	1,700	1,800	2,000	2,400
Good	59.50	57.43	55.80	54.48	53.41	52.52	51.79	51.18	50.66	49.86	48.84
Average	55.30	53.38	51.86	50.63	49.64	48.82	48.13	47.57	47.09	46.34	45.40
Low	49.53	47.81	46.45	45.35	44.46	43.72	43.11	42.60	42.17	41.50	40.66

Service Stations
Ranch or Rustic Type

	Best Quality	Good Quality	Average Quality	Low Quality
Foundation and Floor	Reinforced concrete.	Reinforced concrete.	Reinforced concrete.	Reinforced concrete.
Walls	Steel frame.	Steel frame.	Steel frame, wood frame or masonry.	Steel frame, wood frame or masonry.
Roof Structure	Steel frame, hip or gable type.	Steel frame, hip or gable type.	Steel or wood frame, hip or gable type.	Steel or wood frame, hip or gable type.
Exterior Finish	Natural stone veneer.	Used brick veneer.	Painted steel and masonry veneer.	Painted steel or wood siding.
Roof Cover	Shingle tile or mission tile.	Heavy wood shakes or shingle tile.	Wood shakes or tar and rock.	Composition shingle or tar and gravel.
Glass Area	Large area plate glass in heavy aluminum frame.	Large area plate glass in heavy aluminum frame.	Large area, painted steel frames.	Average area, painted steel frames.
Lube Room Doors	Painted steel or aluminum and glass sectional roll up.	Painted steel or aluminum and glass sectional roll up.	Painted steel sectional roll up.	Painted steel sectional roll up.
Floor Finish	Concrete floors, ceramic tile in office.	Concrete floors, ceramic tile in office.	Concrete floors.	Concrete floors.
Interior Wall Finish	Painted steel panels or gypsum wallboard and paint.	Painted steel panels or gypsum wallboard and paint.	Painted steel panels or gypsum wallboard and paint.	Painted steel panels or gypsum wallboard and paint.
Ceiling Finish	Painted steel panels.	Painted steel panels.	Painted steel panels, gypsum wallboard, or "V" rustic and paint.	Painted steel panels, gypsum wallboard or "V" rustic and paint.
Restroom Finish	Ceramic tile floors, ceramic tile walls, painted steel ceilings.	Ceramic tile floors, ceramic tile walls, painted steel ceilings.	Ceramic tile floors, 5' ceramic tile wainscot, painted steel ceilings.	Ceramic tile floors, 5' ceramic tile wainscot, painted steel ceilings.
Restroom Fixtures	5 good fixtures.	5 good fixtures.	5 good fixtures.	5 good fixtures.
Exterior Appointments	3' to 6' overhang on all sides, 3' raised walk on 3 sides, fluorescent soffit lights on all sides.	3' to 6' overhang on all sides, 3' raised walk on 3 sides, fluorescent soffit lights on all sides.	3' to 6' overhang on 3 sides, 6' x 8' sign pylon, 3' raised walk on 3 sides, fluorescent soffit lights on 3 sides.	2' to 3' overhang on 3 sides, 3' raised walk on 3 sides, fluorescent soffit lights on 3 sides.

Square foot costs include the cost of the following components: Foundations as required for normal soil conditions. Floor, wall and roof structure. Interior floor, wall and ceiling finishes as described above. Interior partitions. Exterior finish and roof cover. A built-in work bench, tire rack and shelving. Electrical services and fixtures contained within the building. Air and water lines within the building. That portion of rough plumbing serving the building and plumbing fixtures within the building. Roof overhangs and raised walks as described above. Lube room doors. Permits and fees. Contractors' mark-up.

The in-place cost of these extra components should be added to the basic building cost to arrive at the total structure cost. See the section "Additional Costs for Service Stations" on page 169. Canopies. Pumps, dispensers and turbines. Air and water services outside the building. Island lighters. Gasoline storage tanks. Hoists. Compressors. Yard lights. Signs. Paving. Curbs and fences. Miscellaneous equipment and accessories. Island office and storage buildings. Site improvements. Heating and cooling systems.

Land improvement costs: Most service station sites require an expenditure of $7,000 or more for items such as leveling, excavation, curbs, driveways, relocation of power poles, replacement of sidewalks with reinforced walks and street paving.

Service Stations
Ranch or Rustic Type

Estimating Procedure

1. Establish the structure quality class by applying the information on page 167.
2. Compute the building floor area.
3. Multiply the square foot cost by the building floor area.
4. Multiply the total cost by the location factor listed on page 7
5. Add the cost of appropriate equipment and fixtures from the section "Additional Costs for Service Stations" beginning on page 169.

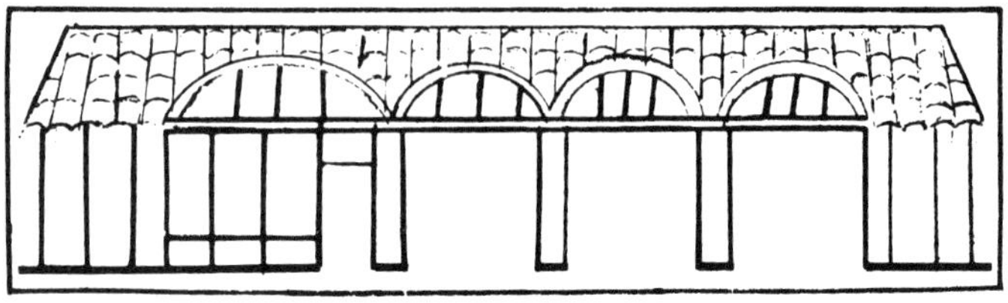

Best Quality

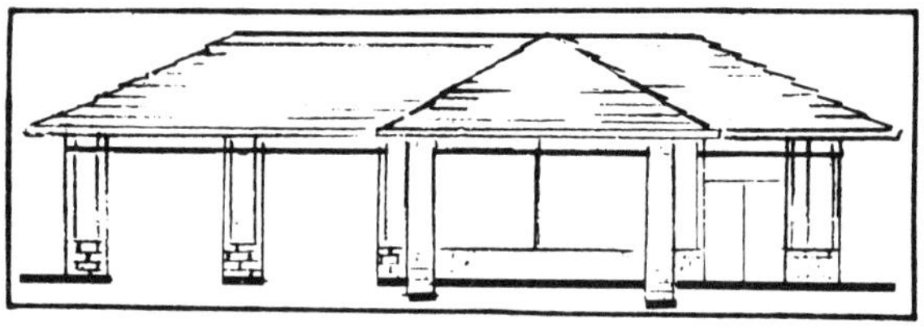

Good Quality

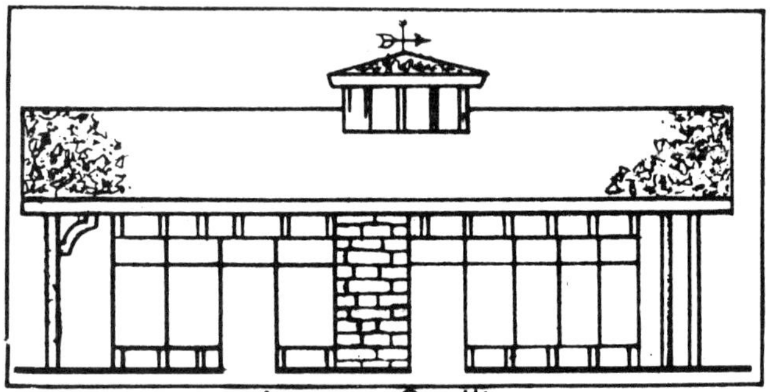

Average Quality

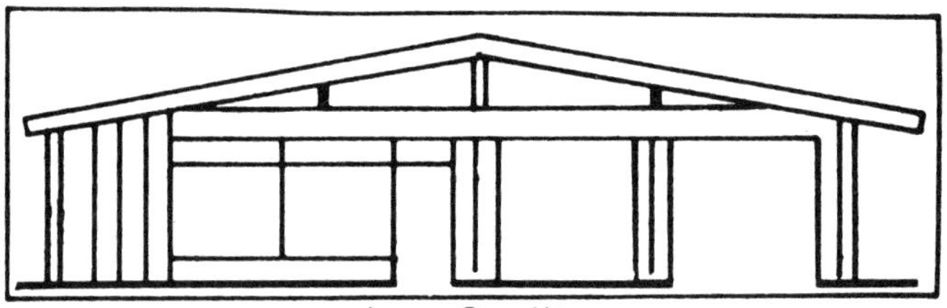

Low Quality

Square Foot Area

Quality Class	1,000	1,100	1,200	1,300	1,400	1,500	1,600	1,700	1,800	2,000	2,400
Best	69.14	66.69	64.76	63.21	61.95	60.92	60.07	59.36	58.77	57.85	56.72
Good	63.98	61.72	59.93	58.50	57.33	56.38	55.59	54.94	54.39	53.54	52.49
Average	59.25	57.16	55.50	54.17	53.10	52.21	51.48	50.88	50.37	49.58	48.61
Low	55.06	53.11	51.57	50.34	49.34	48.52	47.84	47.27	46.80	46.07	45.17

Additional Costs for Service Stations

A portion of the typical plumbing or electrical cost has been added to each item of equipment requiring these services. It will not be necessary except in rare instances to add extra cost for these items.

Canopies cost per square foot

Type	Less than 500 S.F.	500 to 1,000 S.F.	Over 1,000 S.F.
Painted steel	$11.00 to $12.50	$10.00 to $11.00	$ 8.70 to $10.00
Porcelain and steel	12.60 to 13.90	11.00 to 12.50	10.00 to 11.00
Ranch style or gable roof type	15.00 to 16.50	12.70 to 13.90	11.00 to 12.50
Deluxe steel with illuminated plastic signs on sides or in gables. Also includes illuminiated plastic island lighters.	Complete $21.00 to $23.80		
Round type, good steel	$19.80 to $21.00		

Costs include cost of foundation, steel support column or columns, complete canopy, painting or porcelaining, light fixtures, and electrical service. Ranch or gable roof types include the cost of a rock or shake roof cover. Concrete pads under canopies or masonry trim on support columns are not included in costs.

Island Office and Storage Buildings cost per square foot

	Area						
Type	Under 30	31 - 40	41 - 50	51 to 60	61 - 80	81 - 100	101 - 120
Steel and glass or concrete block	$71.50	$64.00	$56.60	$51.00	$46.80	$43.20	$40.30
Wood frame with stucco and glass	58.00	53.80	47.00	40.60	37.90	35.40	34.00

These buildings are usually found at self-service stations. Add $475 per fixture for any plumbing found in these buildings.

Pumps, Dispensers and Turbines

Type	Equipment Cost	Installed Cost		Equipment Cost	Installed Cost
Single pump	$ 900 ea.	$1,870 ea.			
Twin pump	1,840 ea.	3,040 ea.	Add for lighted dispenser	$ 65 ea.	$ 65 ea.
Single dispenser	870 ea.	1,780 ea.	Turbine pump 1/3 HP	550 ea.	1,200 ea.
Twin dispenser	1,600 ea.	2,800 ea.	Turbine pump 3/4 HP	600 ea.	1,350 ea.

Equipment cost includes the cost of the pump or dispenser only. Installed cost includes the cost of the pump or dispenser, installation cost, electrical hookup cost, a portion of the piping cost and a portion of the island block cost.

All of the above pump and dispenser costs are for the computing type. Deduct $75 each for noncomputing single pumps or dispensers and $150 each for twins.

Dispenser costs do not include the cost of the pump. Turbine pump costs must be added. ⅓ HP turbines will serve a single product up to four dispensers. ¾ HP turbines will serve a single product up to eight dispensers.

Additional Costs for Service Stations

Air and Water Services

Type	Air Only Equipment Cost	Air Only Installed Cost	Air and Water Equipment Cost	Air and Water Installed Cost
Underground disappearing hose type	$145.00	$275.00	$180.00	$360.00
Post type with auto inflater	320.00	475.00	360.00	550.00
Post type with auto inflater and disappearing hoses	480.00	650.00	720.00	920.00

Costs include cost of installation and a portion of the cost of air and water lines.

Island Lighters

Width	Length	4 Tubes	6 Tubes
42″	9′-5″	$ 810 ea.	$1,000 ea.
42″	11′-5″	1,025 ea.	1,170 ea.
42″	15′-6″	1,330 ea.	1,480 ea.
42″	19′-6″	1,630 ea.	1,750 ea.
36″	30′-0″	---	2,050 ea.

Cost includes foundation, davit poles or steel support columns and electrical service.

Cash boxes (installed) $115 ea.

Gasoline Storage Tanks:

Capacity in Gallons	Tank Cost	Installed Cost	Capacity in Gallons	Tank Cost	Installed Cost
110	$ 185	$ 730	5300	$1,600	$2,900
150	195	790	6300	1,730	3,100
280	225	1,010	7400	1,960	3,300
550	430	1,250	8400	2,130	3,550
1000	740	1,900	10500	2,530	4,050
2000	1,160	2,340	12600	3,100	4,470
4000	1,460	2,700	--	---	---

Installed cost includes cost of tank, excavation (4′ bury and soil disposal), placing backfill, fill box (concrete slab over tank), tank piping and vent piping.

Miscellaneous Lube Room Equipment

Air hose reels	$325
Water hose reels	325
Grease pit for trucks	$115 to $220 per linear foot

Yard Lights

Type	Height	One Light	Two Lights	Three Lights
Flood lights	24′	$465	$560	$610
Spotlight clusters	24′	420	480	515
Fluorescent - 4 tube	16′	710	640	875

Cost includes cost of installation and usual electrical hookup.

Additional Costs for Service Stations

Hoists

Type	Equipment Cost	Installed Cost
Single plunger auto-rail frame	$1,200	$1,760
Single plunger auto-flat "H" frame type	1,330	1,930
Single plunger auto-flat "H" frame type with adjustable arms	1,630	2,250
Double plunger auto-disappearing type	2,200	2,880
Double plunger-heavy duty rail type	2,760	3,500
Electric lift 150"	2,150	2,780
Electric lift 165"	2,400	3,000

Air Compressors

Horsepower	Equipment Cost	Installed Cost	Horsepower	Equipment Cost	Installed Cost
½	$490	$550	2	$ 980	$1,200
¾	540	610	3	1,220	1,460
1	630	725	5	1,360	1,640
1½	725	850	7½	1,830	2,150

Costs include compressor and tank.

Paving

Asphalt 3"	$ .55 to $ .80 SF
Concrete, 4" with base	1.40 to 1.75 SF
Concrete, 6" with base	1.80 to 2.10 SF
Oil macadam	.35 SF
Pea gravel	.19 SF

Site Improvement

Vertical curb and gutter	$4.90 to $6.50 LF
Concrete approaches	1.95 SF
6" reinforced concrete sidewalks	1.60 to 2.00 SF
Standard 4" sidewalk	1.25 SF

The above costs are normally included in land value.

Fencing and Curbing

Heavy 2 rail fence, 2" x 6"	$4.50 to $5.60 LF
Rails on 4" x 4" posts 6' to 8' o.c.	5.00 to 6.00 LF
Chain link 3' to 4' high	5.40 to 6.50 LF
Solid board 3' to 4' high	4.50 to 5.50 LF
16" bumper fence	4.80 to 5.80 LF
Metal guard rail	8.00 to 10.25 LF
6" x 6" doweled wood bumper strip	2.00 to 2.25
6" x 6" concrete bumper strip, precast	3.30 to 4.00 LF
6" wide x 12" high concrete curb	4.80 to 5.40 LF
6" concrete block walls	2.30 to 2.90 SF plus $4.00/LF for foundation

Service Station Signs

Painted sheet metal with floodlights	$17.60 to $25.00 SF
Porcelain enamel with floodlights	26.50 to 30.00 SF
Plastic with interior lights	40.00 to 56.00
Simple rectangular neon with painted sheet metal faces and a moderate amount of plain letters	38.00 to 46.00 SF
Round or irregular neon with porcelain enamel faces and more elaborate lettering	40.00 to 48.00 SF

Rotators

Small signs	Less than 50 S.F.	$ 650 to $ 750
Medium signs	50 to 100 S.F.	750 to 1,580
Large signs	100 to 200 S.F.	1,600 to 2,900
Extra large signs	Over 200 S.F.	$14.50 per SF of sign area

All of the above sign costs are for double faced signs. Use ⅔ of these costs for single faced signs. Sign costs include costs of installation and normal electrical hookup. They do not include the post cost. These costs are intended for use on **service station signs only** and are based on volume production. Costs of custom built signs will be higher.

Additional Costs for Service Stations

Post Mounting Costs

	Pole Diameter at Base					
Post Height	4"	6"	8"	10"	12"	14"
15	$570	$665	$ 865	$1,150	$1,670	$1,900
20	650	750	1,000	1,200	1,900	2,160
25	675	800	1,020	1,390	2,100	2,300
30	760	880	1,200	1,500	2,200	2,560
35	--	970	1,350	1,700	2,450	2,700
40	--	1,030	1,500	1,800	2,550	3,000
45	--	--	1,670	1,900	2,800	3,100
50	--	--	1,780	2,000	3,000	3,400
55	--	--	--	2,180	3,100	3,700
60	--	--	--	2,300	3,350	3,900
65	--	--	--	--	3,600	4,100

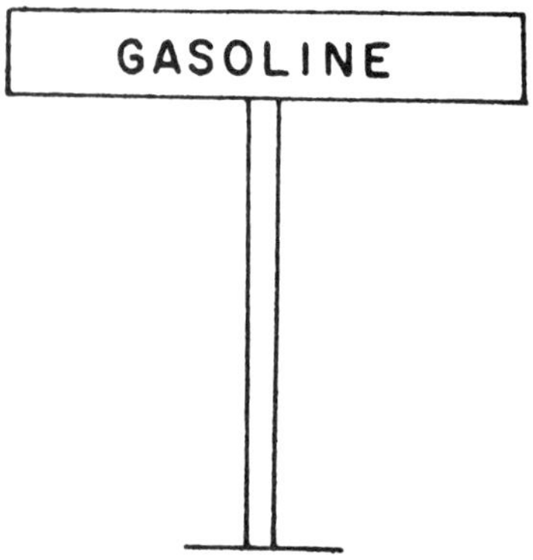

Horizontal Mount

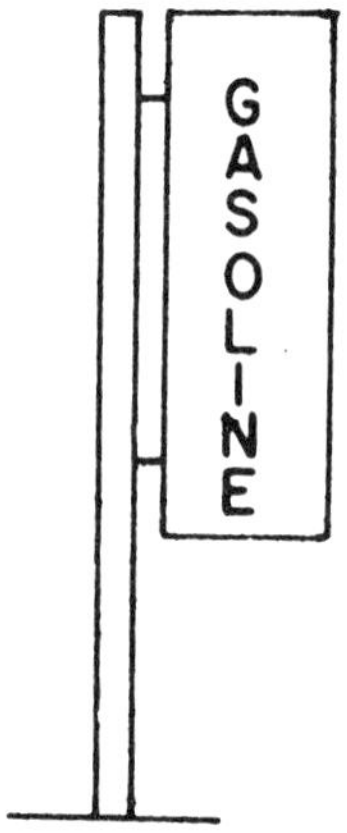

Vertical Mount

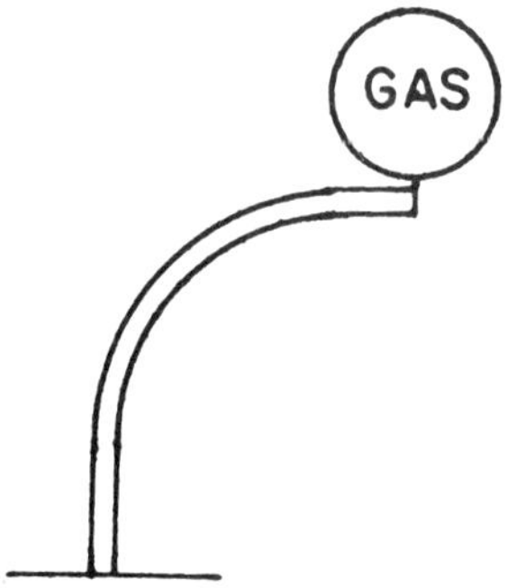

Cantilevered Mount

If signs are mounted on separate posts, post mounting costs must be added. Post mounting costs include the installed cost of the post and foundation. On horizontally mounted signs, post height is the distance from the ground to the bottom of the sign. On vertically mounted signs, post height is the distance to the top of the post.

For cantilevered posts, use one and one-half to two times the conventional post cost.

All of the above post costs are for single posts. Use 90% of the single post costs for each additional post.

If signs are mounted on buildings or canopies and if, because of the extra weight of the sign, extra heavy support posts or foundations are required, a fraction of the post mounting cost should be used.

For example, the cost of a 4' x 25' plastic sign mounted on a 15' by 6'' post that is common to a 12' high canopy might be estimated as follows:

Sign Cost	$5,000
Post Cost $665 x ½ =	333
Total Cost	$5,333

If this sign were mounted on an 8'' post 20' high above this 12' canopy, the cost might be estimated as follows:

Sign Cost	$5,000
Post Cost $1,000 x ¾ =	750
Total Cost	$5,750

Service Garages
Masonry or Concrete
Quality Classification

	Class 1 Best Quality	Class 2 Good Quality	Class 3 Average Quality	Class 4 Low Quality
Foundation	Reinforced concrete or masonry.	Reinforced concrete or masonry.	Reinforced concrete or masonry.	Unreinforced concrete or masonry.
Floor Structure	6" rock fill, 4" concrete with reinforcing mesh.	6" rock fill, 4" concrete with reinforcing mesh.	4" rock fill, 4" concrete with reinforcing mesh.	Unreinforced 4" concrete.
Walls	8" reinforced concrete block, 12" common brick.	8" reinforced concrete block, 6" reinforced concrete.	8" reinforced concrete block, 6" reinforced concrete or 8" common brick.	8" unreinforced concrete block or 8" clay tile.
Roof Structure	Glu-lams or steel trusses on heavy pilasters 20' o.c. 2" x 10" purlins 16" o.c.	Glu-lams or steel trusses on pilasters 20' o.c., 2" x 10" purlins 16" o.c.	Glu-lams or wood trusses with 2" x 8" purlins 16" o.c.	Glu-lams or light wood trusses, 2" x 8" rafters 24" o.c.
Roof Cover	5 ply built-up roof on wood sheathing, with small rock.	4 ply built-up roof on wood sheathing, with small rock.	4 ply built-up roof on wood sheathing.	4 ply built-up roof on wood sheathing.
Restrooms	Two rest rooms with three average fixtures each.	Two rest rooms with two average fixtures each.	One rest room with two low cost fixtures.	One rest room with two low cost fixtures.
Lighting	One incandescent fixture per 300 square feet of floor area.	One incandescent fixture per 300 square feet of floor area.	One incandescent fixture per 300 square feet of floor area.	One incandescent fixture per 300 square feet of floor area.
Windows	3% to 5% of wall area.	3% to 5% of wall area.	3% to 5% of wall area.	3% to 5% of wall area.

Square foot costs include the cost of the following components: Foundations as required for normal soil conditions. Floor, wall and roof structures. Exterior wall finish and roof cover. Entry doors. Basic lighting and electrical systems. Rough and finish plumbing. Permits and fees. Contractors' mark-up.

The in-place cost of these extra components should be added to the basic building cost to arrive at the total structure cost. See page 201. Heating and air conditioning systems. Fire sprinklers. Interior finish costs. Interior partitions. Drive-through doors. Canopies and walks. Exterior signs. Paving and curbing. Miscellaneous yard improvements. Hoists, gas pump and compressor costs are listed in the section ''Additional Costs for Service Stations'' on page 169.

Service Garages
Masonry or Concrete
Length Less Than Twice Width

Estimating Procedure

1. Use these figures to estimate buildings designed primarily for motor vehicle repair. Sales area should be figured separately. Use the costs for urban stores beginning on page 42.
2. Establish the building quality class by applying the information on page 173.
3. Compute the floor area.
4. If the wall height is more or less than 18 feet, add or subtract from the square foot costs below the appropriate amount from the Wall Height Adjustment Table on page 177.
5. Multiply the adjusted square foot cost by the floor area.
6. Deduct for common walls or no wall ownership. Use the figures on page 177.
7. Multiply the total cost by the location factor on page 7.
8. Add the cost of heating and air conditioning systems, fire sprinklers, interior finish and partitions, drive-thru doors, canopies and walks, exterior signs, paving, curbing, and yard improvements. See page 201. Add the cost of hoists, pumps and compressors from pages 169 and 171.

Service Garage (rear portion), Class 2

	Square Foot Area										
Quality Class	2,000	2,500	3,000	4,000	5,000	6,000	7,500	10,000	15,000	20,000	30,000
1, Best	23.30	21.16	19.64	17.60	16.27	15.32	14.30	13.18	11.90	11.17	10.33
1 & 2	22.23	20.19	18.74	16.79	15.52	14.61	13.64	12.57	11.36	10.66	9.86
2, Good	21.12	19.18	17.81	15.96	14.75	13.89	12.96	11.95	10.79	10.13	9.37
2 & 3	19.99	18.15	16.85	15.10	13.96	13.14	12.27	11.31	10.21	9.59	8.87
3, Average	18.90	17.17	15.94	14.28	13.20	12.43	11.60	10.69	9.66	9.07	8.39
3 & 4	17.77	16.14	14.98	13.43	12.41	11.69	10.91	10.05	9.08	8.52	7.88
4, Low	16.68	15.15	14.06	12.60	11.65	10.97	10.24	9.44	8.52	8.00	7.40

Service Garages
Masonry or Concrete
Length Between 2 and 4 Times Width

Estimating Procedure

1. Use these figures to estimate buildings designed primarily for motor vehicle repair. Sales area should be figured separately. Use the costs for urban stores beginning on page 42.
2. Establish the building quality class by applying the information on page 173.
3. Compute the floor area.
4. If the wall height is more or less than 18 feet, add or subtract from the square foot costs below the appropriate amount from the Wall Height Adjustment Table on page 177.
5. Multiply the adjusted square foot cost by the floor area.
6. Deduct for common walls or no wall ownership. Use the figures on page 177.
7. Multiply the total cost by the location factor on page 7.
8. Add the cost of heating and air conditioning systems, fire sprinklers, interior finish and partitions, drive-thru doors, canopies and walks, exterior signs, paving, curbing, and yard improvements. See page 201. Add the cost of hoists, pumps and compressors from pages 169 and 171.

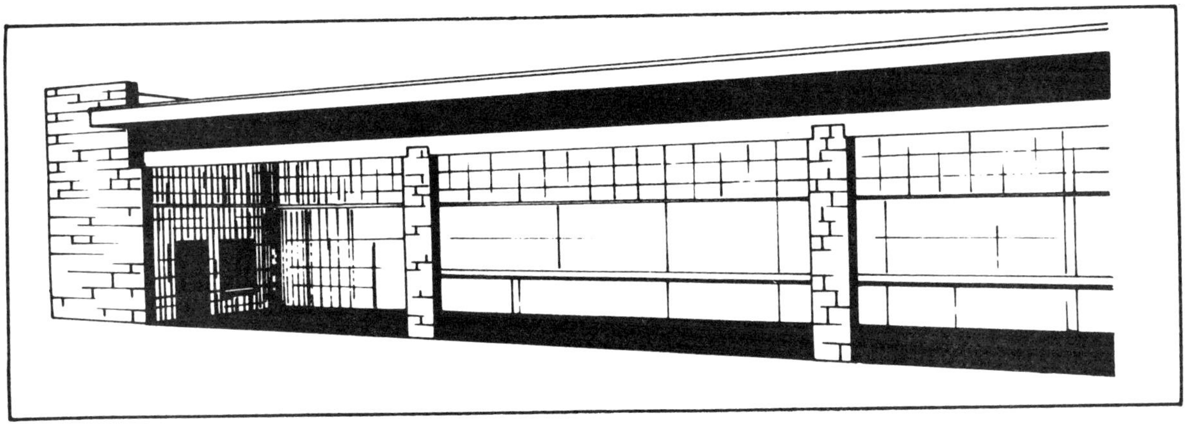

Service Garage, Class 3

	Square Foot Area										
Quality Class	2,000	2,500	3,000	4,000	5,000	6,000	7,500	10,000	15,000	20,000	30,000
1, Best	24.81	22.53	20.91	18.74	17.32	16.31	15.22	14.03	12.67	11.89	10.99
1 & 2	23.62	21.45	19.92	17.85	16.50	15.53	14.49	13.36	12.06	11.32	10.47
2, Good	22.44	20.38	18.92	16.95	15.67	14.75	13.77	12.69	11.46	10.75	9.94
2 & 3	21.21	19.27	17.88	16.03	14.81	13.95	13.02	12.00	10.83	10.17	9.40
3, Average	20.00	18.16	16.86	15.11	13.96	13.15	12.27	11.31	10.21	9.58	8.86
3 & 4	18.86	17.13	15.90	14.25	13.17	12.40	11.57	10.67	9.63	9.04	8.36
4, Low	17.71	16.08	14.93	13.38	12.37	11.64	10.86	10.01	9.04	8.49	7.85

Service Garages
Masonry or Concrete
Length More Than 4 Times Width

Estimating Procedure

1. Use these figures to estimate buildings designed primarily for motor vehicle repair. Sales area should be figured separately. Use the costs for urban stores beginning on page 42.
2. Establish the building quality class by applying the information on page 173.
3. Compute the floor area.
4. If the wall height is more or less than 18 feet, add or subtract from the square foot costs below the appropriate amount from the Wall Height Adjustment Table on page 177.
5. Multiply the adjusted square foot cost by the floor area.
6. Deduct for common walls or no wall ownership. Use the figures on page 177.
7. Multiply the total cost by the location factor on page 7.
8. Add the cost of heating and air conditioning systems, fire sprinklers, interior finish and partitions, drive-thru doors, canopies and walks, exterior signs, paving, curbing, and yard improvements. See page 201. Add the cost of hoists, pumps and compressors from pages 169 and 171.

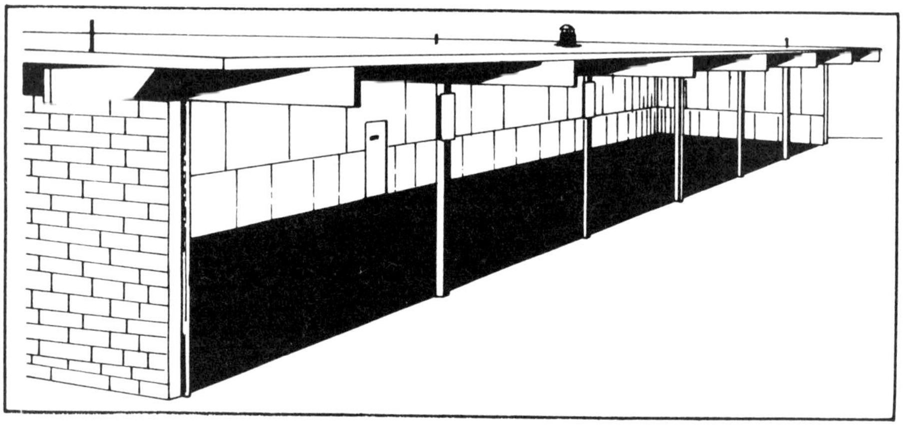

Service Garage, Class 2 & 3

	Square Foot Area										
Quality Class	2,000	2,500	3,000	4,000	5,000	6,000	7,500	10,000	15,000	20,000	30,000
1, Best	26.44	24.00	22.27	19.95	18.44	17.36	16.20	14.94	13.49	12.67	11.72
1 & 2	25.30	22.96	21.31	19.09	17.64	16.61	15.50	14.29	12.91	12.12	11.21
2, Good	23.91	21.70	20.14	18.04	16.68	15.70	14.65	13.51	12.20	11.46	10.60
2 & 3	22.65	20.56	19.08	17.10	15.80	14.88	13.88	12.80	11.56	10.85	10.04
3, Average	21.41	19.44	18.04	16.16	14.93	14.06	13.12	12.09	10.93	10.26	9.49
3 & 4	20.12	18.27	16.95	15.19	14.03	13.21	12.33	11.37	10.27	9.64	8.92
4, Low	18.82	17.08	15.85	14.20	13.13	12.36	11.53	10.63	9.61	9.02	8.34

Service Garages
Masonry or Concrete
Wall Adjustments

Wall Height Adjustment

Add or subtract the amount listed in this table to the square foot of floor cost for each foot of wall height more or less than 18 feet.

Square Foot Area

Area	2,000	2,500	3,000	4,000	5,000	6,000	7,500	10,000	15,000	20,000	30,000
Cost	.24	.22	.20	.17	.15	.13	.11	.10	.07	.06	.04

Perimeter Wall Adjustment

A common wall exists when two buildings share one wall. Adjust for common walls by deducting the linear foot costs below from the total structure cost. In some structures, one or more walls is not owned at all. In this case, deduct the "No Ownership" cost per linear foot of wall not owned.

For a Common Wall, deduct $71.00 per linear foot.
For No Wall Ownership, deduct $116.00 per linear foot.

Service Garages Wood Frame Quality Classification

	Class 1 Best Quality	Class 2 Good Quality	Class 3 Average Quality	Class 4 Low Quality
Foundation	Concrete, heavily reinforced.	Reinforced concrete.	Masonry or reinforced concrete.	Masonry or concrete.
Floor Structure	4" reinforced concrete on 6" rock fill.	4" reinforced concrete on 6" rock fill.	4" concrete on 6" rock fill.	4" concrete on 4" rock fill.
Walls	2" x 4" studs 16" o.c. in walls 14' high; 2" x 6" studs 16" o.c. in walls over 14' high; 3" sill, double plate, adequate blocking and bracing.	2" x 4" studs 16" o.c. in walls 14' high; 2" x 6" studs 16" o.c. in walls over 14' high; 2" sill, double plate, adequate blocking and bracing.	2" x 4" studs 16" o.c. in walls to 14' high; 2" x 6" studs 16" o.c. in walls over 14' high; 2" sill, double plate, minimum blocking and bracing.	2" x 4" studs 24" o.c.; 2" x 4" sill, double 2" x 4" plate, minimum diagonal bracing.
Exterior	Good corrugated iron or board and batt.	Good corrugated iron or board and batt.	Average corrugated iron or board and batt.	Light corrugated iron or board and batt.
Roof Structure	Glu-lams, trusses or tapered steel girders on steel intermediate columns; 2" x 10" rafters 16" o.c.	Glu-lams, average wood trusses, tapered steel girders on steel intermediate columns; 2" x 8" purlins or rafters 16" o.c.	Glu-lams or light wood trusses, on wood posts 18' o.c.; 2" x 8" rafters or purlins 24" o.c.	Light trussed rafters, clear span in small buildings, post and beam support in large buildings.
Roof Cover	Good quality 4 ply composition roofing on wood sheathing.	Average quality 4 ply composition roofing on wood sheathing.	Average quality 3 ply composition roofing on wood sheathing, or good corrugated aluminum.	Light weight 3 ply composition roofing on wood sheathing, or heavy corrugated iron.
Rest Rooms	Two rest rooms with three average fixtures each.	Two rest rooms with three average fixtures each.	One rest room with two low cost fixtures.	One rest room with two low cost fixtures.
Lighting	One incandescent fixture per 300 square feet of floor area.	One incandescent fixture per 300 square feet of floor area.	One incandescent fixture per 300 square feet of floor area.	One incandescent fixture per 300 square feet of floor area.
Windows	3% to 5% of wall area.	3% to 5% of wall area.	3% to 5% of wall area.	3% to 5% of wall area.

Square foot costs include the cost of the following components: Foundations as required for normal soil conditions. Floor, wall and roof structures. Exterior wall finish and roof cover. Entry doors. Basic lighting and electrical systems. Rough and finish plumbing. Permits and fees. Contractors' mark-up.

The in-place cost of these extra components should be added to the basic building cost to arrive at the total structure cost. See page 201. Heating and air conditioning systems. Fire sprinklers. Interior finish costs. Interior partitions. Drive-through doors. Canopies and walks. Exterior signs. Paving and curbing. Miscellaneous yard improvements. Hoists, gas pump and compressor costs are listed in the section ''Additional Costs for Service Stations'' on page 169.

Service Garages
Wood Frame
Length Less Than Twice Width

Estimating Procedure

1. Use these figures to estimate buildings designed primarily for motor vehicle repair. Sales area should be figured separately. Use the costs for urban stores beginning on page 42.
2. Establish the building quality class by applying the information on page 178.
3. Compute the floor area.
4. If the wall height is more or less than 16 feet, add or subtract from the square foot costs below the appropriate amount from the Wall Height Adjustment Table on page 182.
5. Multiply the adjusted square foot cost by the floor area.
6. Deduct for common walls or no wall ownership. Use the figures on page 182.
7. Multiply the total cost by the location factor on page 7.
8. Add the cost of heating and air conditioning systems, fire sprinklers, interior finish and partitions, drive-thru doors, canopies and walks, exterior signs, paving, curbing, and yard improvements. See page 201. Add the cost of hoists, pumps and compressors from pages 169 and 171.

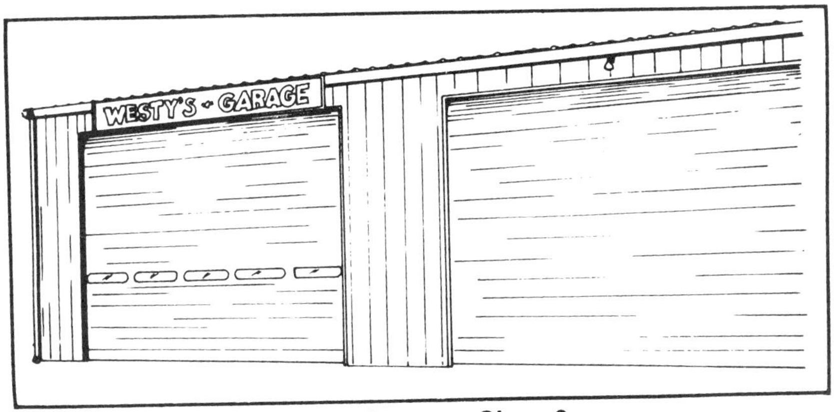

Service Garage, Class 3

	Square Foot Area										
Quality Class	2,000	2,500	3,000	4,000	5,000	6,000	7,500	10,000	15,000	20,000	30,000
1, Best	15.03	13.65	12.67	11.35	10.49	9.88	9.22	8.50	7.69	7.22	6.68
1 & 2	14.18	12.88	11.95	10.71	9.90	9.32	8.70	8.02	7.25	6.81	6.30
2, Good	13.35	12.12	11.25	10.08	9.32	8.77	8.19	7.55	6.82	6.41	5.93
2 & 3	12.48	11.33	10.51	9.42	8.71	8.20	7.65	7.06	6.38	5.99	5.54
3, Average	11.62	10.55	9.80	8.78	8.11	7.64	7.13	6.57	5.94	5.58	5.16
3 & 4	10.85	9.85	9.15	8.20	7.58	7.13	6.66	6.14	5.55	5.21	4.82
4, Low	10.10	9.17	8.51	7.62	7.05	6.64	6.19	5.71	5.16	4.85	4.49

Service Garages
Wood Frame
Length Between 2 and 4 Times Width

Estimating Procedure

1. Use these figures to estimate buildings designed primarily for motor vehicle repair. Sales area should be figured separately. Use the costs for urban stores beginning on page 42.
2. Establish the building quality class by applying the information on page 178.
3. Compute the floor area.
4. If the wall height is more or less than 16 feet, add or subtract from the square foot costs below the appropriate amount from the Wall Height Adjustment Table on page 182.
5. Multiply the adjusted square foot cost by the floor area.
6. Deduct for common walls or no wall ownership. Use the figures on page 182.
7. Multiply the total cost by the location factor on page 7.
8. Add the cost of heating and air conditioning systems, fire sprinklers, interior finish and partitions, drive-thru doors, canopies and walks, exterior signs, paving, curbing, and yard improvements. See page 201. Add the cost of hoists, pumps and compressors from pages 169 and 171.

Service Garage[rear portion] Class 2 & 3

	Square Foot Area										
Quality Class	2,000	2,500	3,000	4,000	5,000	6,000	7,500	10,000	15,000	20,000	30,000
1, Best	15.99	14.52	13.48	12.08	11.17	10.52	9.82	9.05	8.18	7.68	7.10
1 & 2	15.10	13.71	12.73	11.41	10.55	9.93	9.27	8.55	7.72	7.25	6.71
2, Good	14.20	12.90	11.97	10.73	9.92	9.34	8.72	8.04	7.26	6.82	6.31
2 & 3	13.31	12.09	11.22	10.06	9.30	8.76	8.17	7.54	6.81	6.39	5.91
3, Average	12.44	11.30	10.49	9.40	8.69	8.18	7.64	7.04	6.36	5.98	5.53
3 & 4	11.51	10.46	9.71	8.70	8.04	7.57	7.07	6.52	5.89	5.53	5.12
4, Low	10.62	9.64	8.95	8.02	7.42	6.98	6.52	6.01	5.43	5.10	4.72

Service Garages
Wood Frame
Length More Than 4 Times Width

Estimating Procedure

1. Use these figures to estimate buildings designed primarily for motor vehicle repair. Sales area should be figured separately. Use the costs for urban stores beginning on page 42.
2. Establish the building quality class by applying the information on page 178.
3. Compute the floor area.
4. If the wall height is more or less than 16 feet, add or subtract from the square foot costs below the appropriate amount from the Wall Height Adjustment Table on page 182.
5. Multiply the adjusted square foot cost by the floor area.
6. Deduct for common walls or no wall ownership. Use the figures on page 182.
7. Multiply the total cost by the location factor on page 7.
8. Add the cost of heating and air conditioning systems, fire sprinklers, interior finish and partitions, drive-thru doors, canopies and walks, exterior signs, paving, curbing, and yard improvements. See page 201. Add the cost of hoists, pumps and compressors from page 169 and 171.

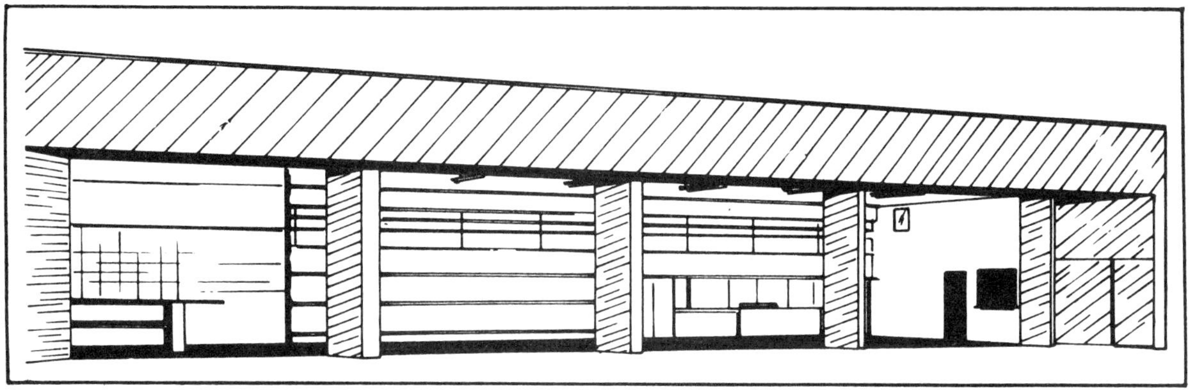

Service Garage, Class 2

	Square Foot Area										
Quality Class	2,000	2,500	3,000	4,000	5,000	6,000	7,500	10,000	15,000	20,000	30,000
1, Best	17.06	15.49	14.38	12.89	11.91	11.22	10.47	9.65	8.72	8.18	7.57
1 & 2	16.11	14.63	13.58	12.17	11.25	10.60	9.89	9.12	8.23	7.73	7.15
2, Good	15.17	13.78	12.79	11.47	10.60	9.98	9.32	8.59	7.76	7.28	6.73
2 & 3	14.21	12.90	11.98	10.74	9.92	9.34	8.72	8.04	7.26	6.81	6.30
3, Average	13.24	12.03	11.16	10.01	9.25	8.71	8.13	7.49	6.77	6.35	5.87
3 & 4	12.28	11.15	10.35	9.28	8.58	8.08	7.54	6.95	6.28	5.89	5.45
4, Low	11.34	10.30	9.56	8.57	7.92	7.46	6.96	6.42	5.80	5.44	5.03

Service Garages
Wood Frame
Wall Adjustments

Wall Height Adjustment

Add or subtract the amount listed in this table to the square foot of floor cost for each foot of wall height more or less than 16 feet.

	Square Foot Area										
Area	**2,000**	**2,500**	**3,000**	**4,000**	**5,000**	**6,000**	**7,500**	**10,000**	**15,000**	**20,000**	**30,000**
Cost	.18	.15	.13	.11	.10	.08	.07	.06	.05	.04	.03

Perimeter Wall Adjustment

A common wall exists when two buildings share one wall. Adjust for common walls by deducting the linear foot costs below from the total structure cost. In some structures one or more walls is not owned at all. In this case, deduct the "No Ownership" cost per linear foot of wall not owned.

For a Common Wall, deduct $23.00 per linear foot.
For No Wall Ownership, deduct $34.50 per linear foot.

Auto Service Centers
Masonry or Concrete
Quality Classification

	Class 1 Best Quality	Class 2 Good Quality	Class 3 Average Quality	Class 4 Low Quality
Foundation	Reinforced concrete.	Reinforced concrete.	Reinforced concrete.	Reinforced concrete.
Floor Structure	4" reinforced concrete on 6" rock fill.	4" reinforced concrete on 6" rock fill.	4" reinforced concrete on 6" rock fill.	4" reinforced concrete on 6" rock fill.
Wall Structure	8" reinforced decorative colored concrete block.	8" reinforced detailed concrete block.	8" reinforced concrete block.	8" reinforced concrete block.
Roof Structure	Steel open web joists, steel deck.	Glu-lams, 3" x 12" purlins 3' o.c. or wood open web joists (truss joists) 4' o.c.	Glu-lams, 3" x 12" purlins 3' o.c., or wood open web joists (truss joists) 4' o.c., ½" plywood sheathing.	Glu-lams, 3" x 12" purlins 3' o.c. or wood open web joists (truss joists) 4' o.c., ½" plywood sheathing.
Floor Finish	Concrete in wood area, vinyl asbestos tile in sales area.	Concrete in wood area, vinyl asbestos tile in sales area.	Concrete in wood area, inexpensive vinyl asbestos tile in sales area.	Concrete.
Interior Wall Finish	Painted concrete block in work area; gypsum wallboard, texture and paint in sales area.	Painted concrete block in work area and sales area.	Unfinished in work area, painted concrete block in sales area.	Unfinished.
Ceiling Finish	Open in work area, acoustical tile suspended in exposed grid in sales area.	Open in work area, acoustical tile suspended in exposed grid in sales area.	Open in work area, Celotex tile in sales area.	Open.
Exterior Finish	Decorative colored concrete block.	Colored or painted detailed block.	Painted concrete block.	Unpainted concrete block.
Display Front (Covers about 25% of the exterior wall)	¼" plate glass in good aluminum frame. Good aluminum and glass doors.	¼" plate glass in average aluminum frame. Aluminum and glass doors.	¼" plate glass in light aluminum frame. Wood and glass door.	Crystal glass in wood frame, wood door.
Roof and Cover	5 ply built-up roofing with insulation.	4 ply built-up roofing.	4 ply built-up roofing.	4 ply built-up roofing.
Plumbing	Two rest rooms with three good fixtures. Metal toilet partitions.	Two rest rooms with two average fixtures. Wood toilet partitions.	One rest room with two fixtures.	One rest room with two fixtures.
Electrical and Wiring	Conduit wiring with triple tube fluorescent strips, 8' o.c.	Conduit wiring with triple tube fluorescent strips, 8' o.c.	Conduit wiring, double tube fluorescent strips, 8' o.c.	Conduit wiring, incandescent fixtures, 10' o.c. or single tube fluorescent strips, 8' o.c.

Square foot costs include the cost of the following components: Foundations as required for normal soil conditions. Floor, wall, and roof structures. Interior floor, wall, and ceiling finishes (in sales area). Exterior wall finish and roof cover. Display windows. Interior partitions. Entry doors. Basic lighting and electrical systems. Rough and finish plumbing. Permits and fees. Contractors' mark-up.

Wall Height Adjustment

Add or subtract the amount listed in this table to the square foot of floor cost for each foot of wall height more or less than 16 feet.

Square Foot Area

Area	1,500	2,000	2,500	3,000	3,500	4,000	5,000	6,000	7,500	10,000	15,000
Cost	.68	.57	.51	.46	.42	.40	.36	.33	.30	.27	.23

Perimeter Wall Adjustment

For a Common Wall, deduct $55.00 per linear foot.
For No Wall Ownership, deduct $110.00 per linear foot.

Auto Service Centers
Masonry or Concrete
Length Less Than Twice Width

Estimating Procedure

1. Use these figures to estimate buildings designed for selling and installing automobile accessories. The square foot costs below allow for a sales area occupying 25% of the building space. The sales area has finished floors, walls and ceiling as described in the quality classification. The remaining 75% of the building is service area and has no interior finish.
2. Establish the building quality class by applying the information on page 183.
3. Compute the floor area. This should include everything within the exterior walls.
4. If the wall height is more or less than 16 feet, add to or subtract from the square foot cost below the appropriate amount from the Wall Height Adjustment Table on page 183.
5. Deduct for common walls or no wall ownership. See page 183.
6. Multiply the total cost by the location factor on page 7.
7. Add the cost of heating and air conditioning systems, fire sprinklers, canopies, walks, exterior signs, paving, curbing, loading docks, ramps, and yard improvements. See page 201. Add the cost of service station equipment from page 169.

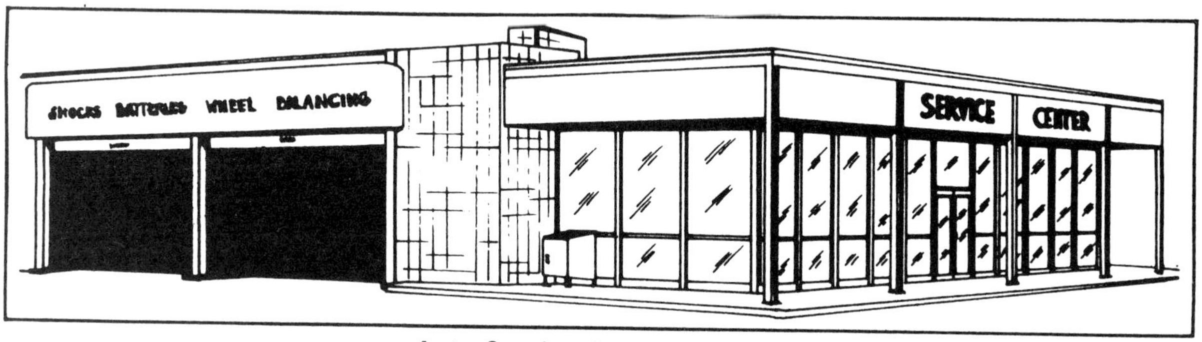

Auto Service Center, Class 2

Quality Class	Square Foot Area 1,500	2,000	2,500	3,000	3,500	4,000	5,000	6,000	7,500	10,000	15,000
Exceptional	44.61	40.82	38.32	36.53	35.16	34.08	32.45	31.28	30.00	28.59	26.95
1, Best	42.14	38.56	36.20	34.51	33.22	32.19	30.66	29.55	28.34	27.00	25.45
1 & 2	40.01	36.61	34.37	32.76	31.54	30.56	29.11	28.05	26.91	25.64	24.17
2, Good	37.85	34.63	32.51	30.99	29.83	28.92	27.54	26.54	25.46	24.25	22.86
2 & 3	35.88	32.83	30.82	29.38	28.28	27.41	26.10	25.16	24.13	22.99	21.67
3, Average	33.99	31.10	29.20	27.83	26.79	25.96	24.73	23.83	22.86	21.78	20.53
3 & 4	32.16	29.42	27.62	26.33	25.35	24.57	23.39	22.55	21.63	20.61	19.42
4, Low	30.50	27.90	26.20	24.97	24.04	23.30	22.19	21.38	20.51	19.54	18.42

Auto Service Centers
Masonry or Concrete
Length Between 2 and 4 Times Width

Estimating Procedure

1. Use these figures to estimate buildings designed for selling and installing automobile accessories. The square foot costs below allow for a sales area occupying 25% of the building space. The sales area has finished floors, walls and ceiling as described in the quality classification. The remaining 75% of the building is service area and has no interior finish.
2. Establish the building quality class by applying the information on page 183.
3. Compute the floor area. This should include everything within the exterior walls.
4. If the wall height is more or less than 16 feet, add to or subtract from the square foot cost below the appropriate amount from the Wall Height Adjustment Table on page 183.
5. Deduct for common walls or no wall ownership. See page 183.
6. Multiply the total cost by the location factor on page 7.
7. Add the cost of heating and air conditioning systems, fire sprinklers, canopies, walks, exterior signs, paving, curbing, loading docks, ramps, and yard improvements. See page 201. Add the cost of service station equipment from page 169.

Auto Service Center, Class 2

Square Foot Area

Quality Class	1,500	2,000	2,500	3,000	3,500	4,000	5,000	6,000	7,500	10,000	15,000
Exceptional	47.41	43.15	40.35	38.35	36.83	35.63	33.83	32.53	31.12	29.56	27.75
1, Best	44.75	40.73	38.09	36.21	34.77	33.63	31.93	30.70	29.37	27.90	26.20
1 & 2	42.48	38.66	36.16	34.36	33.00	31.92	30.31	29.14	27.88	26.48	24.87
2, Good	40.17	36.56	34.20	32.50	31.21	30.19	28.67	27.56	26.37	25.05	23.52
2 & 3	38.12	34.69	32.45	30.84	29.62	28.65	27.20	26.15	25.02	23.76	22.32
3, Average	35.98	32.75	30.63	29.11	27.96	27.04	25.68	24.69	23.62	22.43	21.07
3 & 4	34.18	31.11	29.09	27.65	26.55	25.69	24.39	23.45	22.43	21.31	20.01
4, Low	32.25	29.35	27.45	26.09	25.06	24.24	23.01	22.13	21.17	20.11	18.88

Auto Service Centers
Masonry or Concrete
Length More Than 4 Times Width

Estimating Procedure

1. Use these figures to estimate buildings designed for selling and installing automobile accessories. The square foot costs below allow for a sales area occupying 25% of the building space. The sales area has finished floors, walls and ceiling as described in the quality classification. The remaining 75% of the building is service area and has no interior finish.
2. Establish the building quality class by applying the information on page 183.
3. Compute the floor area. This should include everything within the exterior walls.
4. If the wall height is more or less than 16 feet, add to or subtract from the square foot cost below the appropriate amount from the Wall Height Adjustment Table on page 183.
5. Deduct for common walls or no wall ownership. See page 183.
6. Multiply the total cost by the location factor on page 7.
7. Add the cost of heating and air conditioning systems, fire sprinklers, canopies, walks, exterior signs, paving, curbing, loading docks, ramps, and yard improvements. See page 201. Add the cost of service station equipment from page 169.

Auto Service Center, Class 3

	Square Foot Area										
Quality Class	1,500	2,000	2,500	3,000	3,500	4,000	5,000	6,000	7,500	10,000	15,000
Exceptional	51.11	46.48	43.40	41.15	39.43	38.06	35.99	34.48	32.82	30.97	28.81
1, Best	48.22	43.86	40.94	38.83	37.21	35.91	33.96	32.53	30.97	29.22	27.18
1 & 2	45.75	41.61	38.85	36.84	35.30	34.07	32.22	30.86	29.38	27.72	25.79
2, Good	43.29	39.37	36.76	34.86	33.40	32.24	30.48	29.20	27.80	26.23	24.40
2 & 3	41.04	37.33	34.85	33.05	31.67	30.57	28.90	27.69	26.36	24.87	23.13
3, Average	38.85	35.33	32.99	31.28	29.98	28.93	27.36	26.21	24.95	23.54	21.90
3 & 4	36.86	33.52	31.29	29.68	28.44	27.45	25.95	24.86	23.67	22.33	20.77
4, Low	34.78	31.63	29.53	28.00	26.83	25.90	24.49	23.46	22.33	21.07	19.60

INDUSTRIAL STRUCTURES SECTION

Section Contents

Structure Type	Page
Warehouses	189
Light Industrial Buildings	190
Factory Buildings	191
Internal Offices	192
External Offices	192
Steel Buildings	194
Alternate Costs for Steel Buildings	195
Typical Lives for Commercial and Industrial Buildings	200
Additional Costs for Commercial and Industrial Buildings	201
Display Fronts	208
Signs	211
Yard Improvements	212

Industrial Buildings Quality Classification

	Class 1 Best Quality	Class 2 Good Quality	Class 3 Average Quality	Class 4 Low Quality
Foundations	Continuous reinforced concrete.	Continuous reinforced concrete.	Continuous reinforced concrete.	Reinforced concrete pads under pilasters.
Floor Structure	6" rock base, 6" concrete with reinforcing mesh or bars.	6" rock base, 6" concrete with reinforcing mesh or bars.	6" rock base, 5" concrete with reinforcing mesh or bars.	6" rock base, 4" concrete with reinforcing mesh.
Wall Structure	8" reinforced concrete block or brick with pilasters 20' o.c., painted sides and rear exterior, front wall brick veneer.	8" reinforced concrete block or brick with pilasters 20' o.c., painted sides and rear exterior, stucco and some brick veneer on front.	8" reinforced concrete block or brick, unpainted.	8" reinforced concrete block or brick, unpainted.
Wall Structure	6" concrete tilt-up, ornamental rock on ¼ of exterior walls, remainder painted and patterned panels.	6" concrete tilt-up exposed aggregate on ¼ of exterior walls or patterned and painted panels on ½ of exterior, remainder painted.	6" concrete tilt-up, unpainted.	6" concrete tilt-up, unpainted.
Roof Structure	Glu-lams, wood or steel trusses on steel intermediate columns, span exceeds 70'.	Glu-lams, wood or steel trusses on steel intermediate columns, span exceeds 70'.	Glu-lams or steel beams on steel intermediate columns, short span.	Glu-lams on steel intermediate columns, short span.
Roof Cover	Panelized roof system, ½" plywood sheathing, 5 ply built-up roof.	Panelized roof system, ½" plywood sheathing, 4 ply built-up roof.	Panelized roof system, ½" plywood sheathing, 4 ply built-up roof.	Panelized roof system, ½" plywood sheathing, 4 ply built-up roof.
Skylights	48 S.F. of skylight per 2500 S.F. of floor area (1-6' x 8' skylight 40' to 50' o.c.).	32 S.F. of skylight per 2500 S.F. of floor area (1-4' x 8' skylight 40' to 50' o.c.).	24 S.F. of skylight per 2500 S.F. of floor area (1-4' x 6' skylight 40' to 50' o.c.).	10 S.F. of skylight per 2500 S.F. of floor area (1-2' x 4' skylight 40' to 50' o.c.).
Ventilators	1 large rotary vent per 2500 S.F. of floor area.	1 medium rotary vent per 2500 S.F. of floor area.	1 medium rotary vent per 2500 S.F. of floor area.	1 small rotary vent per 2500 S.F. of floor area.
Rest Rooms, Finish	Good vinyl asbestos tile floors enameled gypsum wallboard partitions.	Vinyl asbestos tile floors, enameled gypsum wallboard partitions.	Concrete floors, painted gypsum wallboard partitions.	Concrete floors, unfinished sheet rock partitions.
Fixtures **Warehouses**	2 rest rooms, 3 good fixtures in each.	2 rest rooms, 3 average fixtures in each.	2 rest rooms, 2 average fixtures in each.	1 rest room, 2 low cost fixtures in each.
Light Industrial	Same as warehouse plus 1 additional fixture per 10,000 S.F. of floor area.	Same as warehouse plus 1 additional fixture per 10,000 S.F. of floor area.	Same as warehouse plus 1 additional fixture per 10,000 S.F. of floor area.	Same as warehouse plus 1 additional fixture per 20,000 S.F. of floor area.
Light Manufacturing	Same as warehouse plus 2 additional fixtures per 10,000 S.F. of floor area.	Same as warehouse plus 2 additional fixtures per 10,000 S.F. of floor area.	Same as warehouse plus 2 additional fixtures per 10,000 S.F. of floor area.	Same as warehouse plus 2 additional fixtures per 20,000 S.F. of floor area.
Lighting **Warehouses**	4' single tube fluorescent fixtures 10' x 12' spacing.	Low cost single tube fluorescent fixtures 12' x 20' spacing.	Low cost 4' single tube fluorescent fixtures 20' x 20' spacing.	Low cost incandescent fixtures 20' x 30' spacing.
Light Industrial	Continuous single tube fluorescent strips 8' o.c.	Continuous single tube fluorescent strips 12' o.c.	Continuous single tube fluorescent strips 16' o.c.	Continuous single tube fluorescent strips 20' o.c.
Light Manufacturing	Continuous triple tube fluorescent strip fixtures 8' o.c.	Continuous triple tube fluorescent strip fixtures 12' o.c.	Continuous double tube fluorescent strips 8' o.c.	Continuous double tube fluorescent strips 12' o.c.

Square foot costs include the following components: Foundations as required for normal soil conditions. Floor, wall, and roof structures. Exterior wall finish and roof cover. Basic lighting and electrical systems. Rough and finish plumbing. A usual or normal parapet wall. Walk-through doors. Contractors' mark-up.

Warehouses

Estimating Procedure

1. Establish the structure quality class by applying the information on page 188.
2. Compute the building floor area.
3. Add to or subtract from the square foot cost below the appropriate amount from the Wall Height Adjustment Table (at the bottom of page 191) if the wall height is more or less than 20 feet.
4. Multiply the adjusted square foot cost by the building area.
5. Deduct, if appropriate, for common walls, using the figures at the bottom of page 191.
6. Multiply the total cost by the location factor listed on page 7.
7. Add the cost of heating and air conditioning equipment, fire sprinklers, interior offices, drive-through or delivery doors, canopies, interior partitions, docks and ramps, paving and curbing and miscellaneous yard improvements. See the section beginning on page 201.

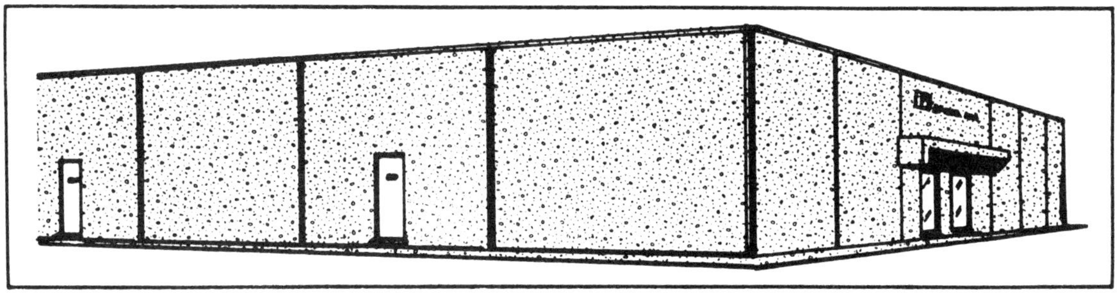

Warehouse, Class 3

Length Less Than Twice Width

Square Foot Area

Quality Class	3,000	4,000	5,000	7,500	10,000	15,000	20,000	30,000	50,000	100,000	200,000
1, Best	31.80	29.10	27.29	24.53	22.90	21.01	19.89	18.58	17.28	15.99	15.08
1 & 2	29.44	26.94	25.27	22.71	21.21	19.45	18.42	17.20	16.00	14.80	13.96
2, Good	27.11	24.81	23.27	20.91	19.53	17.91	16.96	15.84	14.73	13.63	12.86
2 & 3	25.21	23.07	21.63	19.44	18.16	16.65	15.77	14.73	13.70	12.67	11.96
3, Average	23.24	21.27	19.95	17.93	16.74	15.36	14.54	13.58	12.63	11.69	11.02
3 & 4	21.62	19.79	18.56	16.68	15.57	14.28	13.53	12.63	11.75	10.87	10.25
4, Low	19.93	18.24	17.11	15.37	14.36	13.17	12.47	11.65	10.83	10.02	9.45

Length Between 2 and 4 Times Width

Square Foot Area

Quality Class	3,000	4,000	5,000	7,500	10,000	15,000	20,000	30,000	50,000	100,000	200,000
1, Best	33.92	30.89	28.86	25.78	23.98	21.88	20.65	19.22	17.80	16.39	15.40
1 & 2	31.49	28.67	26.79	23.93	22.26	20.32	19.18	17.84	16.52	15.21	14.30
2, Good	28.99	26.40	24.66	22.03	20.49	18.70	17.65	16.42	15.21	14.00	13.16
2 & 3	26.96	24.54	22.93	20.48	19.05	17.39	16.41	15.27	14.14	13.02	12.24
3, Average	24.86	22.63	21.15	18.89	17.57	16.04	15.14	14.08	13.04	12.01	11.29
3 & 4	23.13	21.06	19.68	17.57	16.35	14.92	14.08	13.10	12.13	11.17	10.50
4, Low	21.24	19.34	18.07	16.14	15.02	13.71	12.94	12.04	11.15	10.26	9.65

Length More Than 4 Times Width

Square Foot Area

Quality Class	3,000	4,000	5,000	7,500	10,000	15,000	20,000	30,000	50,000	100,000	200,000
1, Best	36.76	33.39	31.13	27.65	25.61	23.21	21.79	20.13	18.47	16.82	15.66
1 & 2	34.12	31.00	28.89	25.67	23.77	21.54	20.23	18.68	17.14	15.61	14.53
2, Good	31.43	28.56	26.62	23.65	21.90	19.85	18.64	17.21	15.79	14.38	13.39
2 & 3	29.23	26.56	24.76	21.99	20.37	18.46	17.33	16.01	14.69	13.38	12.45
3, Average	26.96	24.50	22.83	20.28	18.78	17.03	15.99	14.76	13.55	12.34	11.48
3 & 4	25.05	22.76	21.21	18.85	17.45	15.82	14.85	13.72	12.59	11.46	10.67
4, Low	23.08	20.97	19.55	17.36	16.08	14.57	13.68	12.64	11.60	10.56	9.83

Light Industrial Buildings

Estimating Procedure

1. Establish the structure quality class by applying the information on page 188.
2. Compute the building floor area.
3. Add to or subtract from the square foot cost below the appropriate amount from the Wall Height Adjustment Table (at the bottom of page 191) if the wall height is more or less than 20 feet.
4. Multiply the adjusted square foot cost by the building area.
5. Deduct, if appropriate, for common walls, using the figures at the bottom of page 191.
6. Multiply the total cost by the location factor listed on page 7.
7. Add the cost of heating and air conditioning equipment, fire sprinklers, interior offices, drive-through or delivery doors, canopies, interior partitions, docks and ramps, paving and curbing and miscellaneous yard improvements. See the section beginning on page 201.

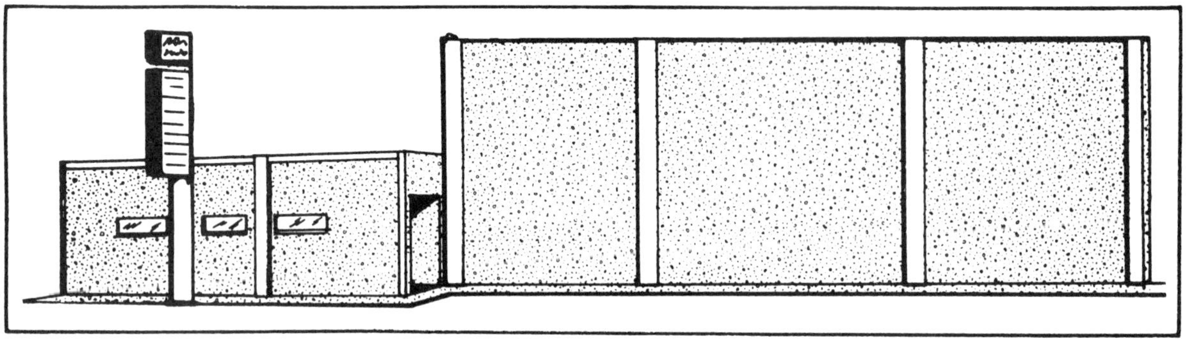

Light Industrial, Class 2

Length Less Than Twice Width

Square Foot Area

Quality Class	3,000	4,000	5,000	7,500	10,000	15,000	20,000	30,000	50,000	100,000	200,000
1, Best	37.54	34.66	32.72	29.75	28.01	25.97	24.76	23.34	21.94	20.53	19.55
1 & 2	34.82	32.15	30.35	27.60	25.98	24.08	22.96	21.65	20.34	19.04	18.13
2, Good	32.07	29.61	27.96	25.42	23.93	22.18	21.15	19.94	18.74	17.54	16.70
2 & 3	30.14	27.82	26.27	23.88	22.48	20.84	19.88	18.74	17.61	16.48	15.69
3, Average	27.64	25.52	24.09	21.90	20.62	19.12	18.23	17.18	16.15	15.12	14.39
3 & 4	25.74	23.77	22.44	20.40	19.21	17.81	16.98	16.01	15.04	14.08	13.40
4, Low	23.82	22.00	20.77	18.88	17.77	16.48	15.71	14.81	13.92	13.03	12.40

Length Between 2 and 4 Times Width

Square Foot Area

Quality Class	3,000	4,000	5,000	7,500	10,000	15,000	20,000	30,000	50,000	100,000	200,000
1, Best	39.93	36.67	34.49	31.16	29.22	26.95	25.62	24.06	22.51	20.98	19.91
1 & 2	37.13	34.10	32.07	28.98	27.17	25.06	23.82	22.37	20.94	19.51	18.51
2, Good	34.31	31.51	29.63	26.78	25.11	23.16	22.01	20.67	19.35	18.03	17.11
2 & 3	31.87	29.26	27.52	24.87	23.32	21.51	20.45	19.20	17.97	16.74	15.89
3, Average	29.41	27.01	25.41	22.96	21.52	19.85	18.87	17.72	16.59	15.46	14.67
3 & 4	27.39	25.16	23.66	21.38	20.05	18.49	17.58	16.51	15.45	14.39	13.66
4, Low	25.34	23.27	21.89	19.78	18.54	17.10	16.26	15.27	14.29	13.32	12.64

Length More Than 4 Times Width

Square Foot Area

Quality Class	3,000	4,000	5,000	7,500	10,000	15,000	20,000	30,000	50,000	100,000	200,000
1, Best	43.01	39.35	36.89	33.12	30.91	28.33	26.80	25.01	23.23	21.46	20.21
1 & 2	40.05	36.64	34.35	30.84	28.78	26.37	24.95	23.29	21.63	19.98	18.82
2, Good	36.92	33.78	31.67	28.44	26.54	24.32	23.01	21.47	19.94	18.42	17.35
2 & 3	34.39	31.46	29.49	26.48	24.72	22.65	21.43	19.99	18.57	17.15	16.16
3, Average	31.76	29.05	27.24	24.46	22.83	20.92	19.79	18.47	17.15	15.84	14.93
3 & 4	29.49	26.98	25.29	22.71	21.19	19.42	18.37	17.14	15.92	14.71	13.86
4, Low	27.33	25.00	23.44	21.05	19.64	18.00	17.03	15.89	14.76	13.63	12.84

Factory Buildings

Estimating Procedure

1. Establish the structure quality class by applying the information on page 188.
2. Compute the building floor area.
3. Add to or subtract from the square foot cost below the appropriate amount from the Wall Height Adjustment Table (at the bottom of this page) if the wall height is more or less than 20 feet.
4. Multiply the adjusted square foot cost by the building area.
5. Deduct, if appropriate, for common walls, using the figures at the bottom of this page.
6. Multiply the total cost by the location factor listed on page 7.
7. Add the cost of heating and air conditioning equipment, fire sprinklers, interior offices, drive-through or delivery doors, canopies, interior partitions, docks and ramps, paving and curbing and miscellaneous yard improvements. See the section beginning on page 201.

Length Less Than Twice Width

Square Foot Area

Quality Class	3,000	4,000	5,000	7,500	10,000	15,000	20,000	30,000	50,000	100,000	200,000
1, Best	40.69	37.79	35.85	32.88	31.14	29.10	27.91	26.50	25.11	23.72	22.76
1 & 2	37.98	35.28	33.46	30.69	29.07	27.17	26.05	24.74	23.44	22.15	21.24
2, Good	35.23	32.72	31.04	28.47	26.96	25.20	24.16	22.95	21.74	20.54	19.70
2 & 3	32.81	30.47	28.90	26.51	25.11	23.47	22.50	21.37	20.25	19.13	18.35
3, Average	30.35	28.19	26.74	24.53	23.23	21.71	20.82	19.77	18.73	17.70	16.98
3 & 4	28.19	26.19	24.84	22.78	21.57	20.17	19.34	18.36	17.40	16.44	15.77
4, Low	26.05	24.20	22.95	21.05	19.94	18.63	17.87	16.97	16.08	15.19	14.57

Length Between 2 and 4 Times Width

Square Foot Area

Quality Class	3,000	4,000	5,000	7,500	10,000	15,000	20,000	30,000	50,000	100,000	200,000
1, Best	42.98	39.75	37.58	34.28	32.35	30.10	28.77	27.22	25.68	24.16	23.09
1 & 2	40.12	37.10	35.08	32.00	30.20	28.10	26.86	25.41	23.98	22.55	21.55
2, Good	37.26	34.46	32.58	29.72	28.05	26.10	24.95	23.60	22.27	20.94	20.02
2 & 3	34.69	32.08	30.33	27.67	26.11	24.29	23.22	21.97	20.73	19.50	18.63
3, Average	32.06	29.65	28.04	25.58	24.14	22.46	21.47	20.31	19.16	18.02	17.23
3 & 4	29.80	27.56	26.06	23.77	22.43	20.87	19.95	18.88	17.81	16.75	16.01
4, Low	27.52	25.45	24.06	21.95	20.71	19.27	18.42	17.43	16.44	15.47	14.78

Length More Than 4 Times Width

Square Foot Area

Quality Class	3,000	4,000	5,000	7,500	10,000	15,000	20,000	30,000	50,000	100,000	200,000
1, Best	46.02	42.41	39.98	36.25	34.06	31.49	29.97	28.18	26.40	24.63	23.39
1 & 2	42.99	39.62	37.35	33.87	31.82	29.42	28.00	26.33	24.67	23.01	21.85
2, Good	39.90	36.77	34.67	31.44	29.53	27.30	25.99	24.44	22.90	21.36	20.28
2 & 3	37.17	34.26	32.30	29.29	27.52	25.44	24.21	22.77	21.33	19.90	18.89
3, Average	34.38	31.68	29.87	27.08	25.45	23.53	22.39	21.05	19.73	18.40	17.47
3 & 4	31.93	29.43	27.74	25.15	23.63	21.85	20.79	19.55	18.32	17.09	16.23
4, Low	29.47	27.16	25.60	23.21	21.81	20.16	19.19	18.05	16.91	15.77	14.98

Wall Height Adjustment

Industrial building costs are based on a 20' wall height, measured from the bottom of the floor slab to the top of the roof cover. Add or subtract the amount listed to the square foot cost for each foot more or less than 20 feet.

Square Foot Area

Area	3,000	4,000	5,000	7,500	10,000	15,000	20,000	30,000	50,000	100,000	200,000
Cost	.30	.27	.25	.21	.19	.15	.13	.10	.07	.04	.02

Perimeter Wall Adjustment

A common wall exists when two buildings share one wall. Adjust for Common Walls by deducting $49.50 per linear foot from the total structure cost. In some structures one or more walls is not owned at all. In this case, deduct $99.00 per linear foot of wall not owned.

Internal Offices

Internal offices are office areas built into the interior area of an industrial building. Add the square foot costs in this section to the basic building cost. The costs include floor finish, partition framing, wall finish, trim and doors, counter, ceiling structure, ceiling finish, and the difference between the cost of lighting and windows in an industrial building and the cost of lighting and windows in an office building.

Plumbing costs are not included in these internal office costs. If a building has fixtures in excess of those listed in the quality classification for the main building, add $430 to $690 for each extra fixture. For two-story internal offices, apply a square foot cost based on the total area of both floor levels.

	Best	Good	Average	Low Cost
Floor Finish	Carpet, some vinyl tile or terrazzo.	Vinyl asbestos tile, some carpet.	Vinyl asbestos tile.	Colored concrete.
Wall Finish	Hardwood veneer, some textured cloth wall cover.	Gypsum board, texture and paint, some hardwood veneer.	Gypsum board, texture and paint.	Plywood and paint.
Ceiling Finish	Illuminated plastic ceilings.	Suspended "T" bar and acoustical tile.	Gypsum board, texture and paint.	Exposed or plywood and paint.
Doors - Interior	Good grade hardwood.	Good grade hardwood.	Low cost hardwood.	Standard paint grade.
Exterior	Pair aluminum entry doors with extensive sidelights.	Aluminum entry door with sidelights.	Wood store door.	Standard paint grade.
Windows	Average amount in good aluminum frame, fixed plate glass in good frame in front.	Average amount in good aluminum frame, some fixed plate glass.	Average amount of average cost aluminum sliding type.	Average amount of low cost aluminum sliding type.
Counters	Good grade hardwood counter, good shelving below.	Good grade plastic counter, average amount of shelving below.	Low cost plastic counter, average amount of shelving.	Paint grade counter, small amount of shelving.
Lighting	Illuminated ceilings.	Recessed fluorescent fixtures.	Average amount of fluorescent fixtures.	One incandescent per each 150 S.F.
Square Foot Costs	$22.60 to $31.30	$17.60 to $22.70	$14.50 to $18.30	$7.00 to $10.25
Wall Height Adjustment*	$.30 to $.45	$.28 to $.38	$.29 to $.36	$.22 to $.32

The range of costs is primarily due to the density of partitions in the office area. Use the lower costs for offices with larger rooms and fewer partitions.

*Wall Height Adjustment. Add or subtract the amount listed in this column to the square foot cost for each foot when the office wall height is more or less than 8 feet.

External Offices

External offices outside the main building walls have one side that is common to the main building. Square foot costs for external office areas should be estimated with the figures for general office buildings with interior suite entrances. See pages 106 to 109 or 114 to 117.

In selecting a square foot cost for the external office area as well as the main building area, the area of each portion should be used separately for area classification. The area and the perimeter formed by these individual areas is used for shape classification of the main area and for calculating the area and perimeter relationship of the office area. A "no wall ownership" deduction based on the office wall cost should be made for the length of the common side. Wall height adjustments should be made for each area individually.

Steel Buildings

Engineered steel buildings are constructed to serve as warehouses, factories, airplane hangars, garages, and stores, among other uses. They are generally either high-profile (4 in 12 rise roof) or low-profile (1 in 12 rise roof).

The square foot costs for the basic building include: Foundations as required for normal soil conditions. A 4 inch concrete floor with reinforcing mesh and a 2 inch sand fill. A steel building made up of steel frames or bents set 20 or 24 feet on centers, steel roof purlins 4½ to 5½ feet on centers, steel wall girts 3½ to 4½ feet on centers, post and beam type end wall frames, 26-gauge galvanized steel on ends, sides, and roof, window area equal to 2% of the floor area. Basic wiring and minimum lighting fixtures. One small or medium gravity vent per 2,500 square feet of floor area.

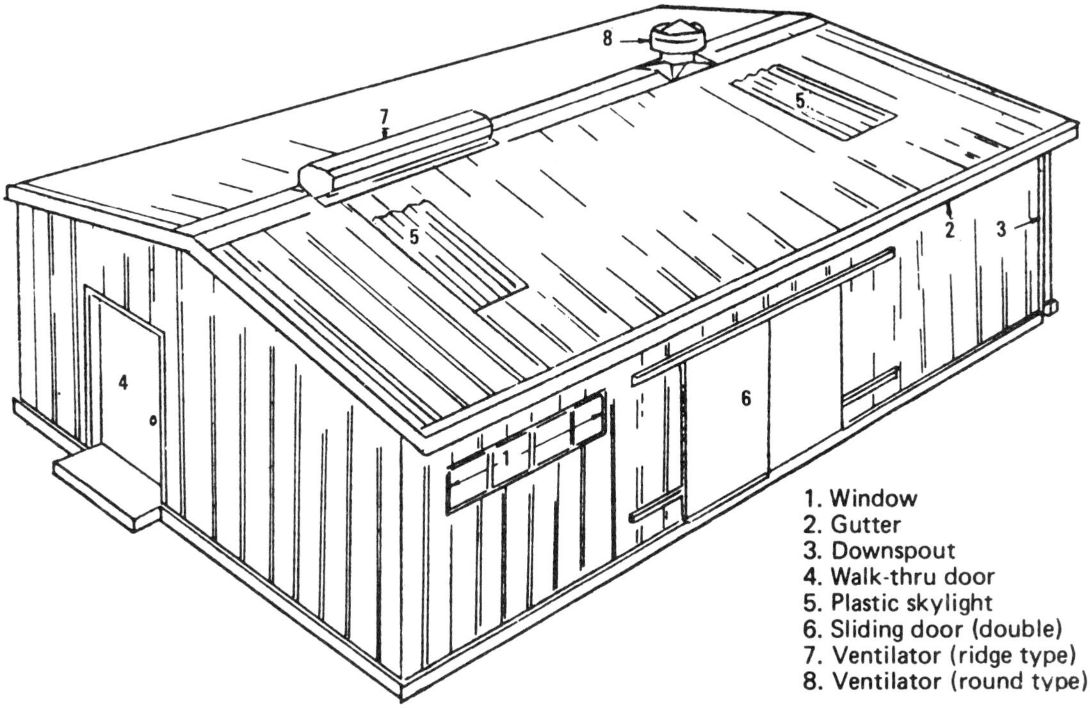

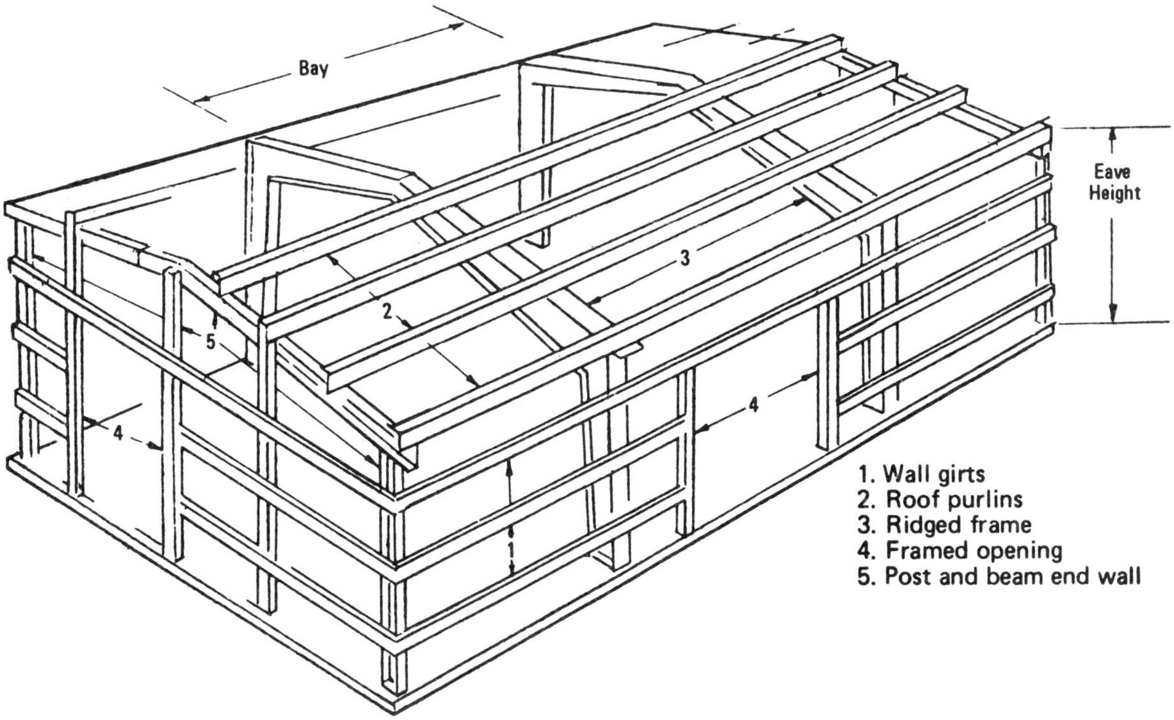

Steel Buildings

Estimating Procedure

1. Compute the building floor area.
2. Add to or subtract from the square foot cost below $.17 for each foot of wall height at the eave more or less than 14 feet. Subtract $.20 from the square foot cost below for low profile buildings.
3. Multiply the adjusted square foot cost by the building area.
4. Multiply the total by the appropriate live load adjustment. The basic building costs are for a 12 pound live load usually built where snow load is not a design factor. A 20 pound live load building is usually found in light snow areas, and a 30 pound live load building is usually found in heavy snow areas.
5. Add or subtract alternate costs from pages 195 and 196. Alternate costs reflect the difference between unit costs of components that are included in the basic building square foot costs and the unit costs of alternate components. Total alternate cost is found by multiplying the alternate cost by the area or number of alternate components.
6. Add the appropriate costs from pages 197 to 199, "Additional Costs for Steel Buildings." These costs reflect the in-place cost of components that are not included in the basic square foot cost but are often found as part of steel buildings. The cost of items that alter or replace a portion of the building reflect the net added cost of the component in-place. The cost of the item that is replaced has been deducted from the total cost of the additive components. No further deduction is necessary.
8. Add the cost of heating and air conditioning systems, fire sprinklers, plumbing, additional electrical work, interior partitions and offices, dock high additive costs, loading docks or ramps, paving and curbing, miscellaneous yard improvements. See the section beginning on page 201.

High Profile Rigid Steel Buildings

Square Foot Cost

Length

Width	60′	80′	120′	Over 140′
20′	$10.60	$9.80	$9.60	$9.20
30′	9.40	8.80	8.20	8.00
40′	8.80	8.00	7.40	7.30
50′ to 70′	8.00	6.90	6.40	6.30
80′ to 140′	---	6.50	6.30	5.20

Deduct .20 cents per square foot for low profile buildings

Live Load Adjustments

	High Profile				Low Profile			
Building	20 Lb. Live Load		30 Lb. Live Load		20 Lb. Live Load		30 Lb. Live Load	
Width	20′ Bays	24′ Bays	20′ Bays	24′ Bays	20′ Bays	24′ Bays	20′ Bays	24′ Bays
20′	1.00	1.01	1.05	1.07	1.00	1.00	1.03	1.04
24′	1.01	1.02	1.06	1.08	1.00	1.00	1.07	1.09
28′	1.01	1.02	1.06	1.08	--	--	--	--
30′	1.01	1.02	1.06	1.08	1.01	1.02	1.09	1.10
32″	1.01	1.02	1.07	1.09	1.01	1.02	1.10	1.11
36′	1.01	1.02	1.08	1.10	1.01	1.02	1.10	1.11
40′	1.02	1.03	1.08	1.11	1.01	1.02	1.10	1.11
50′	1.04	1.06	1.10	1.15	1.02	1.03	1.10	1.12
60′	1.05	1.07	1.12	1.18	1.03	1.04	1.15	1.18
70′	1.07	1.09	1.14	1.20	1.03	1.04	1.15	1.19
80′	1.08	1.10	1.15	1.21	1.04	1.05	1.16	1.20
90′	1.09	1.10	1.16	1.22	1.04	1.05	1.17	1.21
100′	1.09	1.10	1.17	1.23	1.04	1.05	1.17	1.22
110′	1.10	1.11	1.18	1.24	1.05	1.06	1.18	1.24
120′	1.11	1.12	1.19	1.25	1.06	1.07	1.19	1.27
130′	--	--	--	--	1.07	1.08	1.20	1.30
140′	1.13	1.14	1.20	1.27	1.08	1.09	1.22	1.33
150′	--	--	--	--	1.09	1.10	1.24	1.35

Alternate Costs for Steel Buildings

Floors, cost difference per square foot of floor

4" concrete with reinforced mesh and 2" sand fill	0
5" concrete with reinforced mesh and 2" sand fill	+.15
6" concrete with reinforced mesh and 2" sand fill	+.30
6" concrete with ½" rebars, 12" x 12" o.c., and 2" sand fill	+.60
3" plant mix asphalt on 4" untreated rock base	-.43
2" plant mix asphalt on 4" untreated rock base	-.52
Gravel (compacted)	-1.00
Dirt (compacted)	-1.30

Roof Covering, cost difference per square foot of roof

26 gauge galvanized steel	0
24 gauge galvanized steel	+.08
22 gauge galvanized steel	+.28
26 gauge colored steel	+.24
24 gauge colored steel	+.37
.024 aluminum	+.20
.032 aluminum	+.41
.032 colored aluminum	+.60
No roof cover (just purlins)	-.89

Wall Covering, cost difference per square foot of wall

26 gauge colored steel	+.24
24 gauge colored steel	+.37
26 gauge galvanized steel	0
24 gauge galvanized steel	+.08
22 gauge galvanized steel	+.28
26 gauge aluminum and steel	+.19
24 gauge aluminum and steel	+.31
.024 aluminum	+.20
.032 aluminum	+.42
.032 colored aluminum	+.63
26 gauge galvanized steel insulated panel	+.27
26 gauge colored steel insulated panel	+.56
26 gauge aluminum steel insulated panel	+.45
26 gauge galvanized steel mono panel	+.58
26 gauge colored steel mono panel	+.86
No wall cover (just girts)	-.98

The above costs are to be added to or subtracted from basic building costs.

Open Endwall (includes ridged frame in lieu of post and beam), low profile building

Cost Deduction Per Each Endwall

	Eave Heights						
Width	10'	12'	14'	16'	18'	20'	24'
20'	$290	$320	$400	$465	---	---	---
24'	400	430	530	620	---	---	---
30'	475	555	655	765	---	---	---
32'	600	590	710	830	$ 900	$1,050	$1,300
36'	560	630	---	---	---	---	---
40'	600	690	830	945	1,050	1,230	1,490
50'	655	765	955	1,090	1,250	1,450	1,760
60'	730	865	1,050	1,250	1,500	1,760	2,160
70'	---	940	1,160	1,380	1,650	1,900	2,440
80'	----	990	1,200	1,500	1,790	2,050	2,850
90'	---	965	1,300	1,450	1,920	2,150	3,000
100'	---	---	1,090	1,440	1,790	2,140	3,100
110'	---	---	1,020	1,350	1,710	2,100	3,190
120'	---	---	900	1,170	1,650	2,040	3,190
130'	---	---	775	1,130	1,550	1,910	3,100
140'	---	---	565	940	1,500	1,900	3,040
150'	---	---	445	865	1,290	1,760	2,950

These costs are to be subtracted from the basic building cost.

Open Sidewalls, per square foot of wall

Eave Height	10'	12'	14'	16'	18'	20'	24'
	$1.75	$1.73	$1.72	$1.69	$1.72	$1.72	$1.75

These costs are to be subtracted from the basic building cost.

Alternate Costs for Steel Buildings

Open Endwall (includes ridged frame in lieu of post & beam), high profile building

Cost Deduction Per Each Endwall
Eave Heights

Width	10′	12′	14′	16′	18′	20′	24′
20′	$ 345	$ 355	$ 500	---	---	---	---
24′	465	520	655	---	---	---	---
28′	590	640	755	---	---	---	---
30′	645	710	845	---	---	---	---
32′	675	730	865	$1,010	$1,150	$1,300	$1,490
36′	755	820	1,000	---	---	---	---
40′	890	965	1,130	1,290	1,490	1,630	1,900
50′	1,140	1,310	1,490	1,630	1,760	1,900	2,450
60′	1,490	1,630	1,900	2,040	2,290	2,590	3,140
70′	1,760	2,160	2,300	2,590	2,850	3,140	3,790
80′	2,080	2,440	2,590	2,850	3,140	3,530	4,190
90′	---	3,000	3,400	3,650	2,740	4,300	5,230
100′	---	3,130	3,650	3,920	4,320	4,800	5,750
110′	---	---	4,700	5,230	5,750	6,140	7,190

Ridged Frame Endwall in lieu of post and beam, high profile building

Cost Per Each Endwall
Eave Heights

Width	10′	12′	14′	16′	18′	20′	24′
20′	$ 320	$ 345	$ 355	---	---	---	---
24′	345	355	365	---	---	---	---
28′	365	410	420	---	---	---	---
30′	390	420	430	---	---	---	---
32′	420	445	465	$ 520	$ 575	$ 610	$ 720
36′	490	510	545	---	---	---	---
40′	530	575	610	655	710	740	890
50′	730	775	820	865	930	965	1,130
60′	920	975	1,030	1,090	1,140	1,180	1,340
70′	1,270	1,300	1,490	1,430	1,480	1,500	1,630
80′	1,430	1,500	1,630	1,630	1,760	1,900	2,040
90′	---	1,830	2,040	2,160	2,160	2,300	2,590
100′	---	2,290	2,490	2,590	2,720	2,850	3,140
110′	---	---	2,880	3,140	3,140	3,290	3,650

Ridged Frame Endwall in lieu of post and beam, low profile building

Cost Per Each Endwall
Eave Heights

Width	10′	12′	14′	16′	18′	20′	24′
20′	$ 365	$ 400	$ 420	$ 430	---	---	---
24′	400	410	430	445	---	---	---
30′	410	420	445	455	---	---	---
32′	500	510	545	565	$ 590	$ 600	$ 675
36′	545	575	---	---	---	---	---
40′	565	610	645	675	710	730	820
50′	790	840	865	900	955	1,000	1,140
60′	1,020	1,050	1,100	1,140	1,160	1,200	1,350
70′	---	1,350	1,450	1,490	1,490	1,600	1,760
80′	---	1,630	1,760	1,830	1,900	1,960	2,040
90′	---	2,040	2,160	2,290	2,490	2,500	2,600
100′	---	---	2,740	2,800	2,870	3,000	3,100
110′	---	---	3,400	3,400	3,500	3,530	3,790
120′	---	---	3,920	4,050	4,190	4,300	4,450
130′	---	---	4,450	4,700	4,800	4,590	5,300
140′	---	---	5,360	5,500	5,700	5,750	6,000

Alternate Costs for Steel Buildings

Canopies (sidewall location), cost per linear foot

Type and Size	5′	6′	8′	10′	12′	15′
	Width					
High profile 20′ bays	$17.40	$19.70	$24.60	$28.60	$31.70	$36.20
High profile 24′ bays	16.80	19.00	23.70	27.60	30.20	34.70
Low profile 20′ bays	13.65	15.80	20.10	24.00	27.80	32.90
Low profile 24′ bays	13.20	15.10	19.30	23.00	26.30	31.20
	Cost Per Each Canopy					
Basic end cost	$129.00	$157.00	$224.00	$295.00	$395.00	$500.00

The above costs are for 26 gauge galvanized roofing.

For endwall location, add the following to sidewall location costs above.

Eave Height	10′	12′	14′	16′	18′	20′	24′
Cost	$155.00	$190.00	$195.00	$225.00	$245.00	$258.00	$350.00

Sliding Sidewall Doors, cost per opening, including framed opening.

Single Doors	10′	12′	14′	16′	18′	20′	24′
8′ x 9′	$ 570	$ 630	$ 640	$ 650	$ 670	$ 695	$ 695
10′ x 9′	615	660	670	700	715	715	730
10′ x 11′	---	640	705	715	730	730	740
10′ x 13′	----	---	660	730	740	740	760
12′ x 11′	---	740	810	830	840	850	860
12′ x 13′	---	---	785	850	860	890	910
12′ x 15′	---	---	---	850	910	920	930
12′ x 9′	818	850	885	920	930	940	960
12′ x 11′	---	850	930	940	960	975	985
12′ x 13′	---	---	910	960	975	985	1,000
12′ x 15′	---	---	---	950	1,000	1,020	1,040
14′ x 9′	910	985	1,000	1,020	1,040	1,050	1,060
14′ x 11′	---	1,000	930	1,050	1,060	1,070	1,080
14′ x 13′	---	---	975	1,060	1,090	1,100	1,100
14′ x 15′	---	---	---	1,020	1,120	1,160	1,170
16′ x 9′	1,030	1,110	1,150	1,160	1,180	1,180	1,190
16′ x 11′	---	1,050	1,170	1,180	1,190	1,200	1,230
16′ x 13′	---	---	1,110	1,210	1,250	1,280	1,300

The above costs are for 26 gauge colored sliding.
If door is in an endwall, use the lowest cost of size desired and add $115.00.

Hollow Metal Walk-Thru Doors, cost per opening including framed opening.

Door Size	1-3/8″ Thick				1-3/4″ Thick			
	Flush		Half Glass		Flush		Half Glass	
	Galv.	Colored	Galv.	Colored	Galv.	Colored	Galv.	Colored
2′0″ x 6′8″ single	$285	$310	$350	$370	--	--	--	--
2′6″ x 6′8″ single	285	315	350	380	--	--	--	--
3′0″ x 6′8″ single	295	340	360	415	$415	$460	$470	$510
3′0″ x 7′0″ single	310	350	370	425	425	470	480	535
3′4″ x 7′0″ single	---	---	---	---	435	490	500	555
4′0″ x 7′0″ single	---	---	---	---	480	545	575	620
6′0″ x 6′8″ pair	480	580	580	665	665	750	785	880
6′0″ x 7′0″ pair	510	610	610	700	700	785	820	905
8′0″ x 7′0″ pair	---	---	---	---	750	880	905	1,030

Half glass doors include glazing.

Additives:

When walk-thru door is used as a pilot door in a sliding door, add $45.00 to the costs above.

Heavy duty lock set	Add $ 55 each		
Door closer	Add $125 each	Panic hardware pairs	Add $575 pair
Panic hardware single	Add $240 each	Weatherstripping	Add $10.50 each

Additional Costs for Steel Buildings

These costs are to be added to the basic building cost.

Framed Openings, cost per opening

	Eave Height						
Location	10′	12′	14′	16′	18′	20′	24′
Sidewalls	$195	$195	$220	$230	$280	$310	$370
Endwalls	280	280	290	300	350	380	440

These costs are for the standard widths of 8′, 10′, 12′, 14′, 16′, 18′, and 20′. If the width is other than standard **add** $73.00 per opening. If there is only one opening per building, add $95.00.

Gutters and Downspouts, cost per linear foot

Eave gutter (4″)	
24 gauge galvanized steel	$4.75
24 gauge colored steel	5.00
.032 aluminum	5.25
Valley gutter (6″)	
12 gauge galvanized steel	12.50
24 gauge galvanized steel	8.60
Downspouts	
4″ x 4″ 26 gauge galvanized steel	3.20
4″ x 4″ 26 gauge colored steel	3.40
6″ x 6″ 24 gauge galvanized steel	3.55
6″ x 6″ 24 gauge colored steel	4.10

Door Hoods, cost each

Width of Door	Galvanized	Colored
12′	$82.00	$96.00
16′	105.00	117.00
20′	125.00	138.00

Louvers, cost each, including screens

	Fixed		Adjustable	
Size	Galv.	Color	Galv.	Color
3′ x 2′	$175	$185	$240	$250
3′ x 3′	185	200	250	270
5′ x 5′	225	225	355	385

Insulation, cost per square foot of surface area

	Roof Thickness			Wall Thickness		
Size and Type	1″	1½″	2″	1″	1½″	2″
.6 pound density						
White vinyl faced	$.50	$.56	$.63	$.58	$.63	$.72
Colored vinyl faced	.64	.70	.76	.72	.76	.86
Aluminum faced	.72	.76	.86	.79	.87	.93
.75 pound density						
White vinyl faced	.56	.60	.70	.63	.70	.76
Colored vinyl faced	.70	.74	.82	.76	.82	.90
Aluminum faced	.76	.83	.91	.87	.91	1.00

Overhangs (sidewall location), cost per linear foot of overhang, 26 gauge galvanized

	20′ Bays			24′ Bays		
Size and Type	3′	4′	5′	3′	4′	5′
No soffit	$11.00	$13.40	$15.80	$10.30	$12.30	$14.70
Soffit	24.00	28.20	32.00	23.20	27.00	30.80
Color, soffit	24.60	28.80	32.70	23.70	27.80	31.60
Color, no soffit	11.90	14.60	17.00	11.00	13.40	15.90
Color, galvanized soffit	24.90	29.30	33.40	24.00	28.00	32.30
Color, color soffit	25.40	29.90	34.30	37.40	28.70	32.90

Overhangs (endwall location), cost per linear foot of overhang, 26 gauge, galvanized

Size and Type	3′	4′	5′	6′
No soffit	$7.05	$9.40	$11.80	$13.80
Galvanized soffit	17.00	21.60	25.60	29.10
Color, soffit	17.60	21.00	26.30	29.90
Color, no soffit	7.70	10.30	12.90	15.60
Color, galvanized soffit	17.80	22.50	26.80	30.50
Color, color soffit	18.30	22.80	27.30	30.30

Additional Costs for Steel Buildings

These costs are to be added to the basic building cost

Partitions, Interior Steel, cost per square foot

26 gauge galvanized, without flashing	$2.15
26 gauge galvanized, with flashing	2.50

Skylights, Plastic panels, cost each

	3′ x 6′	3′ x 9′	3′ x 11′
Standard			
6 oz. (green or white)	$26.50	$39.00	$45.00
8 oz. (green or white)	32.00	48.00	56.00
Tedlar (self-extinguishing)	53.00	79.00	96.00
Insulite (clear finish)	104.00	153.00	185.00

Ventilators

Round Type, includes screen

Cost Per Each

	Stationary			Rotary		
Diameter	Galv.	Colored	Aluminum	Galv.	Colored	Aluminum
12″	$100	$105	$165	$150	$165	$215
16″	140	145	195	200	220	270
20″	165	175	225	250	260	310
24″	180	185	260	275	280	350

Ventilator-Dampers, cost each

Diameter	12″	16″	20″	24″
Damper only	$36	$49	$54	$59
Damper with cords and pulleys	73	90	105	120

Steel Sliding Windows, includes glass and screens, cost per window

3′ x 2′-6″	$190
6′ x 2′-6″	230
6′ x 3′-8″	270

Ventilators, ridge type, includes screen and damper

Throat Size	4″	9″	12″	14″
Cost per linear foot, galvanized	$24.50	$39.00	$48.00	$55.00
Cost per linear foot, colored	27.00	42.50	52.50	59.40

Continuous Ridge Ventilator, includes screen and damper, cost per 10 foot unit

	First 10 Ft.		Each Add. 10 Ft.	
Size & Type	Galv.	Color	Galv.	Color
9″ throat	$355	$390	$320	$355
10″ throat	390	420	355	390
12″ throat	500	555	455	500

Smoke and Heat Vents, automatic control, cost per 10 foot unit

	First 10 Feet		Each Add. 10 Ft.	
Size & Type	Galv.	Color	Galv.	Color
9″ ridge mounted	$1,340	$1,400	$1,070	$1,120
9″ slope mounted	1,390	1,470	1,120	1,180

Add for operators:
One or two 10 foot sections, add $49.00.
Two to seven 10 foot sections, add $89.00.

Aluminum Industrial Windows, includes glass and screens, cost per window

Size and Type	Project Out	Fixed
3′ x 2′-6″	$178	$125
3′ x 3′-8″	200	145
6′ x 2′-6″	250	185
6′ x 3′-8″	315	215

Aluminum Sliding Windows, includes glass and screens, cost per window

	Height				
Width	2′	2′-6″	3′	3′-6″	4′
2′	$138.00	$150.00	$160.00	$165.00	$175.00
3′	150.00	160.00	165.00	175.00	185.00
4′	---	165.00	180.00	185.00	200.00
5′	---	180.00	195.00	220.00	230.00
6′	---	205.00	220.00	250.00	265.00

If window is fixed, deduct $2.75 per window. For mullions add $5.50 each.

Typical Physical Lives in Years for Commercial and Industrial Structures, by Quality Class

Building Type	Masonry or Concrete 1	2	3	4	5	6	Wood or Wood and Steel Frame 1	2	3	4	5
Urban Stores	70	70	70	60	60	60	70	60	60	60	---
Suburban Stores	70	60	60	60	60	--	60	50	50	45	--
Supermarkets	70	60	60	60	--	--	60	50	50	45	--
Small Food Stores	60	60	60	60	--	--	50	50	45	45	--
Discount Houses	70	60	60	60	--	--	60	50	50	45	--
Banks and Savings Offices	70	70	70	60	60	--	60	60	60	50	50
Department Stores	70	60	60	60	--	--	60	50	50	45	--
General Office Buildings	60	60	60	60	--	--	60	50	50	45	--
Medical-Dental Buildings	60	60	60	50	--	--	60	50	50	45	--
Convalescent Hospitals	60	60	60	50	--	--	55	50	50	45	--
Funeral Homes	70	70	70	60	60	--	60	60	60	50	50
Restaurants	70	60	60	60	--	--	60	50	50	45	--
Theaters	50	50	50	50	--	--	50	45	45	40	--
Service Garages	60	60	50	50	--	--	45	45	40	40	--
Auto Service Centers	50	50	45	45	--	--	--	--	--	--	--
Warehouses	55	55	50	50	--	--	--	--	--	--	--
Light Industrial Buildings	55	55	50	50	--	--	--	--	--	--	--
Factory Buildings	40	40	35	35	--	--	--	--	--	--	--

Service Stations located on main highways or in high land value areas can be expected to become obsolete in 20 years. Other service stations can be expected to become obsolete in 25 years. Reinforced Concrete Department Stores have a typical physical life of 80 years for class one structures and 70 years for lower quality class structures. Steel Buildings have a typical physical life of 50 years.

Normal Percent Good Table for Commercial and Industrial Structures

Average Life In Years

	20 Years		25 Years		30 Years		35 Years	
Age	Rem. Life.	% Good	Rem. Life	% Good	Rem. Life	% Good	Rem. Life	% Good
0	20	100	25	100	30	100	35	100
1	19	95	24	97	29	98	34	99
2	18	90	23	93	28	96	33	97
3	17	85	22	90	27	93	32	95
4	16	79	21	86	26	90	31	93
5	15	73	20	82	25	88	30	91
6	14	67	19	78	24	85	29	89
7	13	61	18	74	23	82	28	87
8	12	56	17	70	22	79	27	85
9	11	51	16	65	21	75	26	83
10	10	49	15	60	20	72	25	80
11	9	48	14	56	19	68	24	78
12	9	46	13	52	18	65	23	75
13	8	44	12	50	17	61	22	72
14	7	43	11	48	16	58	21	69
15	6	43	10	47	15	54	20	66
16	6	41	9	46	14	50	19	63
17	5	39	8	45	13	49	18	60
18	5	38	8	44	12	48	17	57
19	5	37	7	43	12	47	16	54
20	4	35	7	42	11	47	15	51
21	4	34	6	41	11	46	14	50
22	4	33	6	40	10	45	13	49
23	3	32	5	39	10	44	13	48
24	3	30	5	38	9	43	12	47
25	3	29	5	37	9	43	12	47
26	3	28	4	36	8	42	11	46
27	2	27	4	35	8	41	11	45
28	2	25	4	34	7	40	10	44
29	2	24	4	33	7	39	10	43
30	2	22	3	32	6	38	9	43
31	2	21	3	31	6	37	9	42
32	1	20	3	30	5	36	8	42
33	--	--	3	29	5	35	8	41
34	--	--	3	28	5	35	7	40
35	--	--	2	27	5	34	7	39
36	--	--	2	26	4	33	6	38
38	--	--	2	24	4	32	6	37
40	--	--	2	22	3	30	5	36
42	--	--	1	20	3	28	5	34
45	--	--	--	--	2	26	4	32
48	--	--	--	--	2	23	3	30
52	--	--	--	--	1	20	3	27
56	--	--	--	--	--	--	2	24
62	--	--	--	--	--	--	1	20

Average Life In Years

	40 Years		45 Years		50 Years		55 Years	
Age	Rem. Life	% Good	Rem. Life	% Good	Rem. Life	% Good	Rem. Life	% Good
0	40	100	45	100	50	100	55	100
2	38	98	43	99	48	99	53	99
4	36	96	41	97	46	98	51	98
6	34	93	39	95	44	97	49	97
8	32	90	37	93	42	95	47	96
10	30	86	35	90	40	93	45	95
12	28	82	33	87	38	91	43	94
14	26	78	31	84	36	88	41	92
16	24	73	29	81	34	85	39	90
18	22	68	27	77	32	82	37	88
20	20	63	25	73	30	80	35	86
22	18	58	23	69	28	77	33	83
24	17	53	21	65	26	73	31	80
26	15	50	20	60	24	69	29	77
28	14	48	18	55	23	65	27	74
30	13	47	17	50	21	61	26	71
32	11	45	15	49	20	57	24	67
34	10	44	14	48	18	53	22	63
36	9	43	13	47	17	50	21	59
38	8	42	12	46	16	48	19	55
40	8	40	11	44	14	47	18	52
42	7	39	10	43	13	46	17	50
44	6	38	9	42	12	45	16	49
46	6	36	8	41	11	44	15	48
48	5	35	7	40	10	43	14	47
50	5	34	7	38	10	42	13	45
52	4	32	6	37	9	41	12	44
54	4	31	6	36	8	40	11	43
56	3	30	5	35	8	39	10	42
58	3	29	5	34	7	38	9	41
60	3	27	4	32	7	37	9	40
62	2	26	4	31	6	36	8	39
64	2	25	4	30	6	35	8	38
66	2	24	3	29	5	34	7	37
68	2	22	3	28	5	33	7	36
70	2	21	3	27	4	32	6	36
72	1	20	3	25	4	31	6	35
74	--	--	2	24	4	30	5	34
76	--	--	2	23	3	28	5	32
82	--	--	1	20	3	26	4	30
84	--	--	--	--	2	24	4	29
88	--	--	--	--	2	22	3	27
92	--	--	--	--	1	20	2	25
96	--	--	--	--	--	--	2	23
102	--	--	--	--	--	--	1	20

Average Life In Years

	60 Years		70 Years	
Age	Rem. Life	% Good	Rem. Life	% Good
0	60	100	70	100
2	58	99	68	99
4	56	99	66	99
6	54	98	64	99
8	52	97	62	98
10	50	96	60	98
12	48	95	58	97
14	46	94	56	96
16	44	93	54	96
18	42	92	52	95
20	40	89	50	94
22	38	87	48	93
24	36	85	46	92
26	34	83	45	91
28	32	81	42	89
30	30	78	40	87
32	29	75	39	85
34	27	72	37	83
36	25	69	35	81
38	24	66	33	79
40	22	63	31	76
42	21	60	30	73
44	20	56	29	70
46	18	52	27	67
48	17	49	26	64
50	16	48	25	61
52	15	47	23	58
54	14	46	22	56
56	13	46	21	54
58	12	45	20	52
60	11	44	19	50
64	10	42	17	48
68	9	40	15	46
72	8	38	13	44
76	7	36	12	43
80	6	35	11	41
86	5	32	9	39
92	4	29	8	36
100	3	25	6	33
108	2	22	4	29
112	1	20	3	27
122	--	--	2	24
130	--	--	1	20

Additional Costs for Commercial and Industrial Structures

Additional Structure Costs 202
Basements 202
Communications Systems 202
Public Address Systems 202
Burglar Alarms 202
Canopies 202
Docks 202
Doors 203
Draperies 203
Dumbwaiters 203
Elevators 203
Escalators 203
Fill 204
Fire Escapes 204
Fire Extinguishers 204
Fire Sprinklers 204
Fireplaces 204
Heating and Cooling Systems 204
Kitchen Equipment 205
Mezzanines 205
Partitions 205
Pneumatic Tube Systems 205
Seating 205
Skylights 205
Ventilators 206
Walk-in Boxes 207

Display Fronts 207
Display Front Illustrations 208
Display Front Classification and Costs 209
Bulkhead Walls 210
Ceiling 210
Entrances 210
Glass 210
Lighting 210
Platforms 210
Wall Finish 210

Signs 211
Post Mounting Costs 211
Rotators 212

Yard Improvements 212
Asphaltic Concrete Paving 212
Concrete Paving 212
Curbs 212
Gates 212
Striping 212
Lighting 213
Chain Link Fences 213
Wood Fences 213
Drainage Items 213

Additional Structure Costs

Basements

Cost includes concrete floor and walls, open ceiling, minimum lighting, no plumbing, and no wall finish. Cost per square foot of floor at 12′ wall height.

Square Foot Area

Area	500	1,000	1,500	2,000	3,000	4,000	5,000	7,500	10,000	15,000	20,000
Cost	20.40	18.20	15.80	14.40	13.25	12.25	11.50	10.50	10.00	9.60	9.10

Basement Wall Height Adjustment

Add or subtract the amount listed in this table to the square foot of floor cost for each foot of wall height more or less than 12 feet.

Square Foot Area

Area	500	1,000	1,500	2,000	3,000	4,000	5,000	7,500	10,000	15,000	20,000
Cost	1.35	1.00	.86	.71	.56	.51	.47	.36	.31	.23	.21

Communications Systems

Intercommunication Systems, cost per station	
Central control	$155 to $175
Fully intercommunicating	345 to 565
Nurses' call system	175 to 300

Sound Systems, cost per unit	
Voice only (per unit)	$45 to $70
Music (add to above) small units	40 to 50
Music (add to above) large units	60 to 70

Larger installations cost the least per unit

Public Address Systems, cost per speaker

Less than 5 speakers	$400 to $515
5 to 10 speakers	315 to 450
Over 10 speakers	300 to 405

Burglar Alarms

Cost to wire building for service	$ 900 to $1,450
Cost to connect to central service	5,200 to 7,200
Total cost in place	6,100 to 8,600

Canopies, per S.F. of canopy area

Light frame, flat roof underside, plywood and paint or cheap stucco supported by wood or light steel posts, 4″ to 6″ wood fascia.	$5.00 to $7.60
Average frame, underside of good stucco, flat roof, cantilevered from building or supported by steel posts, 6″ to 12″ metal fascia.	7.65 to 10.00
Same as above but with sloping shake or tile roof.	13.50 to 16.00
Corrugated metal on steel frame.	6.75 to 9.00

Canopy Lights, per S.F. based on one row of lights for 5′ canopy

Recessed spots (1 each 6 linear ft.)	$1.45
Single tube fluorescent	2.75
Double tube fluorescent	3.70

Docks for unloading trucks

Cost per S.F. of dock at 4′ height

Depth	10′	20′	30′	50′	100′	200′
5′	$11.70	$10.20	$9.00	$8.10	$8.00	$7.70
10′	9.50	8.20	6.40	6.00	5.60	5.50
15′	9.10	6.50	6.10	5.50	5.00	4.80
20′	8.20	6.30	5.50	5.00	4.80	4.60

Cost includes compacted fill, three concrete walls, concrete floor, and rock base.

Loading Ramps, cost per S.F. of ramp

Size	
Under 300 S.F.	$4.00
Over 300 S.F.	2.80

Mechanical Dockboards, each

Size	
6′ x 6′	$2,500 to $2,900
7′ x 6′	2,800 to 3,000
6′ x 8′	2,900 to 3,400
7′ x 8′	3,000 to 3,800

Additional Structure Costs

Doors

Exterior, commercial, cost per door	
Glass in wood (3' x 7')	$ 680 to $ 790
¼" plate in aluminum (3' x 7')	785 to 880
Automatic, tempered glass (3' x 7')	2,600 to 2,900
Residential type (3' x 7')	155 to 235
Interior, commercial and industrial, cost per S.F.	
Hollow core wood	$5.35 to $6.40
Solid wood	6.35 to 7.40
Hollow core metal	14.00 to 16.00
Fire, cost per S.F.	
Hollow metal, 1¾"	$ 14.20 to 17.50
Metal clad, rolling	13.00 to 19.70
Metal clad, swinging	14.50 to 17.50

Dumbwaiters, includes door

	First Two Stops	Add. Stops
Hand operated	$3,000 to $3,300	$ 525
Electric operated	7,600 to 10,000	1,100

Elevators, Freight, Electric, car and equipment, per shaft, car speed in feet per minute

Capacity	50 to 75	100 to 150	200
2,500 lbs	$21,000	$ ---	$ ---
3,000	23,500	28,000	33,000
3,500	24,400	29,000	36,500
4,000	25,500	31,000	39,000
5,000	30,000	34,500	42,000
6,000	31,000	36,500	46,500
8,000	34,000	41,000	52,000
10,000	39,000	48,000	---

For manual doors, **add** $2,750 for each stop. For power operated doors, **add** $5,000 for each stop. Add $3,700 per car for self-leveling cars.

Elevators, Freight, Hydraulic, per shaft

Capacity	Car and Mach.	Per Stop Manual Door.	Per Stop Auto. Door
2,000 lbs	$19,000	$5,000	$10,000
6,000 lbs	23,000	5,000	10,000

Drive Through Overhead Steel Doors, cost each

10' x 10'	$570
12' x 12'	635
12' x 14'	700
Chain operation (add to above)	65
Motor controlled (add to above)	570

Draperies, cost per square yard

Standard quality	$12.00
Better quality, window	30.00
Better quality, sliding door	26.80
Lining for drapes	2.10

Escalators, cost per flight up or down

Floor Rise	32"	40"	48"
8'	$71,000	$75,000	$81,000
10'	73,000	80,000	83,000
12'	75,000	81,000	84,000
14'	80,000	84,000	87,000
16'	82,000	86,000	92,000
18'	83,000	88,000	94,000
20'	84,000	92,000	98,000
22'	86,000	95,000	100,000
24'	92,000	98,000	107,000

Add for glass side enclosure: $9,000.

Elevators, Hydro-Electric, car and machinery cost, per shaft, car speed in feet per minute

Per Shaft	100 to 150	160 to 210
1,500 lbs.	$15,000	$16,000
2,000	16,000	17,000
2,500	17,000	18,000
3,000	18,000	19,000
3,500	19,000	20,000

Add per stop: $1,800.
Add for double center-opening doors, per stops: $225.
Add for deluxe cab (raised panel interior, drop ceiling) $750 to $2,000.

Elevators, Passenger, Electric, car and machinery cost, per shaft

Capacity	100 F.P.M.	200 F.P.M.	250 F.P.M.	300 F.P.M.	350 F.P.M.
1,200 lbs.	$40,000	$ ---	$ ---	$ ---	$ ---
2,000	42,500	48,800	51,000	53,000	54,000
2,500	44,500	52,000	53,000	56,000	57,000
3,000	45,600	55,000	54,000	58,000	60,000
3,500	47,700	57,000	59,000	60,000	62,500
4,000	49,000	59,000	61,500	64,000	66,000
Hospital Type Car					
4,500 to 5,000 lbs.	51,000	62,500	65,700	69,000	74,000

Add for doors, per stop: baked enamel, $3,700 stainless, $4,800.
Add for deluxe cab: $5,600.
Add for interconnecting control system on 3 or more shafts: $9,500.

Additional Structure Costs

Fill, compacted under raised floor, includes perimeter retaining wall but not slab, per cubic foot.

Up to 10,000 S.F.	$.47 to $.68
Over 10,000 to 50,000 S.F.	.33 to .47

Fire Escapes

Type	Unit	Cost
Second story	Each	$1,300 to $1,900
Additional floors	Per story	1,100 to 1,500

Fire Sprinklers

Area	Standard System		Hydraulically Designed*	
	Single	Double	Single	Double
0 to 10,000	$1.75	$2.75	$2.75	$3.90
10,001 to 20,000	1.45	2.33	2.10	2.75
20,001 to 50,000	1.30	2.10	1.90	2.60
Over 50,000	1.20	1.75	1.60	2.20

Costs include normal installation, service lines and valves.
*Hydraulically designed systems are custom engineered to meet code or insurance requirements and are usually so identified by a metal plate attached to the riser.

Fire Extinguishers, cost

Fire hose and cabinet	$200 to $290
Extinguisher cabinets	100 to 140
Extinguishers	72 to 135

Fireplaces

	1 Story	2 Story
5′ base, common brick or concrete block, low cost Douglas fir or common brick mantle.	$1,030	$1,300
6′ base, common brick, used brick, face brick or natural stone on interior face with average wood mantle.	1,350	1,780
6′ base, common brick, face brick or good wood mantle.	1,620	2,100
6′ base, common brick, used brick or natural stone on interior face, raised hearth.	2,200	2,500
8′ base, common brick, used brick or natural stone on interior face.	2,700	3,100

Heating and Cooling Systems

Type and Use	Cost per S.F. of floor area: Heating Only	Heating and Cooling
Urban stores	$1.40 to $2.15	$3.35 to $5.90
Surban stores	1.40 to 2.10	2.70 to 5.30
Small food stores	.95 to 1.70	2.70 to 5.60
Supermarkets	.90 to 1.70	2.35 to 4.20
Discount houses	.90 to 1.50	2.30 to 4.20
Department stores	1.45 to 2.10	2.30 to 5.90
Reinforced concrete	1.45 to 2.15	4.60 to 7.00
General offices		
Forced air	1.95 to 2.80	5.60 to 7.85
Hot & chilled water	---	6.60 to 10.30
Bank and savings offices	2.40 to 3.65	6.75 to 10.00
Medical-Dental offices		
Forced air	2.25 to 3.50	5.60 to 8.95
Hot & chilled water	---	7.85 to 10.60
Convalescent hospitals		
Forced air	2.25 to 3.35	4.25 to 6.80
Hot & chilled water	---	6.45 to 9.80
Funeral homes	2.00 to 2.70	5.60 to 7.85
Theaters	2.10 to 2.70	6.70 to 9.40
Restaurants	1.65 to 2.45	6.70 to 11.20
Industrial buildings	1.15 to 2.40	2.90 to 5.00
Interior offices	2.00 to 3.75	4.00 to 8.00

Use the higher figures where more heating and cooling density is required.

Overhead Suspended Heaters, per unit

25 M.B.T.U.	$440 to $ 510
50	500 to 640
75	560 to 720
100	640 to 830
150	840 to 890
200	900 to 1,100
250	1,000 to 1,200

Electric Heating Units

Baseboard, per linear foot	$20.00 to $22.00
Add for thermostat	50.00
Cable in ceiling, per S.F.	.65 to 1.60
Wall heaters, per K.W.	90.00 to 110.00

Additional Structure Costs

Kitchen Equipment, cost per linear foot of stainless steel fixture

Work tables	$150 to $175
Serving fixtures	145 to 190

Partitions, cost per S.F. of surface

Gypsum on wood frame (finished both sides)

2" x 4" wood studs, 24" on center with ½" gypsum board, taped, textured and painted. $2.70

Plaster on wood frame (finished both sides)

2" x 4" wood studs, 24" on center with 2 coats plaster over gypsum lath, painted with primer and 1 coat enamel. $3.60

Seating, cost per seat space

Theater, economy	$ 79.00
Theater, lodge	102.00
Pews, bench type	34.50
Pews, seat type	48.00

Mezzanines, cost per S.F. of floor

Unfinished (minimum lighting and plumbing)	$5.00 to $10.20
Store mezzanines	12.00 to 18.40
Office mezzanines (without partitions)	14.00 to 24.70
Office mezzanines (with partitions)	16.20 to 26.50

Costs include floor system, floor finish, stairways, lighting, plumbing, and partitions where applicable.

Pneumatic Tube Systems

Twin tube two station system, 500 to 1,500 feet

2¼" round	$11,300 to $21,700
3" round	13,400 to 24,400
4" round	14,600 to 31,000
4" x 7" oval	20,800 to 36,000

Automatic System

4" round	$12,800 to $14,600
4" x 7" oval	16,600 to 19,500

Skylights

Plastic Rectangular Domes, cost per unit

Size	Single Plastic Panel: Skylight Only	Single Plastic Panel: With 4" or 9" Insulated Curb	Double Plastic Panel: Skylight Only	Double Plastic Panel: With 4" or 9" Insulated Curb
16" x 16"	$110 to $140	$180 to $200	$130 to $140	$190 to $200
16" x 24"	130 to 150	200 to 210	130 to 150	200 to 210
16" x 48"	140 to 180	230 to 260	200 to 210	280 to 290
24" x 24"	140 to 150	210 to 230	180 to 200	210 to 250
24" x 32"	140 to 180	230 to 260	210 to 230	280 to 290
24" x 48"	180 to 200	260 to 300	230 to 260	310 to 350
28" x 92"	300 to 320	420 to 460	450 to 460	530 to 580
32" x 32"	150 to 180	230 to 260	200 to 210	250 to 290
32" x 48"	200 to 210	260 to 300	250 to 290	310 to 350
32" x 72"	260 to 300	380 to 420	410 to 420	480 to 530
39" x 39"	210 to 230	260 to 300	260 to 300	310 to 360
39" x 77"	320 to 360	420 to 470	510 to 520	550 to 630
40" x 61"	290 to 300	380 to 420	410 to 420	480 to 530
48" x 48"	230 to 250	310 to 360	310 to 320	390 to 420
48" x 64"	320 to 360	460 to 510	460 to 470	560 to 620
48" x 72"	360 to 410	510 to 540	540 to 560	630 to 670
48" x 92"	470 to 510	580 to 660	690 to 700	690 to 810
48" x 122"	670 to 700	---	---	---
58" x 58"	360 to 410	470 to 510	560 to 570	640 to 670
60" x 72"	460 to 510	580 to 650	670 to 700	750 to 810
60" x 92"	580 to 610	700 to 790	840 to 980	940 to 1000
64" x 64"	460 to 470	560 to 610	610 to 650	650 to 720
77" x 77"	670 to 690	800 to 870	1020 to 1060	1110 to 1170
94" x 94"	1170 to 1280	---	1840 to 2010	---

The above costs are for three or more skylights. If less than three, add 20% to above cost.

Add:		
	Gravity vent	$240 to $490
	8" exhaust fan	390 to 590
	12" exhaust fan	600 to 820

Inside ceiling dome or egg crate (min. cost $50) $6.50 per square foot.

Additional Structure Costs

Plastic Circular Dome Skylights, cost each

	Single Plastic Panel		Double Plastic Panel		Additives	
Size	4″ Curb	9″ Curb	4″ Curb	9″ Curb	Ceiling Dome	Wall Liner
30″	$ 420	$ 430	$ 495	$ 520	$ 195	$195 to $215
36″	430	450	520	560	250	215 to 250
48″	570	625	725	760	345	250 to 260
60″	720	750	950	980	495	260 to 325
72″	1,090	1,160	1,440	1,500	650	325 to 360
84″	1,490	1,580	2,100	2,200	930	390 to 440
96″	2,200	2,300	3,050	3,200	1,000	450 to 510

The above costs are for single skylights. For three or more, deduct 20%.

Plastic Pyramid Skylights, cost each

		Installed Cost	
Size	Height	2 or Less	3 or More
39″ x 39″	34″	$680	$580
48″ x 48″	42″	970	810
58″ x 58″	49″	1,380	1,200

Plastic Continuous Vaulted Skylights, cost per L.F.

Width	Single Panel	Double Panel
16″	$ 70	$ 97
20″	75	100
24″	83	120
30″	97	135
36″	105	145
42″	115	155
48″	135	190
54″	145	205
60″	155	220
72″	170	240
	210	300

Plastic Ridge Type Skylights, cost per linear foot.

Width*	Single Panel	Double Panel
18″	$140	$205
24″	155	245
30″	190	285
36″	220	330
42″	245	390
48″	265	500

*Width is from ridge to curb following slope of roof.

Wire Glass Skylights, Exterior Aluminum Frame, cost each

24″ x 48″	$195
24″ x 72″	210
24″ x 96″	280
48″ x 48″	290
48″ x 72″	325
48″ x 96″	390

Ventilators, Roof, Gravity Type, cost each

Throat Diameter	2 or Less	3 or More	Add for Insulated Curb
8″	$175	$160	$ 85
12″	185	170	100
18″	285	275	130
24″	350	335	140

Ventilators, Roof, Power Type, cost each

Throat Diameter	2 or Less	3 or More	Add for Insulated Curb
6″	$260	$240	$ 75
8″	390	355	80
10″	475	440	100
12″	530	500	105
18″	575	530	120
24″	650	610	140
30″	1,190	1,080	160
36″	1,250	1,150	170
42″-48″	2,200 to 2,750	2,000 to 2,500	185

Above costs are for a single-speed motor installation. Dampers and bird screens are included.

Add: Explosive-proof units, **add** $250 each. Two-speed motors, **add** $110 to $650 each. Plastic coating, depending on size of unit, **add** $80 to $440 each.

Heat and Smoke Vents, cost each

Size	Plastic Dome Lid	Aluminum Covered Lid
32″ x 32′	$ 590	$ 720
32″ x 48″	750	770
50″ x 50″	860	890
50″ x 62″	970	1,000
50″ x 74″	1,140	1,170
50″ x 92″	1,200	1,300
50″ x 98″	1,370	1,400
62″ x 104″	1,740	1,800
74″ x 104″	1,950	2,100

Additional Structure Costs

Walk-In-Boxes, cost per S.F. of floor area

Temperature Range	Floor Area 50	100	200	300	400	500	600
Over 45°	$58.70	$39.50	$27.50	$24.00	$22.90	$21.40	$21.10
25° to 45°	67.10	44.50	32.40	28.20	27.00	26.00	24.60
0° to 25°	77.90	49.20	36.00	31.40	30.00	28.60	28.60
-25° to 0°	88.20	57.50	43.20	37.20	36.00	34.90	33.90

Cost Include: Painted wood exterior facing, insulation as required for temperature, interior plaster, one 4 x 7 door per 300 S.F. of floor area. Costs are based upon 8' exterior wall height. Costs do not include machinery and wiring. Figure refrigeration machinery at $1,100per ton capacity.

Walk-In Box Equipment

	Installed Cost Painted	Galvanized
Meat tracks, per L.F.		
50 linear feet or less	$7.00	$9.40
50 linear feet or over	6.30	8.80
Track switches, each		
2 way	$45.50	$64.00
3 way	95.00	135.00
Automatic cut-thru	140.00	185.00

	Installed Cost Painted
Track booms 4 foot, each	
Swinging	$355.00
Rigid	175.00
Track scales, each	
Beam type	2,000.00
Dial type	3,500.00
Magnifying lens for dial type	3,100.00

Display Fronts

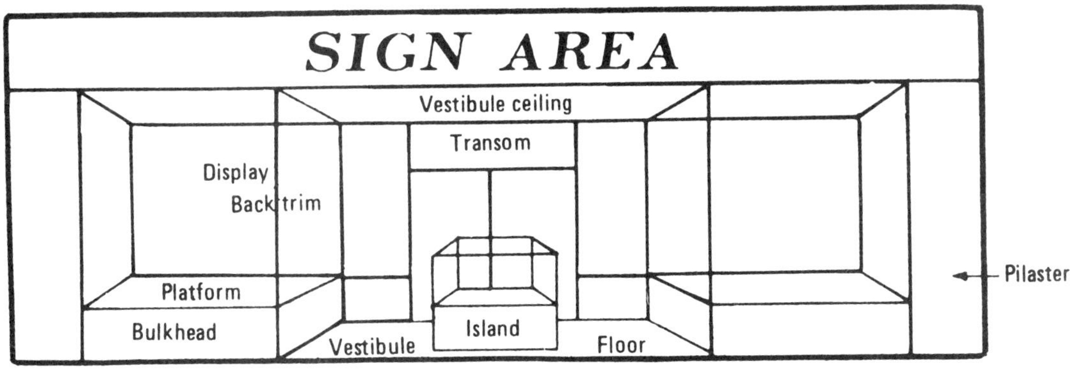

Display front costs may be estimated by calculating the in-place cost of each component or by estimating a cost per linear foot of bulkhead and multiplying by the bulkhead length. This section contains data for both methods. For most fronts, the cost per linear foot method is best suited for rapid preliminary estimates.

Bulkhead length is the distance from the inside of the pilaster and following along the bulkhead or glass to the inside of the opposite pilaster. This measurement **includes** the distance across entryways.

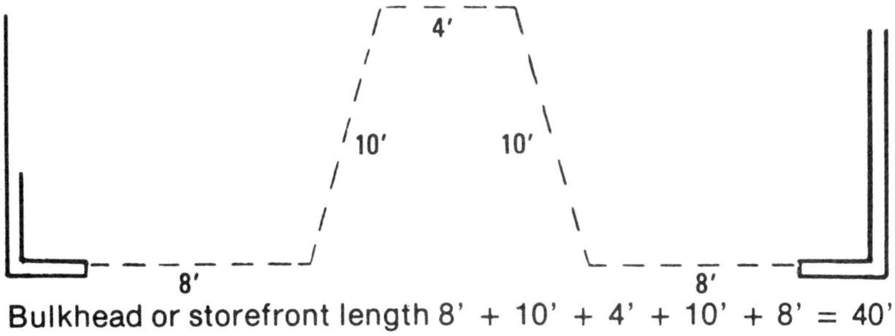

Bulkhead or storefront length 8' + 10' + 4' + 10' + 8' = 40'

The cost per linear foot of bulkhead is estimated using the storefront specifications and costs in this section. This manual suggests linear foot costs for four quality types; low cost, average, good, and very good. Costs are related for each quality type in terms of flat or recessed type fronts.

Recessed type fronts include all components described in the specifications. Flat front costs do not include the following components: Vestibule floor, vestibule and display area ceiling framing, back trim, display platform, lighting. The cost of automatic door openers is not included in front costs.

Display Fronts

Display Front, Best

Display Front, Best

Display Front, Good

Display Front, Good

Display Front, Average

Display Front, Average

Display Front, Low Cost

Display Front, Low Cost

Display Fronts

All front costs are based upon a height of 10 feet from the floor level to the top of the display window. Variations from this standard should be adjusted by using the display window height adjustment costs shown with the front foot costs. These amounts are added or deducted for each foot of variation from the standard.

Bulkhead height variations will not require adjustment. Cost differentials, due to variations in bulkhead height, will be compensated for by equal variations in display window height if overall heights are equal.

Sign areas are based upon 4 foot heights. Cost adjustments are given for flat type and for recessed type fronts. A cost range is given for recessed type fronts because deeply recessed fronts will have lower linear foot costs for sign area components than will moderately recessed fronts because this cost is spread over a longer distance when recesses are deep.

Display island costs are estimated by applying 60 to 80 percent of the applicable linear foot cost to the island bulkhead length. Window height adjustments should be made, but sign height adjustments will not be necessary.

Components	Best	Good	Average	Low Cost
Bulkhead (0 to 4' high)	Vitrolite domestic marble or stainless steel.	Black Carrara flagstone, terrazzo or good ceramic tile.	Average ceramic tile, Roman brick or imitation flagstone.	Stucco, wood or common brick.
Window Area	Bronze or stainless steel. ¼" plate glass with mitered joints.	Heavy aluminum. ¼" plate glass, some mitered joints.	Aluminum. ¼" plate glass.	Light aluminum with wood stops. Crystal or ¼" plate glass.
Sign Area (4' high)	Vitrolite, domestic marble or stainless steel.	Black Carrara flagstone, terrazzo or good ceramic tile.	Average ceramic tile, Roman brick or imitation flagstone.	Stucco.
Pilasters	Vitrolite, domestic marble.	Black Carrara flagstone, terrazzo or good ceramic tile.	Average ceramic tile, Roman brick or imitation flagstone.	Stucco.
Vestibule Floor*	Decorative terrazzo.	Decorative terrazzo.	Plain terrazzo.	Concrete.
Vestibule and* Display Area Ceilings	Stucco or gypsum wallboard and texture.	Stucco or gypsum wallboard and texture.	Stucco or gypsum wallboard and texture.	Gypsum wallboard and texture.
Back* Trim	Hardwood veneer on average frame.	Gypsum wallboard and texture or light frame.	None.	None.
Display Platform Cover*	Excellent carpet.	Good carpet.	Average carpet.	Plywood with linoleum.
Lighting*	1 recessed spot per linear foot of bulkhead.	1 recessed spot per linear foot of bulkhead.	1 exposed spot per 2 linear feet of bulkhead.	1 exposed spot per 4 linear feet of bulkhead.
Doors	¾" glass double doors.	Good aluminum and glass double door or single ¾" glass door.	Average aluminum and glass double door.	Wood and glass.
Costs, Flat Fronts	$405.00/linear foot.	$275.00/linear foot.	$180.00/linear foot.	$135.00/linear foot.
Costs, Recessed Fronts	$440.00/linear foot.	$360.00/linear foot.	$230.00/linear foot.	$160.00/linear foot.
Display Window Adjustments per Foot of Height	$18.20/linear foot.	$16.20/linear foot.	$14.60/linear foot.	$12.90/linear foot.
Sign Area Adjustment per Foot of Height, Flat Joints	$16.10/linear foot.	$10.80/linear foot.	$4.30/linear foot.	$1.60/linear foot.
Sign Area Adjustment per Foot of Height, Flat Fronts	$6.00 to $9.30/linear foot.	$3.80 to $5.50/linear foot.	$1.75 to $2.60/linear foot.	$.82 to $.97/linear foot.

*Not included in flat front costs.

Display Fronts

Bulkhead Walls, cost per S.F. of wall

Up to 5' high	
Concrete	$6.40 to $7.90
Concrete block	3.40 to 6.25
Wood frame	2.20 to 3.30

Ceiling, cost per S.F. of floor

Dropped ceiling framing	$1.50 to $1.75
Acoustical tile on wood strips	1.65 to 1.80
Acoustical plaster including lath	2.20 to 2.65
Gypsum board, texture and paint	1.15 to 1.50
Plaster and paint including lath	1.70 to 1.80

Entrances, cost per entrance

Aluminum and ¼'' plate glass	
Single door 3' x 7'	$ 800 to $ 970
Double door 6' x 7'	1,200 to 1,400
Stainless steel and ¼'' plate glass	
Single door 3' x 7'	1,600 to 1,770
Double door 6' x 7'	2,500 to 2,750
¾'' tempered glass	
Single door 3' x 7'	1,800 to 2,150
Double door 6' x 7'	2,700 to 3,050

Includes door, glass, lock, handles, hinges, sill and frame.

Display Platforms, cost per S.F. of platform area

Framing up to 5' high	$2.50 to $4.50
Hardwood cover	1.50 to 2.65
Plywood cover	1.20 to 1.40

Lighting, cost per fixture

Open incandescent	$31.50 to $49.00
Recessed incandescent	49.00 to 93.00
Fluorescent exposed 4' single	123.00 to 185.00
Fluorescent recessed 4' single	142.00 to 245.00

Exterior Wall Finish, cost per S.F. of wall

Aluminum sheet baked enamel finish	$1.45 to $2.35
Brick veneer	
Common brick	4.25 to 5.50
Roman	6.80 to 7.60
Norman	4.90 to 6.50
Glazed	9.75 to 12.20
Carrara glass	
Black	9.50 to 13.50
Red	12.50 to 16.00
Flagstone veneer	
Imitation	6.60 to 7.85
Natural	9.20 to 12.80
Marble	
Plain colors	15.50 to 22.00
With color variations	23.00 to 31.00
Stucco	1.50 to 2.00
Terrazzo	8.00 to 10.40
Tile - ceramic	6.60 to 9.20

Glass and Window Frames, per S.F. of glass

Glass only installed	
¼'' plate	$3.40 to $4.15
¼'' plate, tempered	7.25 to 8.60
¼'' plate, colored	4.60 to 6.20
Glass including frame (¼'' plate)	
Aluminum frame	9.60 to 14.30
Anodized aluminum frame	11.90 to 17.00
Bronze frame	14.30 to 19.20
Stainless steel frame	14.00 to 19.40
Wood frame	6.10 to 9.60

Signs

Lighted Display Signs

Lighted Display Signs	Cost per S.F. of sign area
Painted sheet metal with floodlights	$20.50 to $26.50
Porcelain enamel with floodlights	24.40 to 31.60
Plastic with interior lights	55.00 to 75.00
Simple rectangular neon with painted sheet metal faces and a moderate amount of plain letters	51.00 to 66.00
Round or irregular neon with porcelain enamel faces and more elaborate lettering	65.00 to 88.00

All of the above sign costs are for double-faced signs. Use ⅔ of these costs for single-faced signs. Sign costs include costs of installation and normal electrical hookup. They do not include costs of posts. If signs are mounted on separate posts, post mounting costs must be added. These costs are for custom-built signs (one-at-a-time orders). Mass-produced signs will have lower costs.

Post Mounting costs for signs

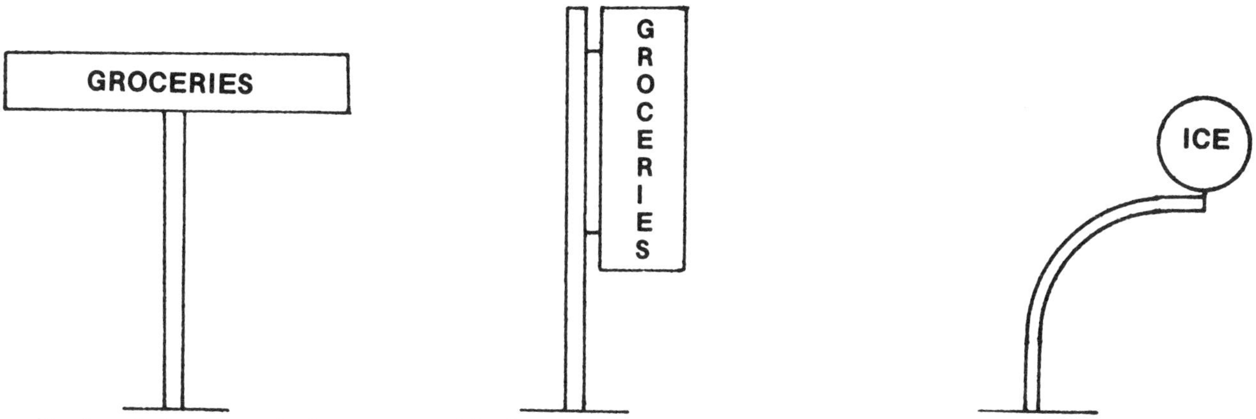

Horizontally Mounted — Vertically Mounted — Cantilever

	Pole Diameter at Base					
Post Height	4″	6″	8″	10″	12″	14″
15	$570	$665	$ 865	$1,150	$1,670	$1,900
20	650	750	1,000	1,200	1,900	2,160
25	675	800	1,020	1,390	2,100	2,300
30	760	880	1,200	1,500	2,200	2,560
35	--	970	1,350	1,700	2,450	2,700
40	--	1,030	1,500	1,800	2,550	3,000
45	--	--	1,670	1,900	2,800	3,100
50	--	--	1,780	2,000	3,000	3,400
55	--	--	--	2,180	3,100	3,700
60	--	--	--	2,300	3,350	3,900
65	--	--	--	--	3,600	4,100

If signs are mounted on separate posts, post mounting costs must be added. Post mounting costs include the installed cost of the post and foundation. On horizontally mounted signs, post height is the distance from the ground to the bottom of the sign. On vertically mounted signs, post height is the distance to the top of the post. For cantilevered posts, use one and one-half to two times the conventional post cost.

All of the above post costs are for single posts. Use 90% of the single post costs for each additional post.

Signs

If signs are mounted on buildings or canopies and if, because of the extra weight of the sign, extra heavy support posts or foundations are required, a fraction of the post mounting cost should be used. For example, the cost of a 4' x 25' plastic sign mounted on a 15' by 6'' post that is common to a 12' high canopy might be estimated as follows:

Sign Cost 100 S.F. at $65.00 =	$6,500
Post Cost $665 x ½ =	333
Total Cost	$6,833

If this sign were mounted on a 8'' post 20' high above this 12' canopy, the cost might be estimated as follows:

Sign Cost 100 S.F. at $65.00 =	$6,500
Post Cost $1,000 =	500
Total Cost	$7,000

Sign Rotators

Small signs	Less than 50 S.F	$ 740 to $ 860
Medium signs	50 to 100 S.F.	850 to 1,840
Large signs	100 to 200 S.F.	1,840 to 3,200
Extra large signs	Over 200 S.F.	$15.50 per S.F. of sign area

Yard Improvements

Bumpers **Pre-Cast Concrete or Good Quality Painted Wood.** Typical 4-foot lengths $3.60 per linear foot

Asphaltic Concrete Paving, cost per square foot

	Under 1,000	1,000 to 2,000	2,000 to 5,000	5,000 to 12,000	12,000 to 25,000	Over 25,000
2'', no base	$1.30	$ .66	$ .53	$ .45	$ .36	$ .30
3'', no base	1.45	.73	.65	.55	.50	.46
4'', no base	1.55	.88	.76	.71	.66	.60
2'', 4'' base	2.35	1.15	. .94	.76	.71	.65
3'', 6'' base	2.50	1.35	1.10	.90	.85	.82

Samples may be adjusted as follows:

1'' thickness of asphaltic paving	= $.17
1'' thickness of base	= .045

Concrete Paving, small quantities, per S.F.

4'' concrete (without wire mesh) no base	$1.55
4'' concrete (without wire mesh) 4'' base	1.75
5'' concrete (with wire mesh) no base	1.85
6'' concrete (with wire mesh) no base	1.95
Add for color	.15

Gates, Coin or Card Operated

Single gate	$4,400 to $4,950
Double gate	5,300 to 6,100

Striping

Parking spaces	
Single line between spaces	$4.50 per space
Double line between spaces	5.40 per space
Directional arrows	6.50 each

Curbs, per linear foot

Asphalt 6'' high	$2.50 to $3.60
Concrete 6'' wide x 12'' high	4.80 to 6.50
Concrete 6'' wide x 18'' high	6.00 to 7.40
Wood bumper rail 6'' x 6''	2.00 to 2.75
Vertical curb and gutters	5.00 to 6.50

Yard Improvements

Lighting

Fluorescent Arm-Type Fixtures	One Arm Unit Pole Height 12 Ft.	16 Ft.	30 Ft.	Two Arm Unit Pole Height 12 Ft.	16 Ft.	30 Ft.	Add For 3-Fixture Unit
2 tube, 48"	$625	$680	$ ---	$790	$840	$ ---	$325
4 tube, 48"	690	740	---	830	930	---	370
4 tube, 72"	735	790	930	910	960	1,100	435
6 tube, 72"	810	875	1,000	985	1,050	1,200	460
4 tube, 96"	840	910	1,050	1,030	1,100	1,280	500
6 tube, 96"	875	940	1,100	1,050	1,150	1,400	550

Mercury Vapor	1-Fixture Unit Pole Height 12 Ft.	16 Ft.	30 Ft.	2-Fixture Unit Pole Height 12 Ft.	16 Ft.	30 Ft.	4-Fixture Unit Pole Height 12 Ft.	16 Ft.	30 Ft.
400 Watt	$1,200	$1,350	$1,500	$1,690	$1,860	$2,070	$2,440	$2,670	$2,900
1,000 Watt	---	1,480	1,640	---	1,990	2,500	---	3,050	---

Incandescent Scoop Reflector	1-Fixture Unit Pole Height 12 Ft.	16 Ft.	30 Ft.	2-Fixture Unit Pole Height 12 Ft.	16 Ft.	30 Ft.	Add For 3-Fixture Unit
500 Watt	$390	$420	$540	$520	$560	$660	$140
1,000 Watt	485	540	650	625	660	775	160

Mall and Garden Lighting	1 Fixture Unit, Pole Height 3 Ft.*	5-7 Ft.	10 Ft.
Incandescent 60-200 watt	$345	$420	$465
Mercury vapor 100-250 watt	365	595	650

*Area lights usually found in service stations

Chain Link Fences, 9 gauge 2" mesh with top rail line posts 10' on center*

Height	Under 150 Linear Feet	150 to 1000 Linear Feet	1000 to 2000 Linear Feet	Over 2000 Linear Feet	Add For 12" Barb Wire	Add For Line Posts 8' or Less O.C.	Deduct For No Top Rail	Deduct For No. 11 Gauge Wire
4'	$6.50	$6.00	$5.50	$5.25	$.75	$.55 to $.80	$1.10	$.71
5'	8.10	6.50	6.00	5.75	.80	.55 to .80	1.10	.75
6'	7.30	7.00	6.60	6.40	.85	.60 to 1.20	1.10	.82
8'	10.20	9.65	9.50	9.25	.90	.65 to 1.25	1.10	1.00
10'	12.20	11.20	11.10	11.00	.95	.65 to 1.35	1.10	1.20
12'	15.50	14.00	13.50	13.40	1.00	.65 to 1.50	1.10	1.60

*Includes typical number of gates.

Wood Fences, per linear foot

5 foot high grapestake	$8.90
6 foot high grapestake	9.40
5 foot basketweave	9.20
6 foot basketweave	9.60
5 foot board and batt (simulated)	9.00
6 foot board and batt (simulated)	10.00
5 foot four rail diamond top	10.00
6 foot four rail diamond top	10.90
3 foot two rail picket	6.50
4 foot two rail picket	7.40
Typical 3' gate	90.00

Drainage, per linear foot

6" unreinforced concrete pipe	$7.10 to $9.00
8" unreinforced concrete pipe	7.75 to 9.30
10" unreinforced concrete pipe	8.45 to 10.20
12" unreinforced concrete pipe	9.30 to 11.40

The cost varies with the depth of the pipe in the ground.

Drainage Items

Size and Type	Cost
Catch basin 18" to 30" x 2' deep	$480 to $560 each
For each additional foot in depth, add	$41.40 per foot
Drop inlets	240 to 480 each
Manholes 4' diameter x 5' deep	720 to 1,140 each
For each additional foot in depth, add	$60.00 per foot

AGRICULTURAL STRUCTURES SECTION

Section Contents

Structure Type	Page
General Purpose Barns	215
Hay Storage Barns	216
Feed Barns	217
Shop Buildings	218
Machinery and Equipment Sheds	219
Small Sheds	220
Pole Barns	221
Low Cost Dairy Barns	222
Stanchion Diary Barns	223
Walk-Through Dairy Barns	224
Modern Herringbone Barns	225
Miscellaneous Dairy Costs	226
Poultry Houses, Conventional	227
Poultry Houses, Modern Type	228
Poultry Houses, High Rise Type	229
Poultry Houses, Deep Pit Type	230
Poultry House Equipment	231
Labor Camps	232
Miscellaneous Agricultural Structures	233
Typical Lives for Agricultural Buildings	233

General Purpose Barns
Quality Classification

General Purpose Barn, Class 1

General Purpose Barn, Class 3

Component	Class 1 Good Quality	Class 2 Average Quality	Class 3 Low Quality
Foundation	Continuous concrete.	Concrete or masonry piers.	Redwood or cedar mudsills.
Floor	Concrete.	Dirt.	Dirt.
Wall Structure	Good wood frame 10′ eave height.	Average wood frame 10′ eave height.	Light wood frame 10′ eave height.
Exterior Wall Cover	Good wood siding painted.	Standard gauge corrugated iron, aluminum or average wood siding.	Light aluminum or low cost boards.
Roof Construction	Medium to high pitch, good wood trusses.	Medium to high pitch, average wood trusses.	Medium to high pitch, 2″ x 4″ rafters 24″ to 36″ o.c. or light wood trusses.
Roof Cover	Wood shingles.	Standard gauge corrugated iron or aluminum.	Light aluminum.
Electrical	Four outlets per 1,000 S.F.	Two outlets per 1,000 S.F.	None.
Plumbing	Two cold water outlets.	One cold water outlet.	None.

	Square Foot Area										
Quality Class	1,000	2,000	3,000	4,000	5,000	6,000	7,000	8,000	9,000	10,000	11,000
1, Good	15.88	13.68	12.75	12.21	11.85	11.58	11.38	11.21	11.08	10.96	10.87
2, Average	10.14	8.74	8.14	7.80	7.56	7.39	7.26	7.16	7.07	7.00	6.94
3, Low	7.47	6.43	6.00	5.74	5.57	5.45	5.35	5.27	5.21	5.16	5.11

Hay Storage Barns
Quality Classification

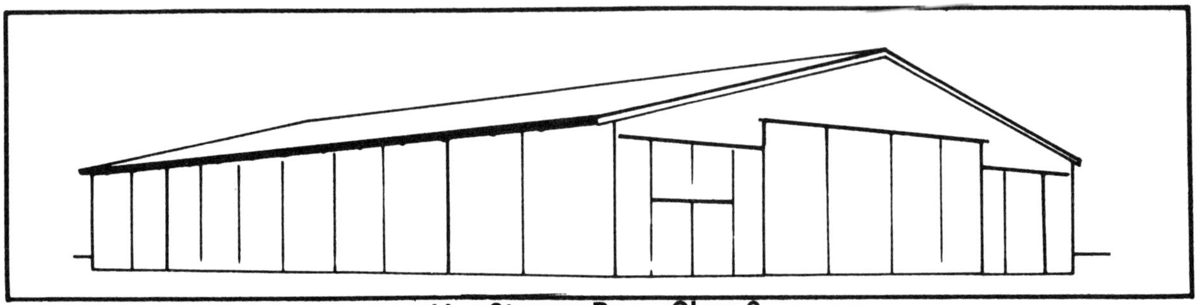

Hay Storage Barn, Class 2

	Class 1 Good Quality	Class 2 Average Quality	Class 3 Low Quality
Foundation	Continuous concrete.	Concrete or masonry piers.	Redwood or cedar mudsills.
Floor	Concrete.	Dirt.	Dirt.
Wall Structure	Good wood frame, 20′ eave height.	Average wood frame, 20′ eave height.	Light wood frame, 20′ eave height.
Exterior Wall Cover	Good wood siding, painted.	Standard gauge corrugated iron or aluminum.	Light aluminum or low cost boards.
Roof Construction	Low to medium pitch, good wood trusses.	Low to medium pitch, average wood trusses.	Low to medium pitch, 2″ x 4″ rafters 24″ to 36″ o.c. or light wood trusses.
Roof Cover	Wood shingles.	Standard gauge corrugated iron or aluminum.	Light aluminum.
Electrical	None.	None.	None.
Plumbing	None.	None.	None.
Shape	Length between one and two times width.	Length between one and two times width.	Length between one and two times width.

Square Foot Area

Quality Class	1,000	2,000	3,000	4,000	5,000	6,000	7,000	8,000	9,000	10,000	11,000
1, Good	12.21	10.96	10.14	9.58	9.16	8.84	8.57	8.36	8.17	8.01	7.87
2, Average	7.21	6.47	5.99	5.65	5.41	5.21	5.06	4.93	4.82	4.73	4.65
3, Low	6.29	5.64	5.22	4.93	4.72	4.55	4.42	4.30	4.21	4.13	4.05

Feed Barns
Quality Classification

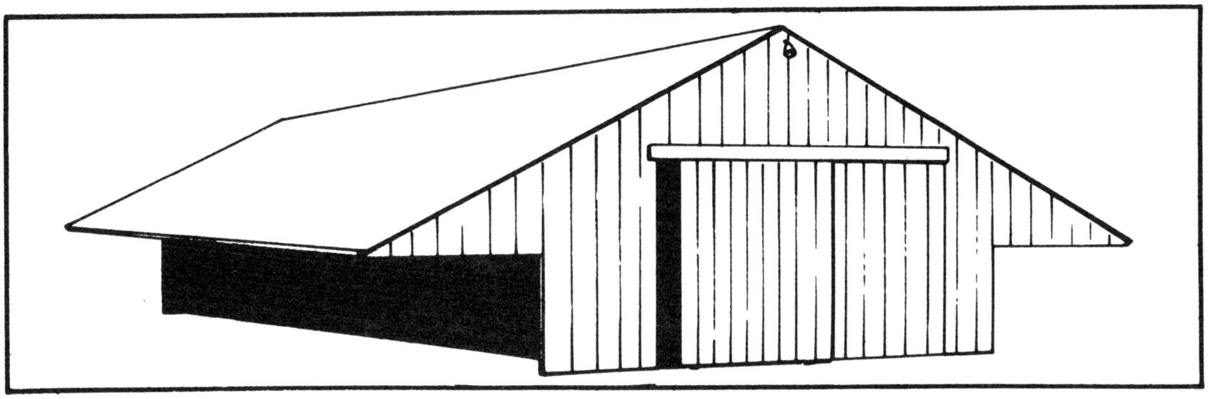

Feed Barn, Class 2

Component	Class 1 Good Quality	Class 2 Average Quality	Class 4 Low Quality
Foundation	Continuous concrete.	Concrete or masonry piers.	Redwood or cedar mudsills.
Floor	Concrete in center section.	Concrete in center section.	Dirt.
Wall Structure	Good wood frame, 8′ eave height at drip line.	Average wood frame, 8′ eave height at drip line.	Light wood frame, 8′ eave height at drip line.
Exterior Wall Cover	Open sides, good siding painted on ends.	Open sides, standard gauge corrugated iron, aluminum or average wood siding on ends.	Open sides and ends.
Roof Construction	Medium to low pitch, good wood trusses.	Medium to low pitch, average wood trusses.	Medium to low pitch, 2″ x 4″ rafters, 24″ to 36″ o.c. or light wood trusses.
Roof Cover	Wood shingles.	Standard gauge corrugated iron or aluminum.	Light aluminum.
Electrical	Four outlets per 1,000 S.F.	Two outlets per 1,000 S.F.	None.
Plumbing	Two cold water outlets.	One cold water outlet.	None.

Square Foot Area

Quality Class	1,000	2,000	3,000	4,000	5,000	6,000	7,000	8,000	9,000	10,000	11,000
1, Good	7.24	6.80	6.62	6.51	6.44	6.39	6.35	6.32	6.29	6.27	6.25
2, Average	6.26	5.89	5.73	5.64	5.57	5.53	5.50	5.47	5.44	5.43	5.41
3, Low	3.96	3.73	3.63	3.57	3.53	3.50	3.48	3.46	3.45	3.44	3.42

Shop Buildings
Quality Classification

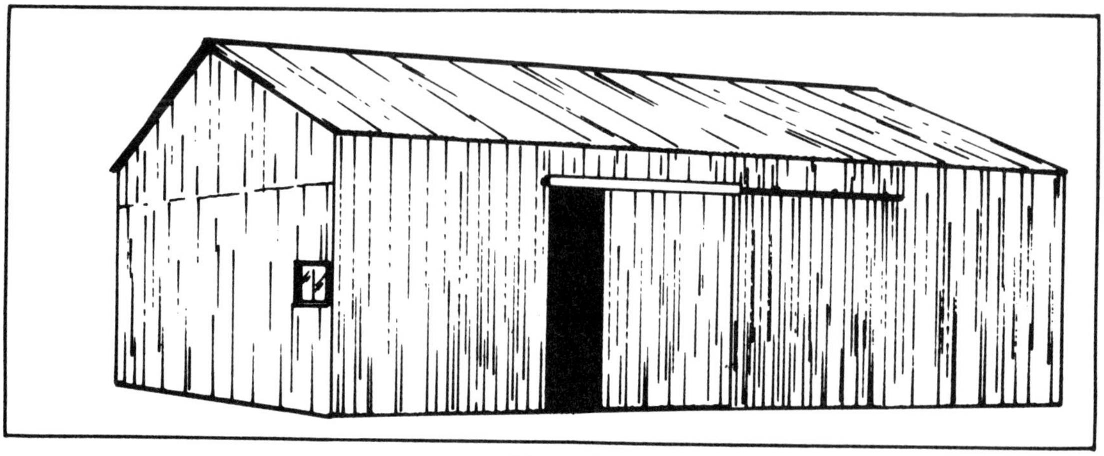

Shop, Class 2

Component	Class 1 Good Quality	Class 2 Average Quality	Class 3 Low Quality
Foundation	Continuous concrete.	Light concrete.	Light concrete.
Floor	Concrete.	Concrete.	Concrete.
Wall Structure	Good wood frame, 15′ eave height.	Average wood frame, 15′ eave height.	Light wood frame, 15′ eave height.
Exterior Wall Cover	Good wood siding, painted.	Standard gauge corrugated iron, aluminum or average wood siding.	Light aluminum or low cost boards.
Roof Construction	Low to medium pitch, good wood trusses.	Low to medium pitch, average wood trusses.	Low to medium pitch, 2″ x 4″ rafters, 24″ to 36″ o.c. or light wood trusses.
Roof Cover	Wood shingles.	Standard gauge corrugated iron or aluminum.	Light aluminum corrugated iron.
Electrical	Four outlets per 1,000 S.F.	Two outlets per 1,000 S.F.	Two outlets per 1,000 S.F.
Plumbing	Two cold water outlets.	One cold water outlet.	None.
Doors	One drive-thru door per 1,000 S.F. plus one walk-thru door.	One average sliding or swinging door per 2,000 S.F.	One light sliding or swinging door per 200 S.F.
Windows	Five percent of floor area.	None or few low cost.	None.
Shape	Length between one and two times width.	Length between one and two times width.	Length between one and two times width.

Square Foot Area

Quality Class	1,000	1,500	2,000	2,500	3,000	4,000	5,000	6,000	8,000	10,000
1, Good	12.76	11.64	10.98	10.53	10.20	9.74	9.42	9.19	8.87	8.65
2, Average	10.12	9.23	8.71	8.35	8.09	7.72	7.47	7.29	7.03	6.86
3, Low	8.01	7.31	6.89	6.61	6.40	6.11	5.92	5.77	5.57	5.43

Machinery and Equipment Sheds Quality Classification

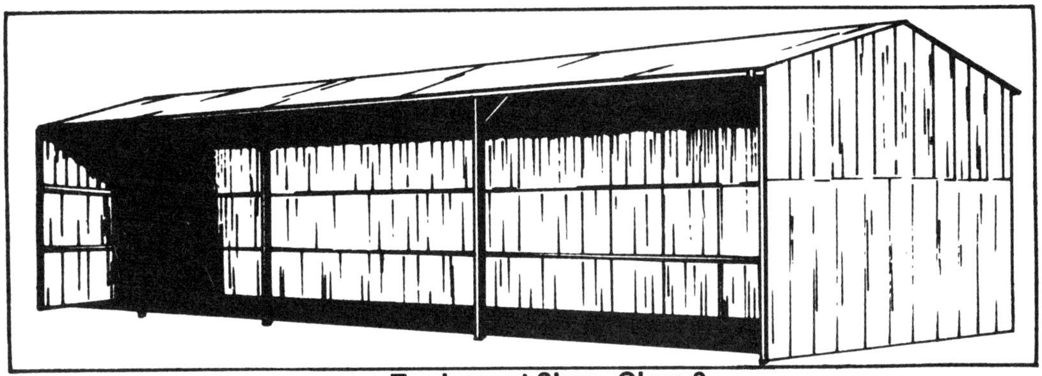

Equipment Shop, Class 3

Component	Class 1 Good Quality	Class 2 Average Quality	Class 3 Low Quality
Foundation	Continuous concrete.	Concrete or masonry piers.	Redwood or cedar mudsills.
Floor	Concrete.	Concrete.	Dirt.
Wall Structure	Good wood frame, 10' eave height.	Average wood frame, 10' eave height.	Light wood frame, 10' eave height.
Exterior Wall Cover	Good wood siding, painted.	Standard gauge corrugated iron, aluminum, or average wood siding.	Light aluminum or low cost boards.
Roof Construction	Low to medium pitch, gable or shed type, good wood framing.	Low to medium pitch, gable or shed type, average wood framing.	Low to medium pitch, shed type, light wood framing.
Roof Cover	Wood shingles.	Standard gauge corrugated iron or aluminum.	Light aluminum.
Electrical	Four outlets per 1,000 S.F.	Two outlets per 1,000 S.F.	None.
Shape	Usually elongated, width between 15 and 30 feet, any length.	Usually elongated, width between 15 and 30 feet, any length.	Usually elongated, width between 15 and 30 feet, any length.

All Sides Closed

Square Foot Area

Quality Class	500	1,000	1,500	2,000	2,500	3,000	3,500	4,000	4,500	5,000	6,000
1, Good	10.00	8.74	8.30	8.07	7.92	7.82	7.75	7.69	7.65	7.61	7.56
2, Average	7.50	6.56	6.23	6.05	5.94	5.87	5.82	5.77	5.74	5.71	5.67
3, Low	4.97	4.34	4.12	4.01	3.94	3.89	3.85	3.82	3.80	3.78	3.75

One Side Open

Square Foot Area

Quality Class	500	1,000	1,500	2,000	2,500	3,000	3,500	4,000	4,500	5,000	6,000
1, Good	9.21	7.77	7.29	7.04	6.89	6.79	6.71	6.66	6.62	6.58	6.53
2, Average	6.85	5.78	5.42	5.24	5.12	5.05	4.99	4.95	4.92	4.90	4.86
3, Low	4.40	3.71	3.48	3.36	3.29	3.24	3.21	3.18	3.16	3.14	3.12

Small Sheds Quality Classification

Component	Class 1 Good Quality	Class 2 Average Quality	Class 3 Low Quality
Foundation	Continuous concrete.	Concrete or masonry piers.	Redwood or cedar mudsills.
Floor	Concrete.	Boards.	Dirt.
Wall Structure	Good wood frame, 8' eave height.	Average wood frame, 8' eave height.	Light wood frame, 8' eave height.
Exterior Wall Cover	Good wood siding, painted.	Standard gauge corrugated iron, aluminum or average wood siding.	Light aluminum or low cost boards.
Roof Construction	Low to medium pitch, gable or shed type, good wood framing.	Low to medium pitch, gable or shed type, average wood framing.	Low to medium pitch, shed type, light wood framing.
Roof Cover	Wood shingles.	Standard gauge corrugated iron or aluminum.	Light aluminum.
Electrical	None.	None.	None.
Shape	Usually elongated, width between 6 and 12 feet, any length.	Usually elongated, width between 6 and 12 feet, any length.	Usually elongated, width between 6 and 12 feet, any length.

All Sides Closed

Square Foot Area

Quality Class	50	60	80	100	120	150	200	250	300	400	500
1, Good	13.00	12.00	10.00	9.00	9.00	8.00	8.00	7.00	7.00	7.00	7.00
2, Average	9.00	8.00	7.00	7.00	6.00	6.00	6.00	5.00	5.00	5.00	5.00
3, Low	6.00	6.00	5.00	4.00	4.00	4.00	4.00	4.00	3.00	3.00	3.00

One Side Open

Square Foot Area

Quality Class	50	60	80	100	120	150	200	250	300	400	500
1, Good	9.45	8.89	8.09	7.55	7.15	6.70	6.18	5.83	5.57	5.20	4.95
2, Average	6.94	6.53	5.95	5.55	5.25	4.92	4.54	4.28	4.09	3.82	3.64
3, Low	4.44	4.18	3.80	3.55	3.36	3.15	2.91	2.74	2.62	2.45	2.33

Pole Barns

These prices are for pole barns with a low pitch corrugated iron or aluminum covered roof supported by light wood trusses and poles 15' to 20' o.c. The gable end is enclosed and the roof overhangs about 2' on two sides. Wall height is 18 feet. Where sides are enclosed, the wall consists of a light wood frame covered with corrugated metal.

All Sides Open

	Side Length									
End Width	34	51	68	85	102	119	136	153	170	187
20	3.58	3.45	3.38	3.35	3.32	3.31	3.30	3.29	3.28	3.28
25	3.36	3.23	3.17	3.14	3.12	3.10	3.09	3.09	3.08	3.08
30	3.23	3.11	3.05	3.02	2.99	2.98	2.97	2.96	2.96	2.95
35	3.11	2.99	2.94	2.91	2.89	2.87	2.87	2.86	2.85	2.85
40	3.08	2.96	2.90	2.87	2.85	2.84	2.83	2.82	2.82	2.81
45	3.02	2.90	2.85	2.82	2.80	2.79	2.78	2.77	2.77	2.76
50	2.96	2.85	2.80	2.77	2.75	2.74	2.73	2.72	2.71	2.71
60	2.94	2.83	2.78	2.75	2.73	2.72	2.71	2.70	2.70	2.69
70	2.94	2.82	2.77	2.74	2.72	2.71	2.70	2.70	2.69	2.69
80	2.92	2.81	2.75	2.73	2.71	2.69	2.69	2.68	2.67	2.67

Ends and One Side Closed, One Side Open

	Side Length									
End Width	34	51	68	85	102	119	136	153	170	187
20	6.45	5.68	5.31	5.09	4.95	4.86	4.78	4.73	4.69	4.66
25	5.90	5.19	4.85	4.66	4.53	4.44	4.37	4.33	4.29	4.26
30	5.56	4.90	4.58	4.39	4.27	4.19	4.12	4.08	4.04	4.01
35	5.29	4.66	4.35	4.17	4.06	3.98	3.92	3.88	3.84	3.82
40	5.13	4.52	4.22	4.05	3.94	3.86	3.80	3.76	3.73	3.70
45	4.97	4.38	4.09	3.93	3.82	3.74	3.69	3.65	3.61	3.59
50	4.85	4.27	3.99	3.83	3.72	3.65	3.60	3.56	3.53	3.50
60	4.72	4.16	3.89	3.73	3.63	3.56	3.50	3.46	3.43	3.41
70	4.64	4.09	3.82	3.66	3.56	3.49	3.44	3.40	3.37	3.35
80	4.54	3.99	3.73	3.58	3.48	3.41	3.36	3.33	3.30	3.27

All Sides Closed

	Side Length									
End Width	34	51	68	85	102	119	136	153	170	187
20	7.72	6.94	6.55	6.32	6.16	6.05	5.97	5.91	5.86	5.82
25	6.89	6.19	5.84	5.63	5.50	5.40	5.33	5.27	5.23	5.19
30	6.36	5.71	5.39	5.20	5.08	4.99	4.92	4.87	4.83	4.79
35	5.97	5.36	5.06	4.88	4.76	4.68	4.62	4.57	4.53	4.50
40	5.70	5.12	4.84	4.66	4.55	4.47	4.41	4.37	4.33	4.30
45	5.47	4.91	4.63	4.47	4.36	4.29	4.23	4.18	4.15	4.12
50	5.28	4.74	4.47	4.32	4.21	4.14	4.08	4.04	4.01	3.98
60	5.04	4.53	4.27	4.12	4.02	3.95	3.90	3.86	3.83	3.80
70	4.90	4.41	4.16	4.01	3.91	3.85	3.79	3.75	3.72	3.70
80	4.73	4.25	4.01	3.87	3.78	3.71	3.66	3.62	3.59	3.57

Side sheds tying into one side of a pole barn are priced as follows. The shed consists of one row of poles 14' to 16' high, spaced 15' to 20' o.c. A light wood truss covered with a low pitch sheet metal roof spans the distance between the poles and the barn side. If the sides are open, the cost will be between $2.25 and $2.40 per square foot of area covered. If all sides are enclosed with sheet metal and a light wood frame, the square foot cost will be $3.50 to $4.50.

Low Cost Dairy Barns

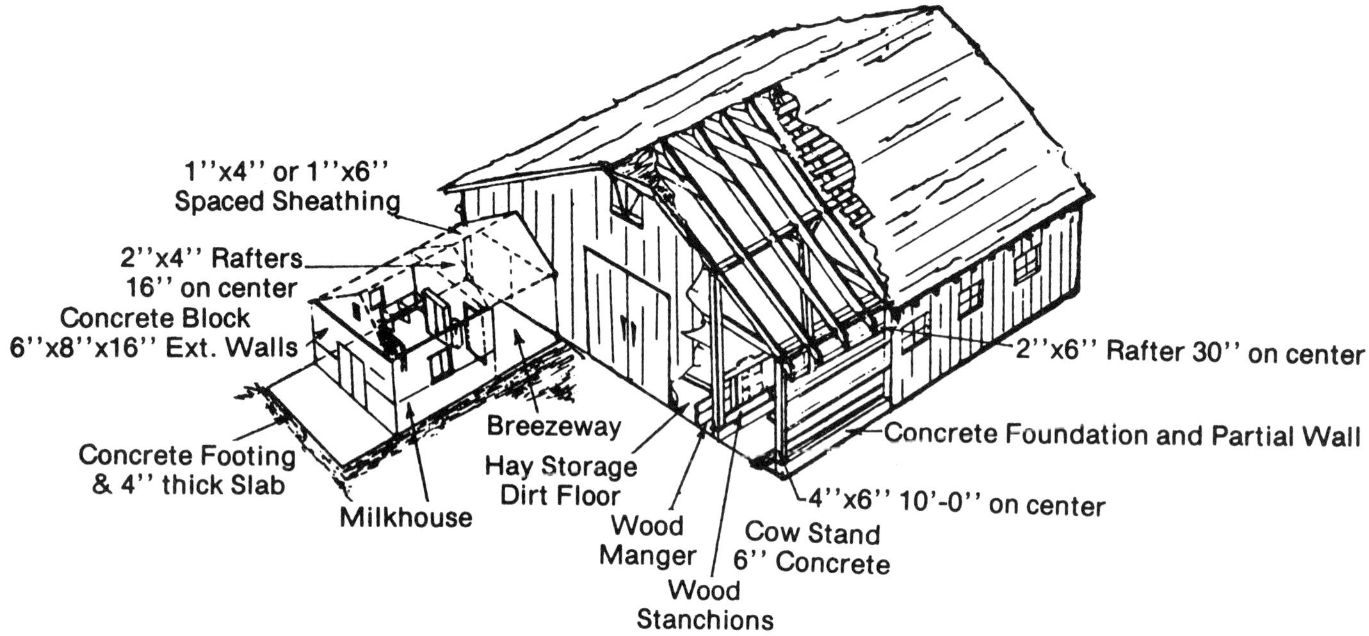

	Milk House	Dairy Barn
Foundation	Concrete.	Light concrete.
Floors	Concrete slab.	Concrete cow stands.
Walls	6" or 8" concrete block 36" high with 2" x 4" @16" o.c. framing above.	Box frame, 4" x 6", 10' o.c.
Roof	Average wood frame, corrugated iron or aluminum cover.	Average wood frame, wood shingles, corrugated iron or aluminum cover.
Windows	Metal sash or metal louvers, on 5% of wall area.	Barn sash.
Interior	Smooth finish plaster.	Unfinished, wood stanchions.
Electrical	Minimum grade, fair fixtures.	None.
Plumbing	One wash basin.	None.
Square Foot Cost	$15.70 to $19.30 (including breezeway).	$6.00 to $7.50 (exclusive of milking equipment).

Stanchion Dairy Barns

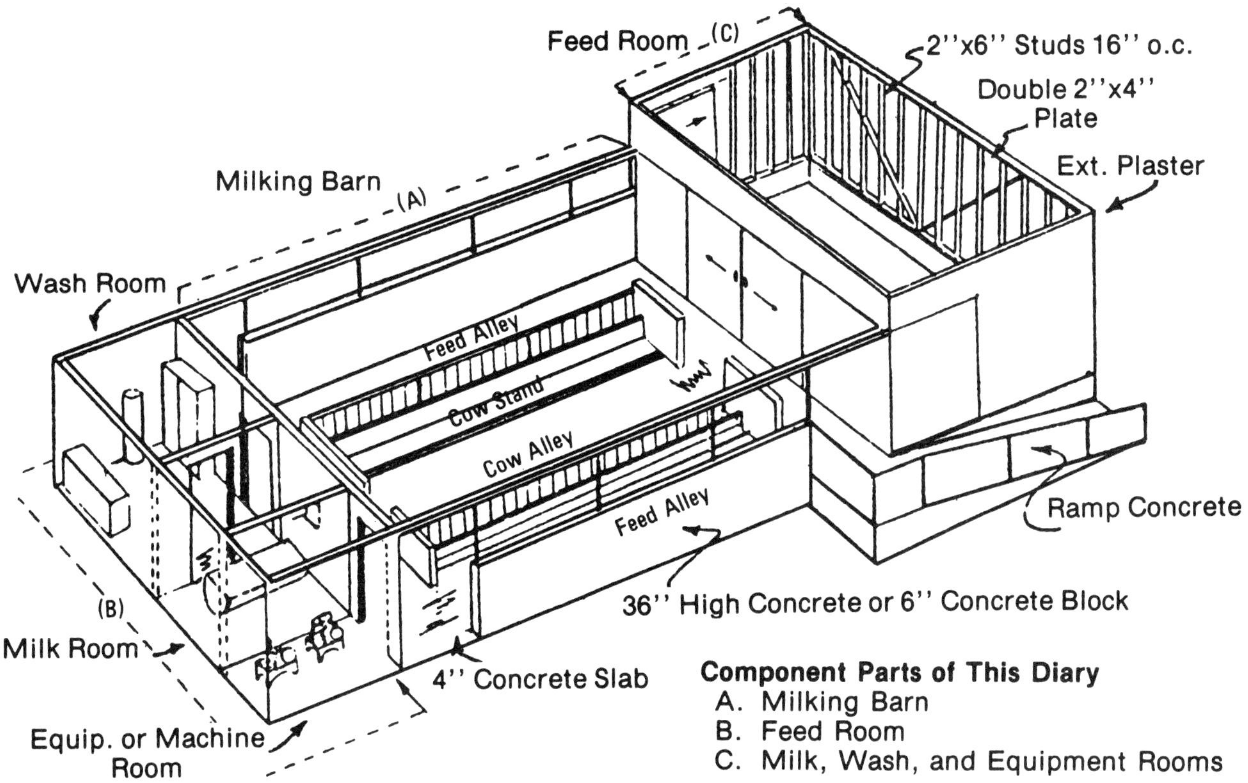

Component Parts of This Diary
A. Milking Barn
B. Feed Room
C. Milk, Wash, and Equipment Rooms

	Feed Room	Milk, Wash and Equipment Room	Milking Barn
Foundation	Reinforced concrete.	Reinforced concrete.	Reinforced concrete.
Floors	Concrete slab.	Concrete slab.	Concrete, well formed gutters and mangers.
Walls	2" x 4" or 2" x 6" @16" o.c. framing.	6" or 8" concrete block 36" high with 2" x 6" @16" o.c. framing above.	6" or 8" concrete block or reinforced concrete 36" high with 2" x 6" @16" o.c. framing above.
Roofs	Average wood frame, corrugated iron or aluminum cover.	Average wood frame, corrugated iron or aluminum cover.	Average wood frame, corrugated iron or aluminum cover.
Windows	None.	Metal sash or metal louvers on 10% of wall area.	Metal sash or metal louvers.
Interior	Unfinished.	Smooth finish plaster, cove base.	Smooth plaster 36" high, metal stanchions.
Electrical	Conduit, average fixtures.	Conduit, average fixtures.	Conduit, average fixtures.
Plumbing	None.	One wash basin with floor drains.	Floor drains and hose bibs.
Square Foot Cost	$6.65 to $10.20.	$22.00 to $27.70 (including breezeway).	$11.40 to $13.30 (exclusive of milking equipment).

Walk-Through Dairy Barns

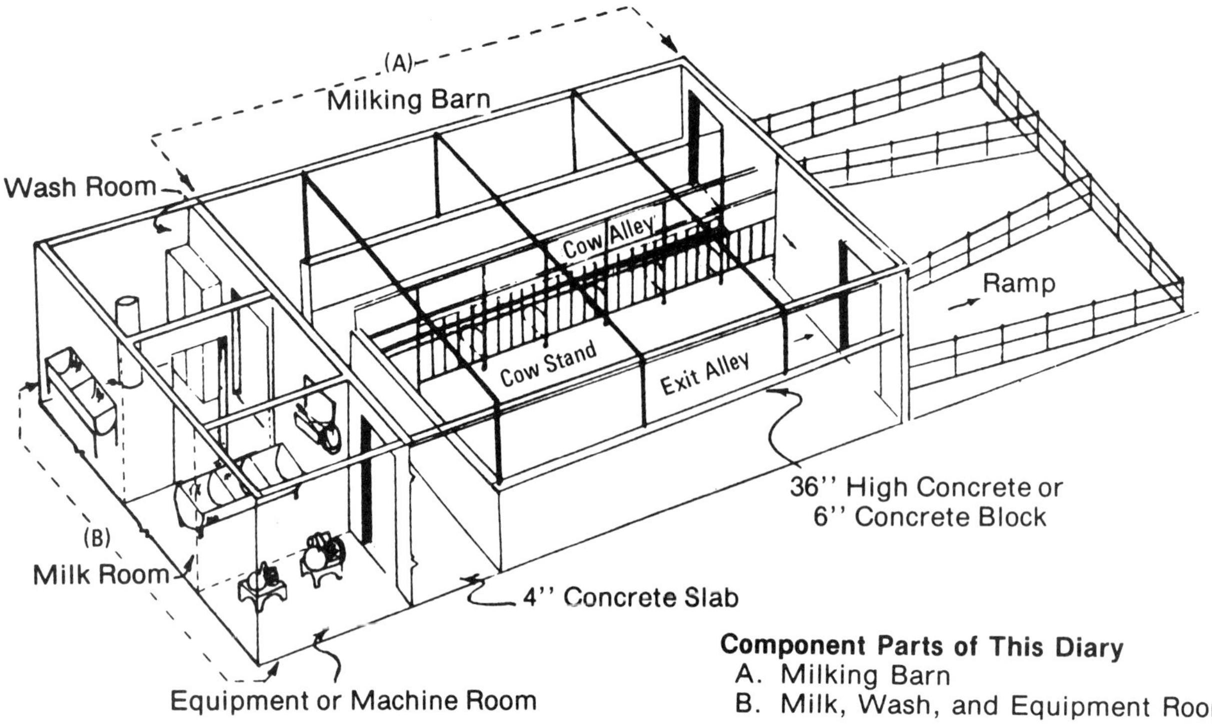

Component Parts of This Diary

A. Milking Barn
B. Milk, Wash, and Equipment Rooms

	Milking Barn	Milk, Wash, and Equipment Room
Foundation	Reinforced concrete.	Reinforced concrete.
Floors	Concrete, well-formed gutters and mangers.	Concrete slab.
Walls	6" to 8" concrete block or reinforced concrete 36" high with 2" x 6" @16" o.c. framing above or all concrete block.	6" or 8" concrete block 36" high with 2" x 6" @16" o.c. framing above, or all concrete block.
Roof	Average wood frame, corrugated iron or aluminum cover.	Average wood frame, corrugated iron or aluminum cover.
Windows	Metal sash or metal louvers.	Metal sash or metal louvers on 10% of wall area.
Interior	Smooth plaster 36" high, metal stanchions.	Smooth finish plaster, cove base.
Electrical	Conduit, average fixtures.	Conduit, average fixtures.
Plumbing	Floor drains and hose bibs.	One wash basin, floor drains.
Square Foot Costs	$13.50 (exclusive of milking equipment).	$13.80 (including breezeway).

Modern Herringbone Barns

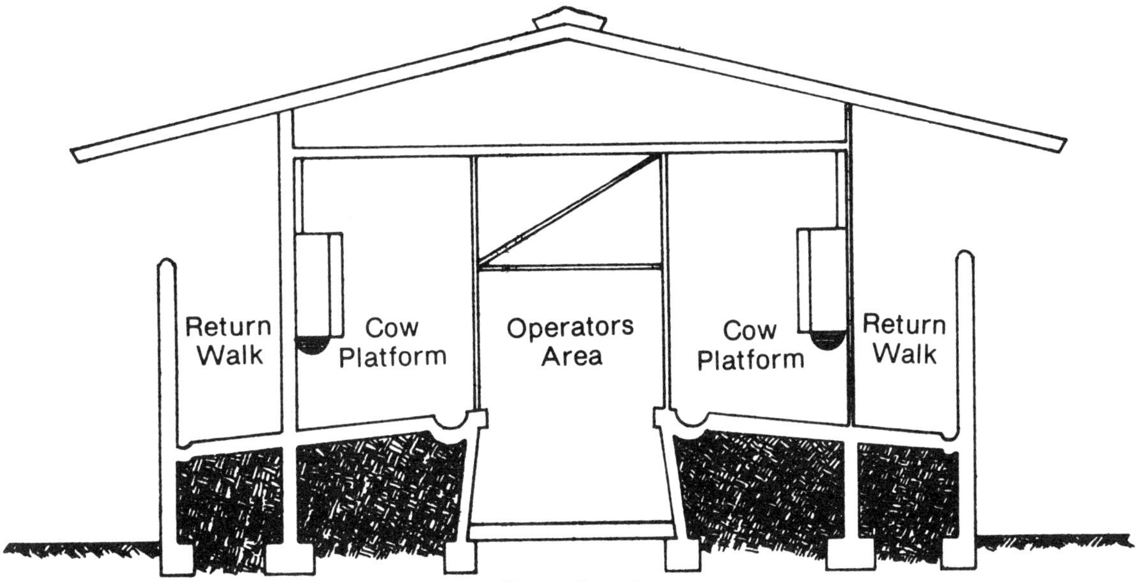

Cross Section

Milking Barn

Foundation	Reinforced concrete.
Floors	Concrete, well-formed gutters and mangers.
Walls	6" or 8" concrete block or reinforced concrete 36" high with 2" x 6" @16" o.c. framing above or all concrete block.
Roof	Average wood frame, corrugated iron or aluminum cover.
Windows	Metal sash or metal louvers.
Interior	Smooth plaster 36" high.
Electrical	Conduit, average fixtures.
Plumbing	Floor drains and hose bibs.
Stanchions	Metal stanchions.
Square Foot Costs	Without stalls, feeding system and milking equipment, $20.40 to $22.30.
Feed System & Stalls	Per double ten barn, $19,000.

Milk, Wash, and Equipment Room

	Average Quality	Good Quality
Foundation	Reinforced concrete.	Reinforced concrete.
Floors	Concrete slab.	Concrete slab.
Walls	8" concrete block with 2" x 6" @16" o.c. framing above or all concrete block.	8" concrete block with 2" x 6" @16" o.c. framing above or all concrete block.
Exterior	Stucco or concrete block.	Stucco and masonry veneer.
Roof	Corrugated iron or aluminum.	Wood shakes or mission tile, 3' to 4' overhang.
Windows	Metal sash on 10% of wall area.	Metal sash on 10% of wall area.
Interior	Smooth finish plaster, cove base.	Smooth finish plaster, cove base.
Electrical	Conduit, average fixtures.	Conduit, average fixtures.
Plumbing	One wash basin, one water closet, one lavatory, floor drains.	One wash basin, one water closet, one lavatory, floor drains.
Square Foot Costs	$19.00 to $20.50.	$20.40 to $21.80.

Miscellaneous Dairy Costs

Holding Corral and Wash Area Costs

Floor or Ramp	Sloping concrete with abrasive finish, $1.60 to $1.90 per square foot.
Walls	5′ to 6′ high plastered interior, $20.40 to $24.00 per linear foot.
Metal Rail Fence	Welded pipe, posts 10′ o.c. in concrete. top rail and 3 cables, $5.20 per linear foot.
Cable Fence	Pipe posts 10′ o.c. set in concrete, $4.40 + .22 per linear foot per cable.
Gates	54″ high, pipe with necessary bracing, $105.00 each.
Sprinklers	Hooded rainbird, $93.00 each, including plumbing and pump.
Roof	Pipe column supports, average wood frame, corrugated iron or aluminum cover, open sides, $2.35 to $3.40 per square foot.
Typical Wash Area Cost	Sloped concrete floors, 5′ concrete block exterior walls, welded pipe interior fences and gates, usual floor drains $11.00 S.F. with roof $7.50 S.F. without sprinkler (Exclusive of sprinkler system)
Sprinkler System	Per double ten barn, $6,000.

Corral Costs

Components	Cost
4″ Concrete Flatwork	$1.35 per square foot.
6″ Concrete Flatwork	$1.55 per square foot.
12″ Curb	$4.30 per linear foot.
Cable Fence	$5.00 per linear foot.
Water Tank, Concrete Typical	$1,350.00 each.
Steel Stanchions	$24.75 to $28.25 each.
Steel Stanchions	$10.50 to $12.30 per linear foot.
Steel Lockable Stanchions	$30.00 to $32.50 each.
Steel Lockable Stanchions	$11.60 to $14.20 per linear foot.
Light Standards	$1,350.00 each.
Sump Pumps 3 HP	$1,400.00
4 HP	$1,900.00
10 HP	$3,000.00
Hay Shelters	$2.45 to $2.85 per square foot.
Loafing Sheds	$2.50 to $3.00 per square foot.
Typical 80-Cow Corral	$16,000.00 to $18,400.00 total.
Typical 80-Cow Corral	$205.00 to $230.00 per cow.

Milk Line, 3″ Stainless Steel

40-Cow Conventional Barn	$10,200 total, $260 per cow.
Double Ten Herringbone Barn.	$6,100 total, $305 per cow.
Dual Milk Pump, Add	$1,170.00.
Incline Filter, Add	$1,200.00.
Incline Cold Plate, Add	$2,900 to $4,000.

Stainless Steel Refrigerated Holding Tanks

5,000 Gallon.	$27,000
4,000 Gallon.	25,000
3,500 Gallon.	24,000
3,000 Gallon.	21,600
2,500 Gallon.	19,600
2,000 Gallon.	18,000
1,500 Gallon.	16,000

Refrigeration Compressors

7½ HP	$3,500.00
10 HP	4,600.00
15 HP	5,650.00

Vacuum Pumps

Units	Cost
8	$3,350.00
12	5,500.00
20	7,500.00

Poultry Houses
Conventional Lay Cage Type

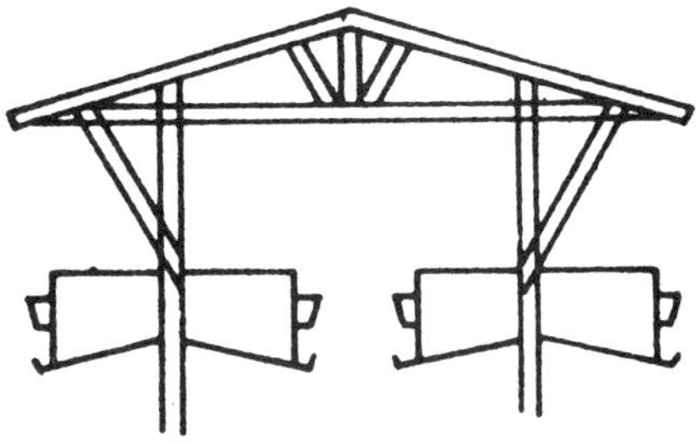

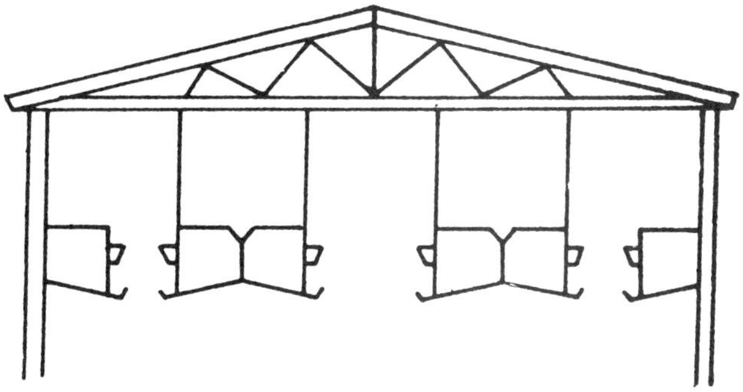

Basic Building

Component	Good Quality	Average Quality	Low Quality
Floors	2" concrete.	Dirt with 4' concrete walkways.	Dirt.
Foundations	Thickened slab.	Concrete piers.	Wood piers.
Frame	Light steel or average wood frame.	Average wood frame.	Light wood frame.
Roof Cover	Aluminum or corrugated iron.	Light aluminum or composition.	Light aluminum or composition.
Exterior	Plywood.	Vinyl curtains.	Wood lath.
Lighting	Good system, automatic controls.	Average system, automatic controls.	Minimum system, manual controls.
Plumbing	Good system.	Average system.	Fair system.
Insulation	Roof only.	None.	None.
Basic Building Cost Per S.F.	$2.60 to $3.25.	$1.75 to $2.40.	$1.50 to $1.75.

Equipment

Component	Best Quality	Good Quality	Average Quality	Low Quality
Cages	12" x 20" double deck.	12" x 20" single deck.	12" x 20" single deck.	12" x 12" single deck.
Water System	Automatic cup system.	Automatic cup system.	Simple "V" trough.	Simple "V" trough.
Feed System	Automatic system.	V trough.	V trough.	V trough.
Egg Gathering	Manual.	Manual.	Manual.	Manual.
Cooling	Pad and fan system.	Pad and fan system.	Simple fogging system.	Simple fogging system.
Cost Per S.F.	$6.90 to $7.85.	$4.15 to $4.85.	$3.15 to $3.80.	$2.65 to $3.45.
Cost Per Bird	$3.90 to $4.90.	$4.15 to $4.90.	$2.85 to $3.80.	$3.40 to $3.40.

Poultry Houses
Modern Controlled Environment Type

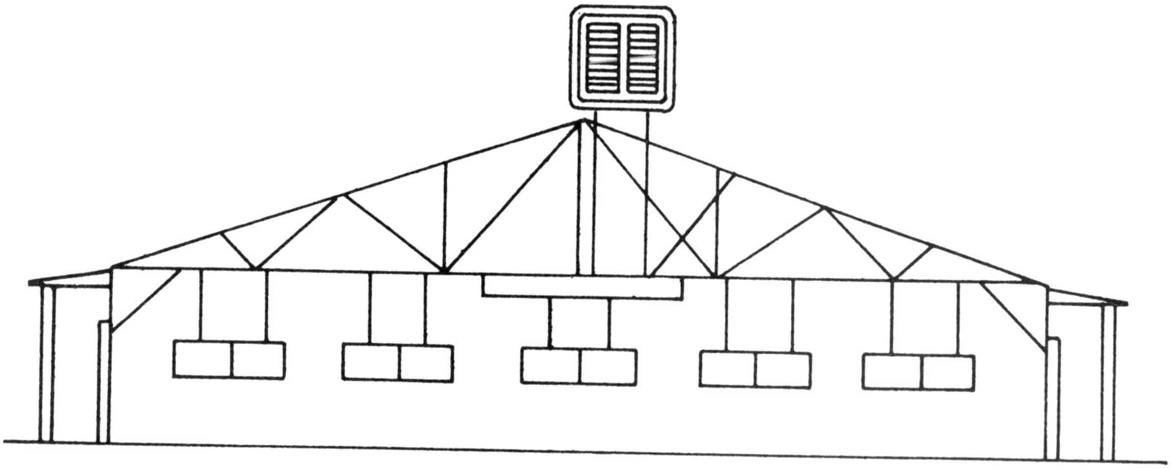

Basic Building

Foundation	Concrete.
Floor	3½" concrete slab.
Wall Frame	2" x 4" @24" o.c.
Roof Frame	Wood trusses with 2" x 4" purlins @24" o.c.
Exterior	Two rib aluminum or corrugated iron.
Interior	4" fiberglass roll with aluminum foil facing or ¾" styrofoam.
Lighting	Fluorescent or good automatic incandescent system.
Plumbing	Good basic system.
Basic Building Cost Per S.F.	$4.25 to $4.90.

Equipment

Component	Single Deck	Stair Step
Cages	12" x 20" single deck.	12" x 20" double deck.
Watering System	Automatic cup system.	Automatic cup system.
Feeding System	Manual.	Manual.
Egg Gathering System	Manual.	Manual.
Cooling	Evaporative coolers.	Evaporative coolers.
Heating	None.	None.
Building and Equipment S.F. Cost	$9.35 to $9.85.	$9.90 to $10.70.
Building and Equipment Cost Per Bird	$5.95 to $6.55.	$6.55 to $7.50.

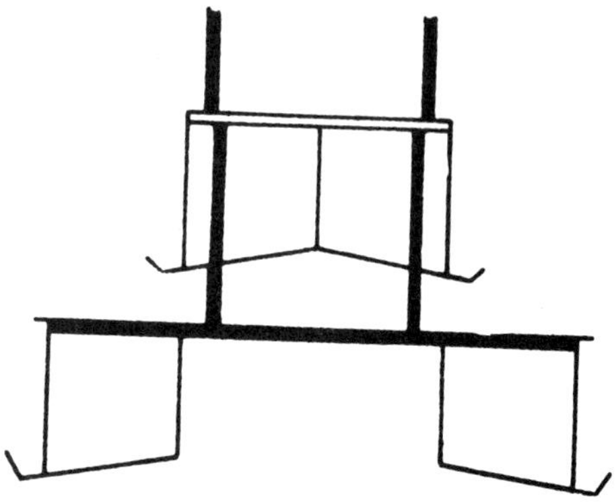

Stair-step cage system

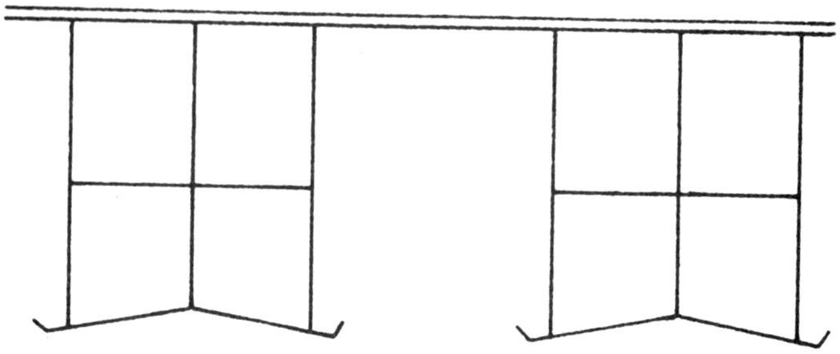

Single-deck cage system

Poultry Houses
High Rise Type

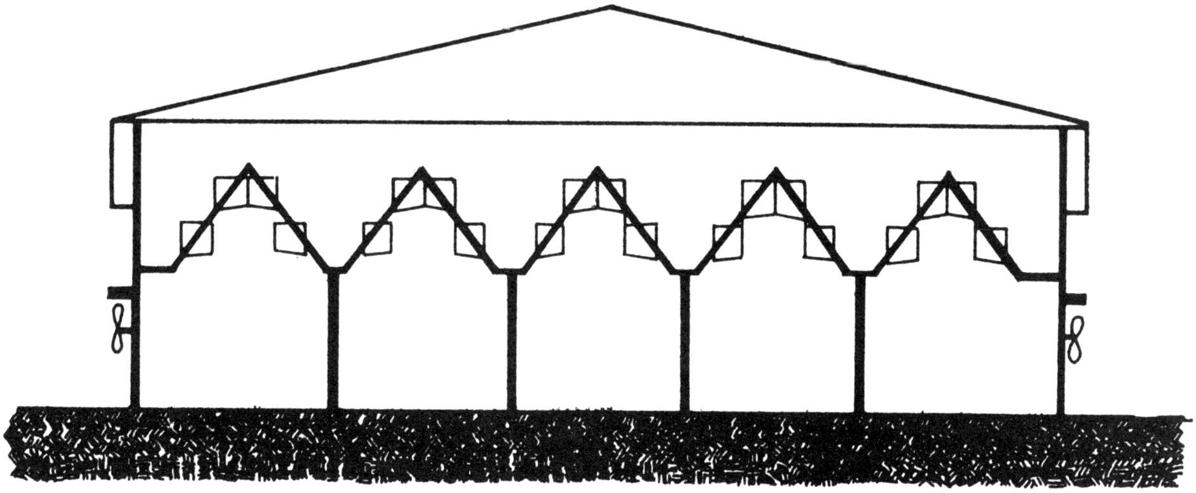

Basic Building

Foundation	Concrete piers.
Floors	Dirt.
Wall Frame	2" x 4" - 24" o.c.
Roof Frame	Wood trusses with 2" x 4" purlins @24" o.c.
Exterior	Two rib aluminum or corrugated iron.
Interior	¾" styrofoam.
Lighting	Fluorescent or good incandescent.
Plumbing	Good basic system.
Basic Building Cost Per S.F.	$4.25 to $4.90

Equipment

Component	Flat Deck	Stair Step
Cages	12" x 20".	12" x 20".
Watering System	Automatic cup system.	Automatic cup system.
Feeding	Automatic system.	Automatic system.
Egg Gathering	Automatic system.	Automatic system.
Cooling	Negative pressure system.	Negative pressure system.
Heating	None.	None.
Building and Equipment S.F. Cost	$8.10 to $9.25.	$9.80 to $10.40 .
Building and Equipment Cost Per Bird	$5.00 to $5.60.	$6.10 to $7.15.

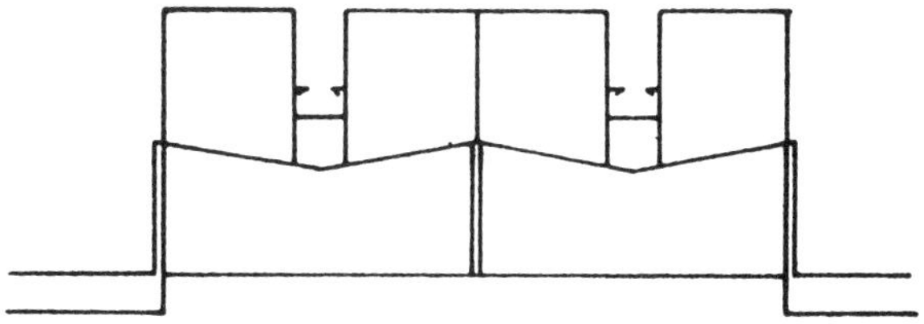

Flat-Deck cage system

Poultry Houses
Deep Pit Type

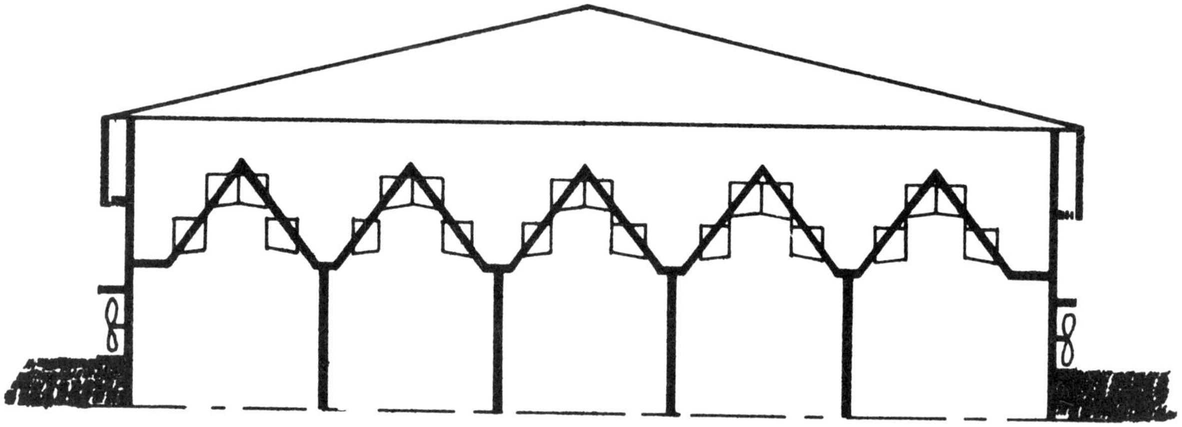

Basic Building

Foundation	Concrete piers.
Floors	Concrete with waterproof membrane.
Wall Frame	2" x 4" - 24" o.c.
Roof Frame	Wood trusses with 2" x 4" purlins @24" o.c.
Exterior	Two rib aluminum or corrugated iron.
Interior	¾" styrofoam.
Lighting	Fluorescent or good incandescent.
Plumbing	Good basic system.
Basic Building Cost Per S.F.	$4.90 to $5.85.

Equipment

Component	Flat Deck	Stair Step
Cages	12" x 20".	12" x 20".
Watering	Automatic cup system.	Automatic cup system.
Feeding	Automatic system.	Automatic system.
Egg Gathering	Automatic system.	Automatic system.
Cooling	Negative pressure system.	Negative pressure system.
Heating	None.	None.
Building and Equipment S.F. Cost	$9.25 to $10.40.	$10.35 to $11.60.
Building and Equipment Cost Per Bird	$6.00 to $7.00.	$7.00 to $9.25.

Poultry Houses
Equipment Costs
To be added to the basic building cost

Component	Serving One Row Of Cages	Serving Two Rows Of Cages
Automatic Feeders	$1.00 per bird.	$.45 per bird.
Automatic Egg Gathering	$ .75 per bird.	$.36 per bird.
Automatic Water Cup System	$ .73 per bird, $2.60 per cup.	$.34 per bird, $2.60 per cup.
"V" Water Trough	$ .14 per bird.	$.09 per bird.
16" Feed Trough	$ .24 per bird.	$.14 per bird.

Foggers	½" galvanized pipe. ¾" galvanized pipe. 1" galvanized pipe.	$ .90 linear foot. .98 linear foot. 1.07 linear foot.
Roof Sprinklers		$4.00 linear foot.
Evaporative Coolers		$820.00 each. $.87 per square foot of building.
Fans 30" 36" 48"		 $295.00 each. 405.00 each. 435.00 each.
Negative Pressure Air Conditioning System		$.55 to $.65 per square foot of building.
Cooling Pads in Wall		$.37 per square foot of surface area.
Heating Systems		$845.00 per unit.
Cages, 12" x 20" or 18"		$3.70 each. $1.00 per bird.

Labor Camps
Quality Classification

	Class 1 Best Quality	Class 2 Prefabricated	Class 3 Good Quality	Class 4 Average Quality	Class 5 Low Quality
Slab Foundation	Spread footing around perimeter and thickened slab at partitions.	Spread footing around perimeter.	Thickened around perimeter.	Thickened around perimeter.	Thickened around perimeter.
Floor	4" concrete slab reinforced.	4" concrete slab reinforced.	4" concrete slab.	4" concrete slab.	4" concrete slab.
Walls	Masonry exterior walls, wood frame interior partitions and ceiling.	Metal building, prefabricated.	2" x 4" studs @16" o.c., 2" x 4" stud partitions.	Box construction 4" x 4" at 48" o.c.	Box construction 2" x 4" at 48" o.c.
Exterior Cover	Natural blocks.	Prefabricated metal building.	Average grade redwood board and batt or horizontal siding or stucco finish.	Fair grade redwood or fir board and batt or horizontal board.	Poor grade of redwood or fir, vertical or horizontal.
Interior Finish	Gypsum wallboard.	None.	Gypsum wallboard.	Plywood partitions.	None.
Roof Framing	Rafters, collar beams and ceiling joists.	Prefabricated metal building.	Rafters, collar beams and ceiling joists.	Very simple truss.	Rafters and tie at plate line.
Roofing	Composition shingles.	Metal.	Aluminum or wood.	Composition or sheet metal.	Composition or used sheet metal.
Doors	1 metal door and metal frame each room.	Metal doors.	1 average door each room.	3 or 4 average doors.	2 or 3 cheap doors.
Windows	1 steel sash or aluminum window in each room.	Approximately 20% of floor area.	1 steel or aluminum window in each room.	1 window each room.	Few and small.
Electrical	1 good light and plug in each room.	1 light, 1 plug for each 300 square feet.	1 light, 1 plug in each room.	1 pull chain light and plug for each 400 square feet.	1 pull chain light for each 500 square feet.

Labor Camp, Class 1

Labor Camp, Class 4

Square Foot Area

Quality Class	400	600	800	1,000	1,200	1,500	2,000	2,500	3,000
1, Best	20.41	18.85	18.01	17.47	17.10	16.70	16.28	16.01	15.82
2, Prefabricated	18.82	17.38	16.60	16.11	15.76	15.40	15.01	14.76	14.58
3, Good	17.25	15.93	15.22	14.77	14.45	14.12	13.76	13.53	13.37
4, Average	14.40	13.29	12.70	12.32	12.06	11.78	11.48	11.29	11.16
5, Low	11.76	10.86	10.37	10.06	9.85	9.62	9.38	9.22	9.11

Costs do not include any plumbing. Add $350 to $435 per fixture.

Miscellaneous Agricultural Structures

Livestock Scales

Type	Size	Capacity	In-Place
Full-capacity beam	16′ x 8′	5 Ton	$5,560
Printing beam	16′ x 8′	5 Ton	6,760
Full-capacity beam	22′ x 8′	10 Ton	7,300
Printing beam	22′ x 8′	10 Ton	8,800

Additional Costs For Livestock Scales

Type and Size	Cost
Each foot arm is removed from scale	$ 65 L.F.
Angle iron stock rack for 16′x8′ scale	1,825 ea.
Angle iron stock rack (wood) for 16′x8′	2,700 ea.
Angle iron stock rack (wood) for 22′x8′	3,170 ea.

Scale pit has 4″ concrete walls and slab poured in place. May be poured in or on top of the ground. If on top, compacted ramps and steps to the scale beam are included.

Motor Truck Scales

Five inch reinforced concrete platform. All-steel structure and scale mechanism. Reinforced concrete pit. Motor truck scales are of two general types: the beam type (either manual or type registering) and the full-automatic dial type. The construction of both, insofar as the weight-carrying mechanism is concerned, is very similar. The method of recording the weights makes the cost vary.

Capacity	Platform Size	Total Cost
20 Tons	24′ x 10′	$12,000
30 Tons	34′ x 10′	18,500
40 Tons	40′ x 10′	21,000
50 Tons	45′ x 10′	22,000
50 Tons	50′ x 10′	23,500
50 Tons	60′ x 10′	26,500
50 Tons	70′ x 10′	29,000

Above costs are for full-capacity beam scales. Add $600 for the registering type beam.

Septic Tanks

1 bath home with 1,000 gallon tank	$1,400
2 bath home with 1,250 gallon tank	1,750
3 bath home with 1,500 gallon tank	2,130

Bulk Feed Tanks

Size and Type	Cost
3 to 4 ton	$465
5 to 6 ton	685
6 to 7 ton	765
7 to 8 ton	855
9 to 10 ton	1,000
10 to 12 ton	1,100
3 to 4 ton twin	590
4 to 5 ton twin	650
8 ton dairy feed	1,200
12 ton dairy feed	1,650
15 ton dairy feed	1,860

Tanks are equipped with a scissor-type opening chute

Domestic Water Systems

Submersible Pump, Installed at 105′ Depth Excluding Well and Casing

Typical Installation

	½ H.P.	¾ H.P.	1 H.P.
Total Cost	$900	$955	$1,140
Pressure Tank Size	82 Gal.	82 Gal.	120 Gal.
Cost Per Ft. Above Or Below 105′ Depth	$2.85	$2.95	$3.80

	1½ H.P.	2 H.P.	3 H.P.
Total Cost	$1,570	$1,820	$2,200
Pressure Tank Size	220 Gal.	220 Gal.	315 Gal.
Cost Per Ft. Above Or Below 105′ Depth	$4.65	$5.75	$6.65

Pressure Tank Sizes and Installed Costs

42 Gal.	16″ Dia. x	48″ Depth	50″	$ 110 to	$ 145
82 Gal.	20″ Dia. x	60″ Depth	63″	205 to	225
120 Gal.	24″ Dia. x	60″ Depth	75″	260 to	310
220 Gal.	30″ Dia. x	72″ Depth	94″	800 to	965
315 Gal.	36″ Dia. x	72″ Depth	113″	900 to	1,000
525 Gal.	36″ Dia. x	120″ Depth	113″	1,200 to	1,400

Typical Physical Lives In Years For Agricultural Structures

Building Type	Quality: Good	Average	Low
Barns	40	30	20
Dairy barns, modern	25	25	--
Dairy barns, low cost	--	20	20
Storage sheds	40	30	20
Poultry houses, modern	30	25	--
Poultry houses, coventional	--	25	20

To determine the useful life remaining, use the percent good table for residential structures on page 40.

MILITARY CONSTRUCTION COSTS

The Office of the Secretary of Defense has prepared the following square foot guidelines to reflect the cost of military construction as of June 1982. The costs are based on repetitive type of permanent facilities built in Washington D.C. Use the "Construction Cost Indices" at the end of this section to adapt the Washington D.C. cost to any other area. The "Size/Unit Cost Adjustment Chart" should be used to determine the approximate cost of a structure larger or smaller than the typical size shown.

Included in these costs are all items of equipment which are permanently built-in or attached to the structure, including items with fixed utility connections.

The following items are typical examples:

- •Furniture, cabinets and shelving, built-in.
- •Venetian blinds and shades.
- •Window screens and screen doors.
- •Elevators and escalators.
- •Drinking water coolers.
- •Telephone, fire alarm and intercom systems.
- •Theater seats.
- •Pneumatic tube systems.
- •Heating, ventilating and air conditioning installations.
- •Electrical generators and auxiliary gear.
- •Waste disposers such as incinerators.
- •Food preparation and serving equipment, built-in.
- •Hoods and vents.
- •Chapel pews and pulpit.
- •Refrigerators, built-in.
- •Laboratory furniture, built-in.
- •Cranes and hoists, built-in.
- •Dishwashers.

The costs listed are the estimated contract award costs, excluding contingencies, supervision, and administration. They include construction to the five-foot line only, but do not include the cost of outside utilities or other site improvements. Figures listed do not include the cost of piles or other special foundations which are considered as an additional supporting item. The cost of air conditioning is included to the extent authorized by the Construction Criteria Manual.

The costs of equipment such as furniture and furnishings which are loose, portable, or can be detached from the structure without tools are excluded from the unit costs. The costs of permanently attached equipment related directly to the operating function for which the structure is being provided, such as technical, scientific, production and processing equipment, is normally excluded from these costs. The following items are typical examples:

- •Furniture, loose.
- •Furnishings, including rugs, loose.
- •Filing cabinets and portable safes.
- •Office machines, portable.
- •Wall clocks, plug-in.
- •Food preparation and serving equipment, including appliances, portable.
- •Training aids and equipment, including simulators.
- •Shop equipment.
- •Bowling lanes, including automatic pin spotting equipment, score table and players' seating.
- •Automatic data processing equipment.
- •Communications equipment.
- •Photographic equipment, portable.
- •Any operational equipment for which installation, mounting and connections are provided in building design and which are detachable without damage to the building or equipment.

Estimating Procedure

Determine the area relationship of the proposed building by dividing the gross area by the typical size as shown in the Square Foot Cost Table. Locate the quotient on the Area Relationship scale and trace vertically to the Factor Line, then trace horizontally to the Cost Relationship scale. This value is then multiplied by the unit cost in the Square Foot Cost Table and factored by the Construction Cost Index to determine the adjusted unit cost for the building.

Square Foot Costs

Facilities	Typical Size Gross S.F.	Unit Costs Per S.F.
Academic Facilities	25,000	62.00
Administrative Facilities		
Up to 3 stories	15,000	55.00
Data processing area facility		73.00
Air Terminal Buildings		
Passenger	17,500	56.00
Freight	45,000	42.00
Unmarried Officer Quarters		
Fiscal 1982		60.00
Barracks, Dormitories		
Fiscal 1982		55.00
Bowling Facilities		
Building only, (8 lanes)	7,800	61.00
Bowling lane, including automatic pin spotting equipment, score table and players' seating		25,000.00
Chapels	8,000	98.00
Clubs (Open Messes)		
Enlisted service	16,000	68.00
Noncommissioned Officer	12,000	75.00
Officer	10,000	81.00
Commissaries	88,500	55.00
Exchange Stores		
Without cafeteria	12,000	65.00
Family Housing		
Fiscal 1982		44.00
Fire and Crash Stations	8,000	98.00
Flight Simulator Buildings	12,000	100.00
Guard Houses, Brigs	6,000	75.00
Gymnasiums	20,000	75.00
Hangers, General Purpose or Maintenance	50,000	65.00

Facilities	Typical Size Gross S.F.	Unit Costs Per S.F.
Libraries	4,800	63.00
Medical Facilities		
Hospitals		123.00
Dispensaries	8,000	85.00
Dental Clinics	10,600	128.00
Messhalls, Dining Facilities, Enlisted	16,000	115.00
Operations Building, Aircraft Without Tower	10,000	66.00
Photographic Laboratories	2,400	99.00
Post Offices	3,750	67.00
Recreation Workshops		
Arts and Crafts	3,400	71.00
Automative	2,000	61.00
Reserve Armory Type Facilities	20.000	62.00
Shops		
Automotive Maintenance	5,000	61.00
Aircraft	12,000	78.00
Installations Maintenance	15,000	54.00
Parachute and Dinghy	8,000	76.00
Weapons Release Systems	5,000	53.00
Squadron Operations Building	12,000	65.00
Storage Facilities		
Auto Garages	20,000	33.00
Cold Storage Facilities	6,000	103.00
General Purpose Warehouses	50,000	30.00
Telephone Exchanges	6,000	115.00
Theaters		
With Stage and Seats	10,000	100.00

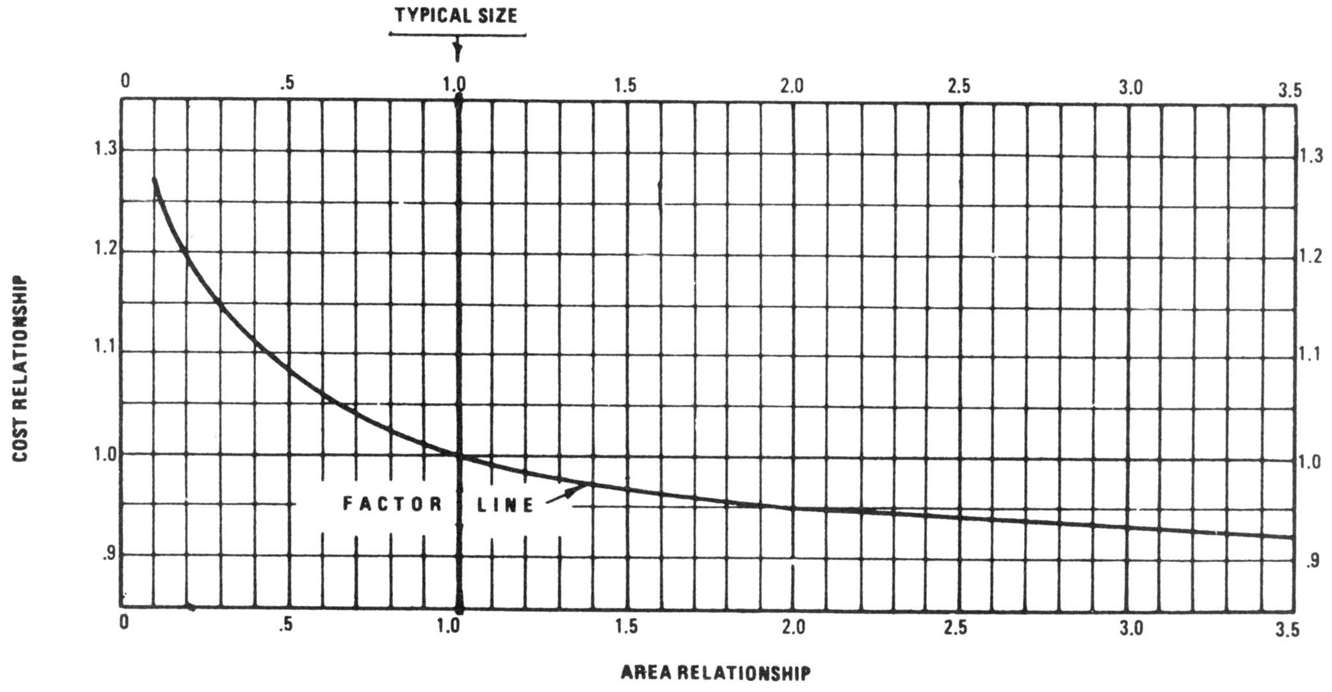

Construction Cost Indices

State	Exceptions	Index
Alabama		.89
	Gulf Coast Area	.97
Alaska		1.40
Aleutian Islands		3.80
Clear AFS & Ft. Greely (Big Delta)		2.20
Cold Bay, Ft. Yukon		3.00
Eielson AFB, Ft. Wainwright & Whittier		2.10
Elmendorf AFB & Ft. Richardson		1.90
Kodiak		2.50
King Salmon		2.60
Nome		2.30
Point Barrow		3.50
Arizona		1.01
	Yuma P.G. & Davis-Monthan AFB	1.10
	Ft. Huachuca	1.20
	Gila Bend AFS	1.15
Arkansas		.87
California		1.11
	Beale, Castle, March, Mather, San Francisco Bay Areas, Sierra Army Depot, McClellan	1.15
Colorado		.98
Connecticut		1.03
New London		1.20
Delaware		.99
District of Columbia		1.00
Florida		.90
	Patrick, Cape Canaveral	1.15
	Key West, Miami and Gulf Coast	1.00
Georgia		.86
Hawaii:		1.20
Kauai		1.80
Idaho		.96
	Mt. Home AFB	1.20
Illinois		1.00
	Scott AFB & Granite City Army Depot	1.20
Indiana		.97
	Grissom AFB	1.10
Iowa		.97
Kansas		.96
Kentucky		.94
Louisiana		.92
	Ft. Polk	1.05
	New Orleans	1.00
Maine		.90
Maryland		.96
	Ft. Richie & Patuxent River ATC	1.10
	D.C. Area	1.00
Massachusetts		1.02
	Ft. Devens	1.15
Michigan		1.02
	Northern Area	1.15

State	Exceptions	Index
Minnestoa		.99
	Northern Area	1.15
Mississippi		.99
	Keesler AFB	1.00
Missouri		.98
	St. Louis	1.02
	Ft. Leonard Wood	1.20
Montana		.95
	Northern Area, Marlmstrom	1.15
Nebraska		.94
Nevada		1.05
Tonopah		1.60
New Hampshire		.92
New Jersey		1.04
New Mexico		.96
Holloman		1.05
New York		1.03
	New York City & Long Island	1.17
North Carolina		.84
	Cherry Point, Camp Lejeune	.95
	Bragg & Pope AFB, Seymour Johnson	1.00
North Dakota		.91
	Northern Area	1.15
Ohio		1.02
	Clifton County AFB	1.02
	Wright-Patterson AFB	1.10
Oklahoma		.94
Tinker AFB		1.00
Oregon		1.02
Pennsylvania		1.01
Rhode Island		1.00
South Carolina		1.00
South Dakota		.92
	Ellsworth AFB	1.15
Tennessee		.88
	NAS Memphis	1.00
Texas		.98
	Carswell	1.10
Utah		.98
	Dugway Proving Ground & Hill AFB	1.30
Vermont		.91
Virginia		.91
	Area Adjacent to D.C., Norfolk-Newport News Area, Dismal Swamp Area, Ft. Lee and Dahgren	1.00
Washington		1.01
	Puget Sound Area	1.15
West Virginia		.96
Wisconsin		.98
Wyoming		.91

Index

A
Academic Facilities, Military, 235
Administrative Facilities, Military, 235
"A-Frame" Cabins, 29, 35-38
"A-Frame" Restaurants, 148-149
Agricultural Labor Camps, 232
Agricultural Structures, 214-233
Air Compressors, 171
Air Conditioning, 39, 204, 231
Air Hose Reels, 170
Air Terminals, Military, 235
Aircraft Towers, Military, 235
Airplane Hangars, 193
Alarms, Burglar, 202
Apartment Buildings, 16-19
Appliances, Residential, 26
Arbors, 24
Armory Facilities, 235
Asphalt Paving, 39, 212
Asphalt Shingles, 24
Auto Repair Garages, 173-182
  Military, 235
Auto Service Centers, 183-186
Automobile Hoists, 171
Average Building Life
  Agricultural, 233
  Commercial, 200
  Industrial, 200
  Residential, 40

B
Bachelor Officer Quarters, 235
Balconies, 25
Banks, 82-92
  Additional Costs, 92
  Concrete, 82-86
  Masonry, 82-86
  Wood Frame, 87-91
Barns, 215-217, 221-226
  Dairy, 222-226
  Feed, 217
  General Purpose, 215
  Hay Storage, 216
  Herringbone, 225
  Pole, 221
Baseboard Heaters, 25, 204
Basements, 24, 202
Basin, Catch, 213
Bathrooms, 27
Blendomatic Pump, 169
Bowling Facilities, Military, 235
Boxes, Safe Deposit, 92
Brigs, Military, 235
Building Cost Index, 9
Building Life
  Agricultural, 233
  Commercial, 200
  Industrial, 200
  Residential, 40
Bulk Feed Tanks, 233
Bulkhead Walls, 210
Bumpers, Parking, 212
Burglar Alarms, 202

C
Cabins, 29-38
Cages, Poultry, 231
Camps, Labor, 232
Canopies, 169, 202
Cantilevered Poles, 172, 211
Carports, 26, 39
Cash Boxes, 170
Catch Basins, 213
Ceiling, Display Fronts, 210
Chain Link Fences, 213
Chapels
  Funeral, 138, 139
  Military, 235
Chicken Houses, 227-231
Cinemas, 150-160
Clay Tile Roofing, 24
Clubs, Military, 235
Coffee Shops, 143-145
Cold Storage Facilities, 235
Commercial Structures, 41-186
Commissaries, Military, 235
Communications Systems, 202
Compressors
  Air, 171
  Refrigeration, 226
Concrete, Decks, 24 39, 212
Concrete Tile Roofing, 24
Construction Cost Indices
  Military, 236
  Non-Military, 7, 8
Convalescent Hospitals, 134-137
"Conventional" Cabins, 29-43
Conventional Poultry Houses, 227
Conventional Restaurants, 146-147
Cooling Systems, 39, 204, 231
Corrals, 226
Cost Index, Building, 8, 236
Counters, 92
Crash Stations, Military, 235
Curbs, 171, 212

D
Dairy Barns, 222-226
  Low Cost, 222
  Modern Herringbone, 225
  Stanchion, 223
  Walk-Through, 224
Davit Poles, 170
Decks, 24, 39
Dental Clinics, Military, 235
Dental-Medical Buildings, 118-133
  Concrete, 118-125
  Masonry, 118-125
  Wood Frame, 126-133
Department Stores, 93-101
  Basements, 95, 98, 101
  Concrete, 96-98
  Masonry, 96-98
  Reinforced Concrete, 93-95
  Upper Floors, 95, 98, 101
  Wood Frame, 99-101
Depreciation
  Agricultural Structures, 233
  Commercial Structures, 200
  Industrial Structures, 200
  Residential Structures, 40
Discount Houses, 78-81
  Concrete and Masonry, 78-79
  Wood and Steel, 80-81
Dishwashers, 26
Dispensaries, Military, 235
Dispensers, Gasoline, 169
Display Fronts, 56, 207-210
Disposers, 26
Dockboards, Mechanical, 202
Docks, 202
Domestic Water Systems, 233
Door Hoods, 198
Doors, 203
  Vault, 92
Downspouts, 198
Drainage, Storm, 213
Draperies, 203
Drive-In Restaurants, 140-142
Drop Inlets, 213
Dumbwaiters, 203
Dwellings, 10-19

E
Egg Gathering Equipment, 231
Electric Baseboard Heating, 25
Elevators
  Freight, 203
  Hydraulic, 27
  Hydro-Electric, 203
  Passenger, 203
Entrances, Display Fronts, 210
Escalators, 203
Evaporative Coolers, Poultry House, 231
Exchange Stores, Military, 235
External Offices, Industrial, 192
External Windows, 92
Extinguishers, Fire, 204

F
Factory Buildings, 191
Family Housing, Military, 235
Fans, Poultry House, 231
Farm Buildings, 214-233
Feed Barns, 217
Feed Tanks, 233
Feeders, Poultry, 231
Fences, 171, 213
Fill, 204
Fire Doors, 203
Fire Escapes, 204
Fire Extinguishers, 204
Fire Sprinklers, 204
Fire Stations, Military, 235
Fireplaces, 26, 39, 204
Flatwork, 39, 212
Flight Simulator Buildings, 235
Flood Lights, 170
Foggers, Poultry House, 231
Food Stores, 70-77
  Small, 74-77
  Supermarkets, 70-73
Forced Air Heating, 25, 204
Freight Elevators, 203
Funeral Homes, 138-139
Furnaces, 25, 204

G
Gable Roofs, 24
Garages, 26, 28, 39
  Service, 173-182
Garbage Disposers, 26
Gas Stations, 163-172
Gasoline Pumps, 169
Gasoline Storage Tanks, 170
Gates, 212
General Office Buildings, 102-117
  Concrete, 102-109
  Masonry, 102-109
  Wood Frame, 110-117
General Purpose Barns, 215
Glass, Display Fronts, 210
Grease Pits, 170
Guard Houses, Military, 235
Gutters, 198
Gymnasiums, Military, 235

H
"Half Classes", 10
Half Story Areas, 27
Hangars, Airplane, 193, 235
Hay Storage Barns, 216
Heat Vents, 199, 206
Heating Systems, 25, 39, 204, 231
Heating Units, Electric, 204
Herringbone Barns, 225
Hoists, Automobile, 171
Holding Tanks, Milk, 226
Homes, Single Family, 10-15
Hoods, Door, 198
Hose Reels, 170
Hospitals
  Convalescent, 134-137
  Military, 235

Hydraulic Elevators, 27, 203

I
Industrial Structures, 187-200
Industrial Windows, 199
Inflators, Tire, 170
Insulation, Steel Buildings, 198
Intercommunications Systems, 202
Internal Offices, Industrial, 192

K
Kitchen Equipment, 205

L
Labor Camps, 232
Lavatory, 39
Lay Cage Poultry Houses, 227
Libraries, Military, 235
Light Industrial Buildings, 190
Lighting
  Display Fronts, 210
  Service Station, 170
  Yard Improvements, 213
Livestock Scales, 233
Loading Ramps, 202
Louvers, 198
Lube Rooms, 170

M
Manholes, 213
Mechanical Dockboards, 202
Medical-Dental Buildings, 118-133
  Concrete, 118-125
  Masonry, 118-125
  Wood-Frame, 126-133
Mess Halls, Military, 235
Mezzanines, 92, 205
Military Structures, 234-236
Milk Holding Tanks, 226
Milk Line, 226
Milk Vacuum Pumps, 226
Milking Barn, 223, 224
Mobile Home Parks, 161-162
Mortuaries, 138-139
Motels, 20-23
Motor Truck Scales, 233
Mounts, Sign Post, 172
Movie Houses, 150-160
Multi-Family Residences, 16-19

N
Nursing Homes, 134-137

O
Office Buildings, 102-117
  Concrete, 102-109
  Masonry, 102-109
  Wood Frame, 110-117
Office, Service Station, 169
Officer Quarters, Military, 235
Offices, Industrial, 192
Operations Building, Aircraft, 235
Oven Units, 26
Overhangs, 198
Overhead Heating, 204

P
Pads, Cooling, Poultry House, 231
Parks, Mobile Home, 161-162
Partitions, 205
  Steel Buildings, 199
Passenger Elevators, 203
Paving
  Asphalt, 39, 212
  Concrete, 39, 212
  Service Stations, 171
Percent Good Tables
  Commercial Structures, 200
  Industrial Structures, 200
  Residential Structures, 40
Pergolas, 24
Pews, 205
Photographic Laboratories, 235
Pits, Grease, 170
Platforms, Display Front, 210
Pneumatic Tube Systems, 205
Pole Barns, 221
Porch Roofs, 24
Porches, 39
  Covered, 24
Post Mounts, Sign, 172
Post Offices, Military, 235
Poultry Cages, 231
Poultry Feeders, 231
Poultry Houses, 227-231
  Controlled Environment, 228
  Conventional, 227
  Deep Pit, 230
  Equipment Costs, 231
  High Rise, 229
Pressure Tanks, 233
Public Address Systems, 202
Pumps
  Milk Vacuum, 226
  Gasoline, 169

R
Ramps, Loading, 202
Ranges, Kitchen, 26
Readiness Crew Buildings, 235
Recreation Workshops, Military, 235
Recreational Dwellings, 29-38
Reels, Hose, 170
Refrigerated Room Coolers, 25
Refrigerators, 26
Residential Structures,10-40
Rest Homes, 134-137
Restaurants, 140-149
  "A-Frame", 148-149
  Coffee Shop, 143-145
  Conventional, 146-147
  Self Service, 140-142
Roll Asphalt Roofing, 24
Roof Sprinklers, Poultry House, 231
Roofs, Porch, 24
Rotators, 171
Rural Buildings, 214-233

S
Safe Deposit Boxes, 92
Savings Offices, 82-92
  Additional Costs, 92
  Concrete, 82-86
  Masonry, 82-86
  Wood Frame, 87-91
Scales, 233
Seating, 205
Self Service Restaurants, 140-142
Septic Tanks, 233
Service Centers, Auto, 183-186
Service Garages, 173-182
  Concrete, 173-176
  Masonry, 173-176
  Wood Frame, 178-182
Service Stations, 163-172
  Additional Costs, 169-172
  Masonry, 163-164
  Porcelain, 165-166
  Ranch Type, 167-168
  Rustic Type, 167-168
  Steel, 163-166
Shed Roof, 24
Sheds, 219-220
  Machinery, 219
  Small, 220
Shop Buildings, Agricultural, 218
Shops, Military, 235
Showers, 39
Sign Post Mounts, 172
Sign Rotators, 212
Signs, 211-212
  Service Stations, 171
Single Family Residences, 10-15
  Four Corners, 10, 12
  Six Corners, 10, 13
  Eight Corners, 10, 14
  Ten Corners, 10, 15
Sinks, 39
Skylights, 199, 205-206
Slate Roofing, 24
Sleeping Porches, 24
Small Food Stores, 74-77
  Masonry, 74-75
  Wood Frame, 76-77
Smoke Vents, 199, 206
Spotlights, 170
Sprinklers
  Fire, 204
  Roof, Poultry House, 231
Stairways, 25
Stanchion Barns, 223
Steel Buildings, Industrial, 193-199
Storage Barns, Hay, 216
Storage Buildings, 169
Storage Tanks, Gasoline, 170
Stores, 42-81
Storm Drainage, 213
Striping, Parking Lot, 212
Suburban Stores, 55-69
Supermarkets, 70-73
  Concrete, 70-72
  Masonry, 70-72
  Wood Frame, 72-73
Supporting Walls, 39

T
Tanks
  Bulk Feed, 233
  Gasoline Storage, 170
  Milk Holding, 226
  Pressure, 233
  Septic, 233
Telephone Exchanges, Military, 235
Theater Seating, 205
Theaters
  Concrete, 150-155
  Masonry, 150-155
  Military, 235
  With Balcony, 151
  Without Balcony, 152-154
  Wood Frame, 156-160
Tile Roofing, 24
Trailer Parks, 161-162
Troughs, Poultry, 231
Truck Scales, 233
Tubs, Bath, 39
Tube Systems, Pneumatic, 205

U
Urban Stores, 42-54
  Concrete, 43-48
  Masonry, 43-48
  Wood Frame, 49-53

V
Vacuum Pumps, Milk, 226
Vault Doors, 92
Ventilators, 199, 206
Vents, 25, 206

W
Walk-In-Boxes, 207
Wall Heaters, 25
Walls, Supporting, 39
Warehouses, 189
  Military, 235
Water Closet, 39
Water Hose Reels, 170
Water Lines, 170
Water Trough, Poultry, 231
Water Systems, 233
Window Frames, 210
Windows
  External, 92
  Industrial, 199
Wood Decks, Uncovered, 24
Wood Fences, 213
Wood Posts, 39
Wood Shakes, 24

Y
Yard Improvements, 212-213
Yard Lights, 170